Power Systems Harmonics: Modeling and Analysis

Power Systems Harmonics: Modeling and Analysis

Edited by Jacob Hunter

STATES
ACADEMIC PRESS
www.statesacademicpress.com

States Academic Press,
109 South 5th Street,
Brooklyn, NY 11249, USA

ISBN: 978-1-63989-431-4

Cataloging-in-Publication Data

Power systems harmonics : modeling and analysis / edited by Jacob Hunter.
p. cm.
Includes bibliographical references and index.
ISBN 978-1-63989-431-4
1. Electric power systems. 2. Harmonics (Electric waves). 3. Electric power production.
4. Electrical engineering. I. Hunter, Jacob.
TK3226 .P69 2022
621.319 21--dc23

For information on all States Academic Press publications
visit our website at www.statesacademicpress.com

STATES
ACADEMIC PRESS

Contents

Preface

The purpose of the book is to provide a glimpse into the dynamics and to present opinions and studies of some of the scientists engaged in the development of new ideas in the field from very different standpoints. This book will prove useful to students and researchers owing to its high content quality.

Electric voltages and currents that can cause power quality issues on an electric power system are called harmonics. It can be produced by electronic devices with nonlinear loads drawing current in sudden short pulses. They are prevalent when multiple computers, printers, fax machines, copiers, UPSs, variable speed drives, etc., are all on the same electrical system. Harmonics specifically lowers the power quality and its efficiency in a commercial building or industrial facility. Overloading neutral conductors, overheating transformers and increased associated losses, and nuisance tripping of circuit breakers are some common problems caused by harmonics. The various studies that are constantly contributing towards advancing technologies and evolution of this field are examined in detail. This book elucidates the concepts and innovative models around prospective developments with respect to this field. A number of latest researches have been included herein to keep the readers up-to-date with the global concepts in this area of study.

At the end, I would like to appreciate all the efforts made by the authors in completing their chapters professionally. I express my deepest gratitude to all of them for contributing to this book by sharing their valuable works. A special thanks to my family and friends for their constant support in this journey.

Editor

1

Materials Used in High-Speed Electrical Machines

Flyur R. Ismagilov, Viacheslav Ye. Vavilov and
Valentina V. Ayguzina

Abstract

The high-speed electrical machines are widely used in different industries, such as machine tools, aerospace engineering, autonomous power engineering, etc. This chapter devoted to materials used in high-speed electrical machines with high-coercitivity permanent magnets (HCPMs). It is considered to be materials of rotor sleeve, shaft, stator magnetic core, and permanent magnet. Material selection methods are presented. In addition, mechanical strength calculation of the rotor sleeve is shown. The obtained results can be used in the design of high-speed electrical machines with high-coercivity permanent magnets and in their future development.

Keywords: high-speed electrical machine, high-coercivity permanent magnets, rotor magnetic system, amorphous magnetic material, eddy-current losses, windage losses

1. Introduction

Mass and size and energy parameters, reliability, and efficiency of high-speed and ultra-high-speed electric machines (EMs) are determined and ensured by the properties of the materials used in them. Therefore, special attention should be given to the selection of materials with the necessary properties in the high-speed EM design process.

Materials used in the static components of high-speed EM (housing, bearing shields, cooling elements, etc.) are no different from the materials of the traditional industrial EMs. But materials used in rotating parts (shaft, rotor shroud, rotor back, etc.) of high-speed EMs with high-coercivity permanent magnet (HCPM) should have increased strength characteristics, from which the EM operating conditions, limiting rotational speed, and power are determined. For example, the air gap value is determined by the rotor sleeve strength of the high-speed EM with the HCPM. Therefore, materials used in rotating parts of the high-speed EM with HCPM are discussed in the following sections in more detail.

2. Rotor sleeve materials of the high-speed EM

At high rotor rotational speeds, the centrifugal forces acting on the HCPM tend to tear them away from the shaft. To counteract these forces and to ensure the rotor mechanical strength, the rotor sleeve is used. The rotor sleeve is installed over the HCPM with a certain shrink fit and can be made as a single nonmagnetic sleeve or wound on the HCPM with a composite material filament glued with epoxy resin. **Figure 1a** shows a rotor with a rotor sleeve made of composite material, and **Figure 1b** shows the rotor sleeve made of composite material. Properties of composite materials used in the rotor sleeve [1] are presented in **Table 1**.

The use of composite materials in the high-speed EMs is more technological than the use of solid non-magnetic sleeves, but this technology requires specialized winding equipment, which limits its wide application. Therefore, in a number of cases, one-piece sleeves are used that are installed with a certain shrink fit on the rotor. Titanium or Inconel 718 is used as a material of the rotor sleeve. **Table 2** shows the main characteristics of these materials.

In addition, to select the rotor sleeve material, it is necessary to take into account that the outer surface of the rotor sleeve determines the friction coefficient, which affects the friction losses. To minimize these losses, special antifriction coatings can be used of which the application increases the efficiency of high-speed EM to 0.5% according to [2].

In the rotor sleeve design, it is necessary to take into account the thermal and mechanical extensions of the rotor sleeve during operation. These extensions in some cases can reach 0.3–0.5 mm. If the air gap is incorrectly selected, the expansion can lead to a mechanical contact between the stator and the rotor, which is unacceptable. That is, this criterion also plays an important role in the material selection of the rotor sleeve.

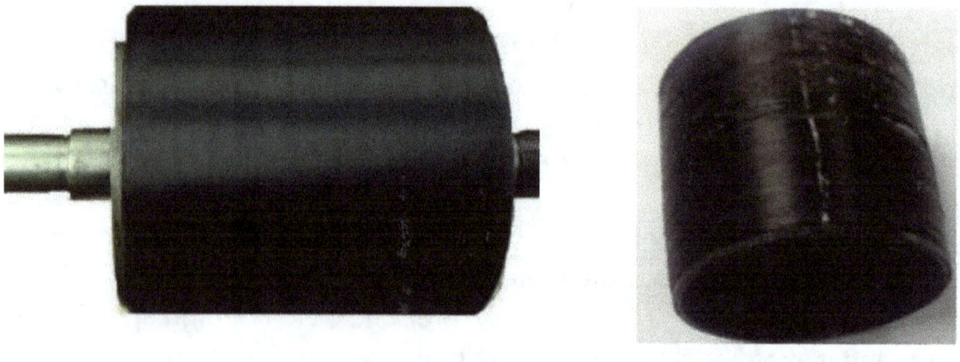

a) b)

Figure 1. Rotor (a) with a rotor sleeve made of composite material (b).

Property	GFRP DW 210	AFRP DW 152	CFRP DW 231	CFRP DW260
Tensile strength (MPa)	1440	1880	2420	2420
Density (kg/m^3)	2100	1330	1520	1520
Maximum operating temperature (°C)	<110	<140	<140	>300
Thermal expansion coefficient	7	0.2	0.2	0.2

Table 1. Properties of composite materials used in the rotor sleeve [1].

Characteristics	Titanium	Inconel 718
Tensile strength (MPa)	1191–1230 (depending on the treatment type)	1443–1553 (depending on the treatment type)
Density (kg/m^3)	4600	8190
Thermal expansion coefficient	7.7–10.4 (depending on the treatment type)	9.5–20.5 (depending on the treatment type)

Table 2. The main characteristics of Titanium and Inconel 718.

It is also important to notice that almost all the materials that are used in the rotor sleeve have electrical conductivity, which leads to eddy currents and heating of the rotor sleeve and HCPM. Therefore, the rotor sleeve should have a minimum electrical conductivity.

2.1. Mechanical strength calculation of the rotor sleeve

In [3], equations for the rotor sleeve thickness calculations are given. These equations do not take into account thermal expansions of the rotor sleeve and shrink fit. Therefore, they are approximate and do not satisfy the specified design accuracy. It seems expedient to consider in more detail the design process of the rotor sleeve of high-speed EM.

In the rotor sleeve calculation, the nominal and maximum rotational speeds are assumed to be set. The material properties, the shrink fit value, and the main rotor geometric dimensions are considered to be known.

It is important to notice that the selection of the rotor sleeve thickness is an important component of the high-speed EM design process. In **Figure 2**, the stress dependencies of the HCPM and the rotor sleeve made of titanium and Inconel 718 obtained in [4] for the EM with a rotational speed of 30,000 rpm are given as an example. It can be seen that an increase in the rotor sleeve shrink fit by 100% leads to a reduction in mechanical stresses in the HCPM by 260% and an increase in mechanical stresses in the rotor sleeve by 60%. The shrink fit value should be selected that the maximum stresses in the rotor sleeve are below the maximum permissible stresses of the material used by 20–30%, and stresses in the HCPM are below the maximum permissible stresses by 20–30%.

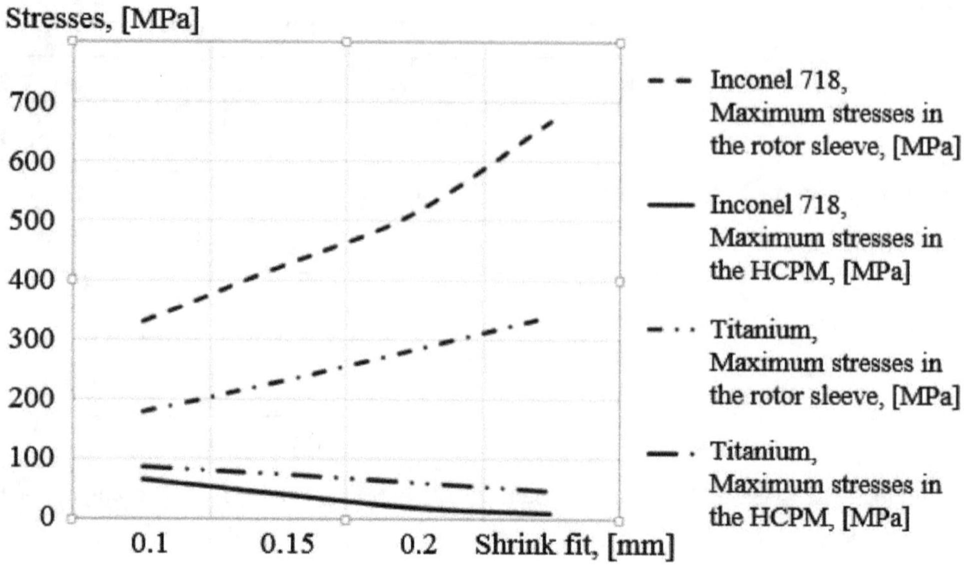

Figure 2. The stress dependences of the HCPM and the rotor sleeve made of titanium and Inconel 718 [4].

The centrifugal forces of the rotor sleeve are determined by the following:

$$F_{r.s} = 11.1 m_{r.s} r_r \left(\frac{n}{1000}\right)^2, \tag{1}$$

where $F_{r.s}$ is the centrifugal force of the rotor sleeve; $m_{r.s}$ is the rotor sleeve mass; n is a rotational speed; and r_r is a rotor radius.

The inertia radius is determined by the known equation: $r_r = \frac{2}{3} r_2 \frac{1+\alpha_r+\alpha^2_r}{1+\alpha_r}$, where $\alpha_r = \frac{r_1}{r_2}$ is a rotor radius ratio.

The centrifugal force of the rotor: $F_r = \frac{2,93}{8} \rho_r l r_2{}^3 \left(1 - \alpha^3{}_r\right) \left(\frac{n}{1000}\right)^2$, where $\rho_r = \frac{\rho_1+\rho_2}{2}$ is the average density of rotor materials, ρ_1 and ρ_2 are, respectively, the density of the HCPM and shaft material.

The preliminary thickness of the rotor sleeve is determined without taking into account the rotor sleeve mass (**Figure 3**):

$$b = \frac{\Omega^2 \rho_1 (r_1 + r_2)\left(D^2{}_4 - D^2{}_3\right)}{16\sigma_M} k_\sigma, \tag{2}$$

where Ω is a rotor rotational speed, (rad/s); σ_M is the tensile strength of the rotor sleeve material; D_3, D_4 are, respectively, the inner and outer HCPM diameters; k_σ is the rotor sleeve safety factor, which is taken in the range from 1.2 to 1.5.

Radial sectional area of the rotor sleeve: $S_{r.s} = \frac{\pi\left(8r_2 b + 4b^2\right)}{4}$.

Tangential stresses on the inner surface of the rotor sleeve: $\sigma = \frac{\eta_1 F_{s.r}}{2\pi S_6} \left(\frac{n}{1000}\right)^2$, where η_1 is the coefficient of tangential forces on the inner surface of the rotor sleeve, which is determined from **Figure 4**.

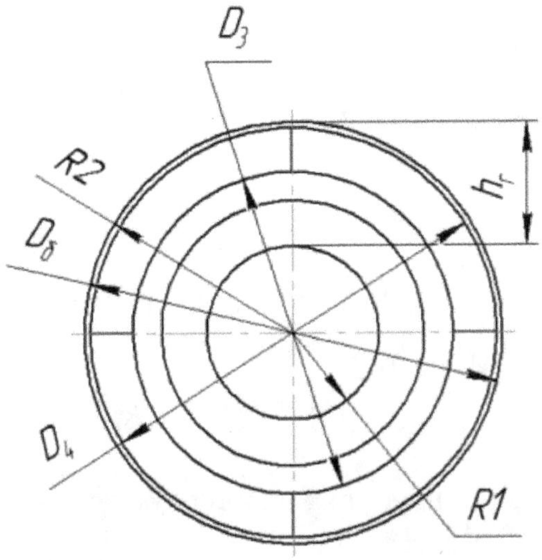

Figure 3. Calculation scheme of the high-speed EM.

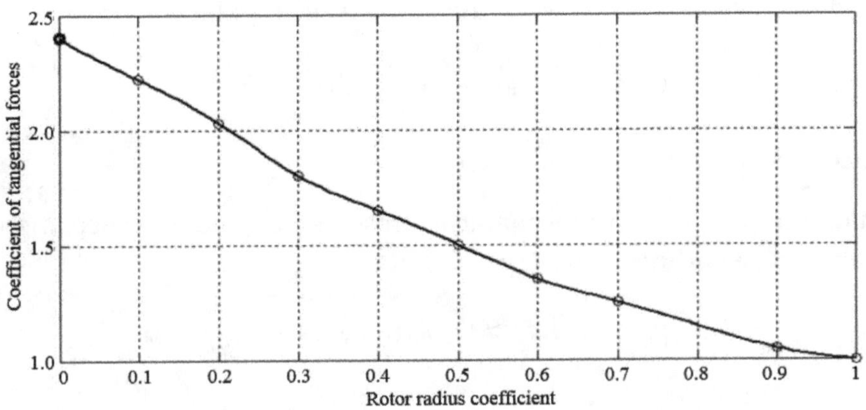

Figure 4. Dependence of the coefficient of tangential forces on the rotor radius coefficient.

Tangential stresses on the outer surface of the rotor sleeve: $\sigma' = \frac{\eta_2 F_{s.r}}{2\pi S_{r.s}}\left(\frac{n}{1000}\right)^2$, where η_2 is the coefficient of tangential forces on the outer surface of the rotor sleeve.

Then calculations are carried out according to the calculation scheme shown in **Figure 3** and the method presented in [5].

Stresses in the rotor sleeve should be less than the tensile strength of the rotor sleeve material, and the rotor sleeve should create a pressure for the HCPM [5]:

$$p_N = p_p + p_t - p_z > 0, \ \ k_\sigma(\sigma_p + \sigma_t + \sigma_z) < \sigma_M; \tag{3}$$

where p_N is the residual rotor sleeve pressure on the permanent magnets; p_p, σ_p are, respectively, the pressure and mechanical stress produced by the rotor sleeve preload; p_t, σ_t are, respectively, the pressure and mechanical stress produced by the thermal expansion of the rotor and permanent magnets; p_z, σ_z are, respectively, the pressure and mechanical stress produced by centrifugal forces.

Pressure and mechanical stress produced by the preload are defined by [5]:

$$\sigma_{\mathrm{p}} = \frac{2\Delta D_{\mathrm{r.s}}}{D} E_{\mathrm{r.s}}, \; p_{\mathrm{p}} = \sigma_{\mathrm{p}} \left[\frac{r_1^2}{r_1^2 - r_2^2} \left(1 + \frac{r_2^2}{r_3^2} \right) \right]^{-1}; \tag{4}$$

where $\Delta D_{\mathrm{r.s}}$ is the rotor sleeve preload; $r_1 = \frac{D_{\mathrm{r.s}}}{2}$; $r_2 = \frac{D}{2}$; $r_3 = \frac{D_{\mathrm{m}}}{2}$.

The pressure and mechanical stress produced by the rotor thermal expansion are determined [5]:

$$\sigma_{\mathrm{t}} = \frac{2\Delta D_{\mathrm{T}}}{D} E_{\mathrm{r.s}}, \; p_{\mathrm{t}} = \sigma_{\mathrm{t}} \frac{b}{R_{\mathrm{r.s}}}; \tag{5}$$

where $\Delta D_{\mathrm{T}} = \alpha_{\mathrm{Bt}} D (T_2 - T_1)$ is the linear expansion of permanent magnets under the temperature; α_{Bt} is the linear expansion coefficient of permanent magnets; $T_2, \; T_1$ are the temperatures of permanent magnets at the ultra-high-speed electrical machine (UHSEM) operation beginning and the steady UHSEM temperature, respectively.

The pressure and the mechanical stress produced by centrifugal forces [5]:

$$\sigma_{\mathrm{z}} = \Omega^2 r_3^2 \rho_{\mathrm{r.s}}^2, \; p_{\mathrm{z}} = \rho_{\mathrm{r.s}} \Omega^2 r_3 b; \tag{6}$$

where $\rho_{\mathrm{r.s}}$ is the density of the bandage material.

After determining the pressure and mechanical stress in the bandage, checking its mechanical strength is performed according to Eq. [5]:

$$p_{\mathrm{N}} = p_{\mathrm{p}} + p_{\mathrm{t}} - p_{\mathrm{z}} > 0, \; k_{\sigma} \left(\sigma_{\mathrm{p}} + \sigma_{\mathrm{t}} + \sigma_{\mathrm{z}} \right) < \sigma_{\mathrm{M}}. \tag{7}$$

3. Shaft materials of the high-speed EMs

The shaft of the high-speed EM provides the necessary rotor stiffness, its mechanical strength, and magnetic properties. The magnetic field of the HCPM should not be closed through bearings to avoid the occurrence of bearing currents. Therefore, it is advisable to use a prefabricated shaft with a back made of magnetic material and a shaft made of a nonmagnetic material. The use of fully nonmagnetic shafts is recommended only for the combined and Halbach magnetic systems. In all other cases, the use of a nonmagnetic shaft will reduce the magnetic flux density in the air gap by 15–25%. For fully magnetic shafts, it is necessary to insulate the bearings. For the nonmagnetic part of the shaft, either titanium or Inconel 718 is usually used.

4. Stator magnetic core materials of the high-speed EMs

Prospects for the use of a particular material of the stator magnetic core in high-speed EMs are determined by the minimum losses, mass and size parameters of the stator magnetic core and EM, maximum EM power and maximum magnetic flux density in the air gap, cost of the stator magnetic core, and manufacturability. Thus, the selection task of the optimal stator magnetic core material is the task of multicriteria optimization. As a selection criterion, the mass and size EM parameters, magnetic flux density, efficiency, manufacturability, and cost of the of the stator magnetic core are used. Since each of these criteria can be represented as a function of magnetic flux density in the stator back and teeth, it is possible to formulate a mathematical function of the stator magnetic core material selection. By using this function for the specified maximum values of the specific losses in the stator back and teeth, the cost of the stator magnetic core, production technology, and effectiveness of a given material can be numerically justified. Taking into account that this is an optimization task, it should be solved after preliminary EM calculations, which will determine the geometric dimensions of the stator magnetic core and electromagnetic loads (linear current load, magnetic flux density in the air gap, etc.). All the aforementioned criteria should be presented as functions of the steel properties. For example, such dependence for magnetic flux density in the air gap is determined from the magnetic circuit calculation:

$$F_M = 2F_\delta + 2F_z + F_j + F_r, \tag{8}$$

where $F_M = H_M l_M$ is the magnetomotive force (MMF) of the HCPM; $H_M l_M$ is a product of the magnetic field strength in the HCPM by the length of the average magnetic line in the HCPM; F_δ is the MMF of the air gap; F_j is the MMF of the stator back; F_z is the MMF of the stator teeth; and F_r is the MMF of the rotor back and shaft.

Eq. (15) is represented in the form of the MMF dependences on the magnetic flux density and the magnetic field strength:

$$F_M = 2B_\delta k_\delta \delta + 2H_z h_z + H_j h_j + H_r h_r, \tag{9}$$

where B_δ is a magnetic flux density in the air gap; δ is an air gap; $H_z = B_z/\mu_s\mu_0$ is the magnetic field strength in the stator teeth; $H_j = B_j/\mu_s\mu_0$ is the magnetic field strength in the stator back; $H_r = B_r/\mu_r\mu_0$ is the magnetic field strength in the rotor back; h_z, h_j, h_r are the height of the stator tooth, stator back, and rotor back, respectively; and k_δ is an air gap coefficient.

One of the features of the EM with HCPM is that it is necessary to perform a check of the magnetic circuit saturation in the idle mode, since under load, the armature reaction magnetic field and the HCPM heating reduce the magnetic flux density in the air gap but do not demagnetize the HCPM. Therefore, in selection of the stator magnetic core material of the high-speed EM, the MMF of the HCPM can be considered as constant. The dependence of the magnetic flux density in the air gap can be represented as a function of magnetic flux density in the stator back and in the stator teeth according to Eq. (16).

The mass of the stator magnetic core is determined by the magnetic flux density in the stator back and teeth, as well as the density of the stator magnetic core material, and can be represented as a function of these parameters. The efficiency is determined by the specific losses of the stator magnetic core material, which can be approximated as a function of frequency, empirical coefficients, and magnetic flux density in the stator back and teeth. The cost of the stator magnetic core is determined by the total mass of the material used in its manufacture (taking into account the mass of waste during stamping or cutting), as well as the manufacture complexity. The task of the stator magnetic core material selection can be represented as a mathematical function:

$$\begin{cases} B_\delta(B_z, B_j) \rightarrow \max \\ M_M(B_z, B_j) \rightarrow \min \\ \eta(B_z, B_j) \rightarrow \max \\ C(B_z, B_j) \rightarrow \min \\ M_{waste}(B_z, B_j) \rightarrow \min \end{cases}, \tag{10}$$

where $B_\delta(B_z, B_j)$ is a magnetic flux density in the air gap; $M_M(B_z, B_j)$ is the EM mass; $\eta(B_z, B_j)$ is the EM efficiency; $C(B_z, B_j)$ is the cost of the stator magnetic core; and $M_{waste}(B_z, B_j)$ is a mass of the waste.

Thus, after determining the maximum permissible values of the magnetic flux density for the stator teeth and back for a specific stator magnetic core material and approximating the specific loss function, the value of each criterion included in Eq. (17) can be determined. Then, a comparative multicriteria analysis of the material expediency and the optimal selection of the stator magnetic core material can be made.

To reduce the number of compared criteria, Eq. (17) can be reduced to a task with one criterion by determining the weight of each function in Eq. (17):

$$f(B_z, B_j) = \lambda_1 B_\delta(B_z, B_j) - \lambda_2 M_M(B_z, B_j) + \lambda_3 \eta(B_z, B_j) - \lambda_4 C(B_z, B_j) - \lambda_4 M_{waste}(B_z, B_j) \rightarrow \max, \tag{11}$$

where $\lambda_1 \ldots \lambda_5$ is, respectively, the weight of each criterion in the overall efficiency function of using a particular stator magnetic core material.

In conversion from the multicriteria comparison task to the function of one variable, which characterizes the effectiveness of the material application, it is necessary to replace the dimensional units in Eq. (17) by the relative units:

$$y^* = \frac{Y(B_z, B_s)}{Y_m(B_z, B_s)}, \tag{12}$$

where y^* is a criterion in relative units; $Y(B_z, B_s)$ is a criterion in dimensional units; and $Y_m(B_z, B_s)$ is a base value in dimensional units.

Since the efficiency of using a known electrotechnical steel or a precision soft magnetic alloy is determined, the base values are parameters in Eq. (18) at a saturation magnetic flux density for a given electrical steel or a precision soft magnetic alloy:

$$\begin{gathered} Ef(B_z, B_j) = -\lambda_1 \frac{B_\delta(B_z, B_j)}{B_\delta(B_{zn}, B_{jn})} + \lambda_2 \frac{M_M(B_z, B_j)}{M_M(B_{zn}, B_{jn})} - \\ -\lambda_3 \frac{\eta(B_z, B_j)}{\eta(B_{zn}, B_{jn})} + \lambda_4 \frac{C(B_z, B_j)}{C(B_{zn}, B_{jn})} + \lambda_4 \frac{M_{waste}(B_z, B_j)}{M_{waste}(B_{zn}, B_{jn})} \rightarrow \min, \end{gathered} \tag{13}$$

where $Ef(B_z, B_j)$ is the efficiency of stator material use.

Thus, the general condition for the stator magnetic core material selection can be formulated as follows. From two stator magnetic core materials, more effective is one with less $Ef(B_z, B_j)$ for the same $\lambda_1 \ldots \lambda_4$. For practical application of the proposed technique, different stator magnetic core material types of the high-speed EMs are considered.

4.1. Electrical steels and soft magnetic alloys

Figure 5 shows the dependences of the specific losses on frequency for various electrical steels. In the high-speed EM, silicon electrical steels are widely used. The main advantages of electrical steels are the simple production technology of the stator magnetic core of the high-speed EM. The disadvantage is a low-saturation magnetic flux density, which leads to an increase in the mass and size parameters of the EM. Higher saturation magnetic flux density has soft magnetic alloys based on cobalt or nickel, for example, VACOFLUX 50, VACOFLUX 48, VACODUR S Plus, VACODUR 50, VACODUR 49, VACOFLUX 18 HR, and VACOFLUX 17. The saturation magnetic flux density for these types of alloys varies from 2.14 to 2.35 T. **Figure 6** shows the magnetization curves for VACOFLUX 48 and VACODUR 49 alloys. The soft magnetic alloys with high saturation induction have significant specific losses that can be reduced by reducing the sheet thickness. In this case, even with a sheet thickness of 0.1 mm, it is inferior to electrical steels. In addition, soft magnetic alloys have a complex production technology and require special treatment. Therefore, these alloys predominantly find wide application in EM with magnetization frequencies below 400 Hz. At higher frequencies, the use of this alloy type requires justification.

4.2. Amorphous magnetic materials (AMMs)

AMM is a kind of alloy, which lacks the periodicity in the atom arrangement. AMMs are produced in the form of thin tape. The thickness of the tape is 25–30 μm.

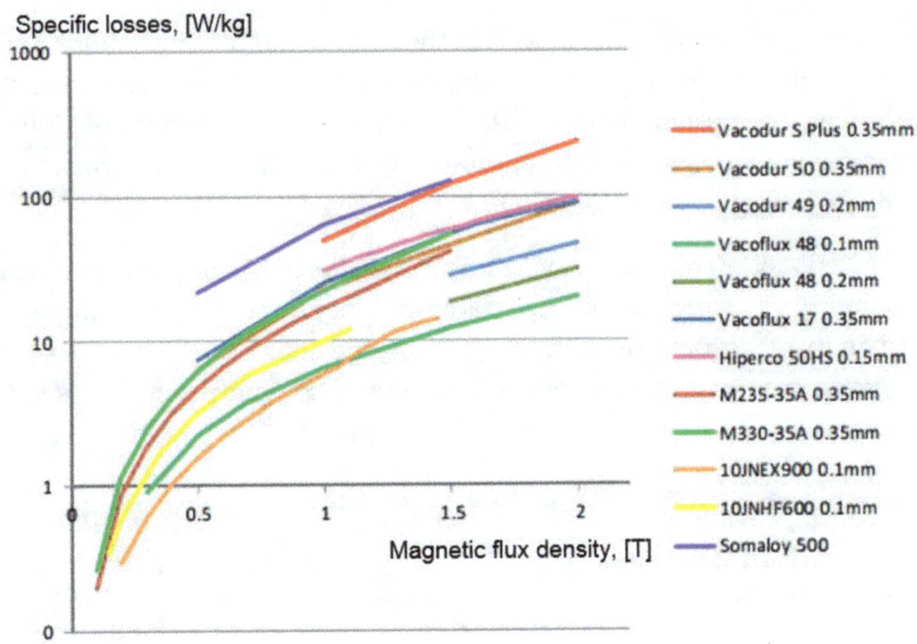

Figure 5. The dependences of the specific losses on frequency for various electrical steels.

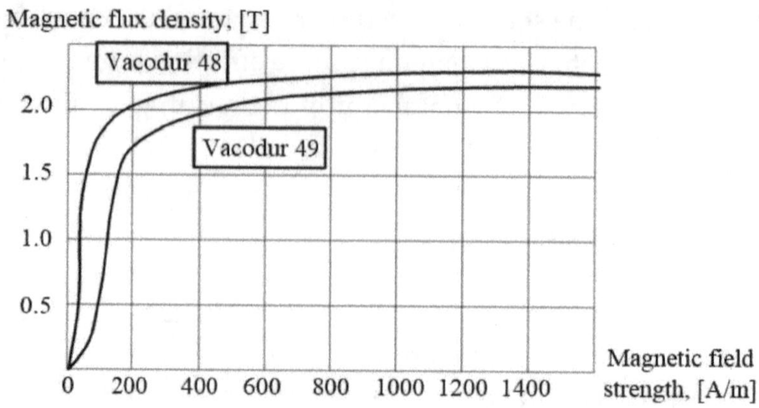

Figure 6. Magnetization curves for VACOFLUX 48 and VACODUR 49 alloys.

The AMM technology was developed in 1960 by a research group of the California Institute of Technology. In 1988, Hitachi Metals first developed a nanocrystalline alloy (metal glass).

The composition of the AMM includes two groups of elements: transition metals (Fe and Co) and amorphous elements (B, C, and Si). The amorphous alloy structure is obtained only at a certain cooling rate. In the AMM structure, there is no periodicity in the atom arrangement. Despite the fact that the AMM density is lower than the density of crystalline analogs, their strength is higher by 400–900% [6]. The higher strength is because the AMMs have no defects such as dislocations and grain boundaries inherent in the crystalline state. The presence of such defects greatly complicates the diffusion (penetration of atoms) through amorphous metallic layers.

The AMM advantages are high wear resistance, high corrosion resistance, high electrical resistivity (ϱ = 1–1.5 $\Omega\cdot mm^2$), low coercive force Hc (less than 8 A/m), high values of magnetic permeability (initial magnetic permeability μ_{in} is up to 8000, and maximum magnetic permeability μmax is up to 300,000–700,000), and low specific losses.

The AMM disadvantage is that it is produced in the form of a thin tape. And it is impossible to make a stator magnetic core by using traditional design technologies. Therefore, slotless stator magnetic core designs or designs with a large area and a small number of slots are used. In addition, AMMs has low saturation induction. For example, the maximum saturation induction value is 1.8 T for Metglas Alloy 2605CO.

Nevertheless, the development of the AMM production technologies makes it possible to talk about their prospects of using in the EM stator magnetic cores. For example, Hitachi has already created an induction motor with the energy efficiency class IE 4 with an AMM stator magnetic core. Radam and Celeroton also actively create the high-speed EMs with the AMM stator magnetic core.

The main AMM features are associated with several facts:

- The fill factor of the AMM stator magnetic core varies from 0.7 to 0.8 because in the AMM production, various defects are formed on the tape.
- AMM tape thickness does not exceed 25–30 μm; therefore, AMM magnetic cores are usually wound.

- AMM has a low saturation magnetic flux density, which usually varies from 1.3 to 1.6 T. For the cobalt-doped AMM, the magnetic flux density increases to 1.8 T. However, the specific losses also increase due to the increase in electrical conductivity.
- The AMM layers in the magnetic cores are not isolated from each other.

These features do not allow the use of known designs in the EMs with the AMM magnetic cores. Therefore, further, it seems appropriate to consider in more detail the AMM magnetic core designs and their technological approaches.

It is promising to use AMM in the high-speed EMs with a rotational speed of below 120,000 rpm and in the ultra-high-speed EMs with a rotational speed above 1,200,000 rpm. For the high magnetization reversal frequency, AMMs are practically the only correct technical solution. In micro-EM with rotational speeds above 700,000 rpm and magnetic flux density in the air gap of 0.2–0.25 T, the use of ferrites is effective, but a low saturation magnetic flux density is limited only by this narrow application field.

Due to the technological complexity of the AMM stator cores, several designs are used:

- EM with a tooth-coil stator magnetic core made of AMM, which forms from U-shaped AMM cores (**Figures 7–9**). The disadvantage of these cores is the technological complexity, and they are mainly performed with a large area and a small number of slots. The problem solution of a small number of slots can be the use of a sectional design of the EM with a stator magnetic core made of AMM (**Figure 10**).
- EM with prefabricated designs of the AMM stator magnetic cores. These designs are characterized by the fact that the magnetic core parts are made separately from simple geometric shapes, after which they are assembled. This technology is of particular interest for axial EM. Various designs of these magnetic cores are shown in **Figure 11**.

In addition, individual geometric figures can be performed not wound but by typing from plates. Further, these sets are inserted into the stator. For example, in [8], a radial EM for this technology is proposed (**Figure 12**).

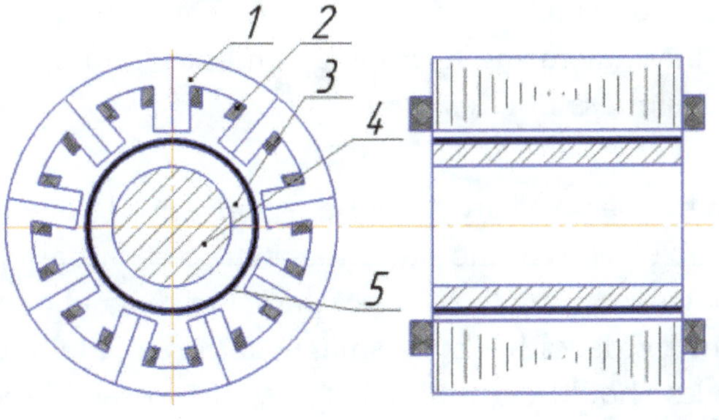

Figure 7. EM with a tooth-coil winding and a stator magnetic core made of AMM: 1 is a stator magnetic core; 2 is a tooth-coil winding; 3 is the HCPM; 4 is a shaft; and 5 is a rotor sleeve.

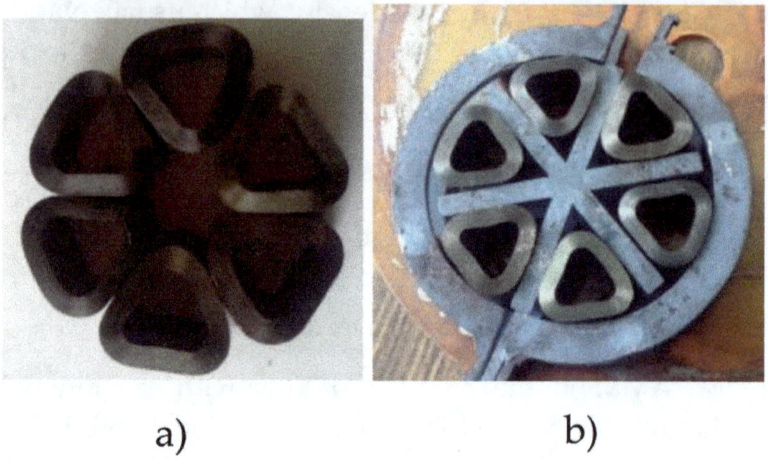

Figure 8. Stator magnetic core made of AMM (a) and its formation technology (b).

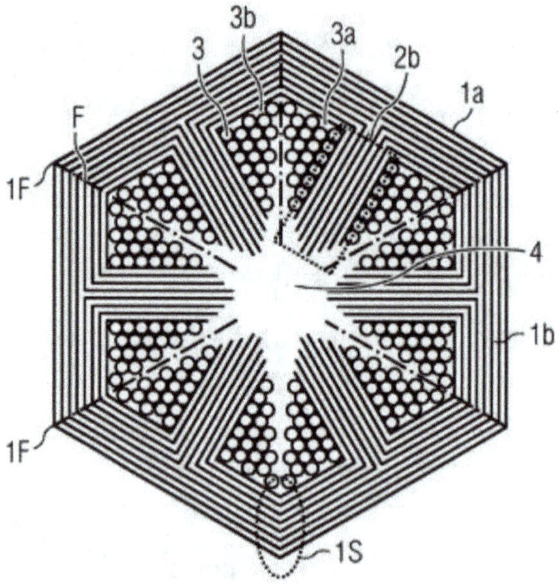

Figure 9. EM with a tooth-coil winding and a stator magnetic core made of AMM [7].

- Slotless EM with the AMM stator magnetic core. This type of AMM magnetic cores is the simplest in the manufacturing technology and has a minimum manufacturing cost. **Figure 13** shows the slotless EM designs.

Special advantages of the slotless EM are for various high-speed systems. The use of the AMM allows not only to minimize losses in the stator magnetic core but also to increase the number of poles to 8–10. This allows reducing the mass of the stator magnetic core of the slotless EM by 20–25%. In the high-speed EMs with a stator magnetic core made of electrical steel or Co-Fe alloy, the number of poles should be minimal. To reduce losses in the stator magnetic core, 2-or 4-pole EMs are used, whose stator magnetic core mass is higher than the stator magnetic core mass of multipolar EMs.

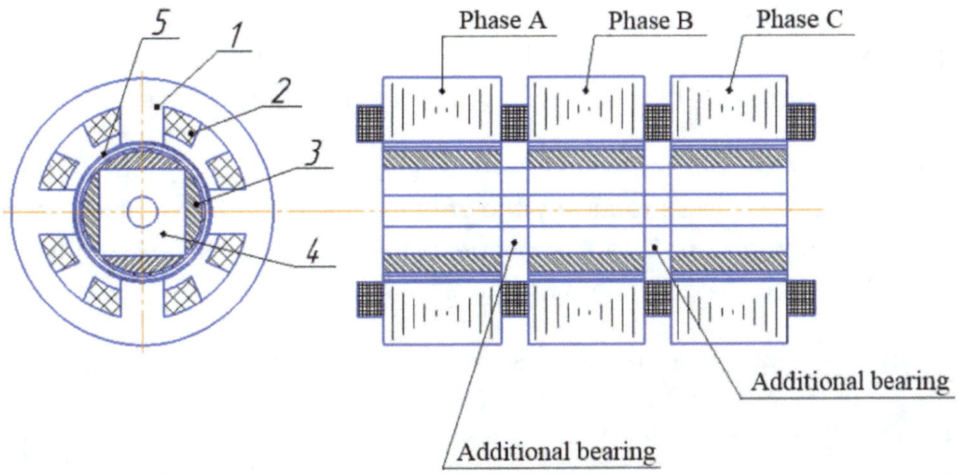

Figure 10. Sectional EM with a tooth-coil winding and a stator core made of AMM: 1 is a stator magnetic core; 2 is a tooth-coil winding; 3 is the HCPM; 4 is a shaft; and 5 is a rotor sleeve.

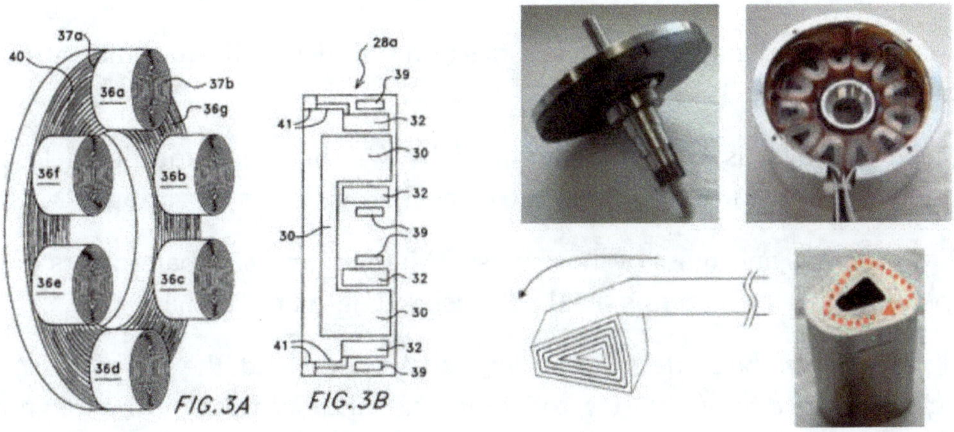

Figure 11. Prefabricated design of the AMM stator magnetic core.

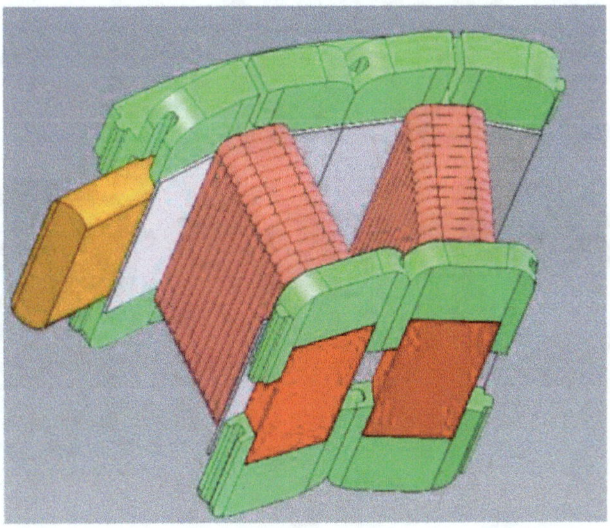

Figure 12. Assembly of the stator magnetic core made of AMM [8]

Another advantage of the slotless EM is the use of magnetic bearings in Lorentz forces, which are integrated directly into the air gap. This reduces the axial length and, thus, provides the necessary rotor dynamics and stiffness in comparison with other designs.

In addition, the slotless EM has the simple technology of producing a slotless stator magnetic core, which makes this EM class practically the only technical solution that allows the creation of ultra-high-speed micromotors and microgenerators. The slot absence results in minimization losses in the HCPM caused by tooth harmonics.

The main disadvantage of the slotless EM is that the entire winding is located in the air gap, and this leads to its significant increase. At the same time, losses to the skin effect and eddy currents in strands appear in the winding, as the entire magnetic field in the air gap crosses the winding.

The efficiency of the slotless EM with HCPM is largely determined by the number of poles, winding type, and also manufacturing method of the slotless stator magnetic core (wound or stamped). The winding type of the slotless EM with HCPM determines:

- the air gap value and energy characteristics of the slotless EM, since the entire winding is located in the air gap;
- the efficiency, which is largely determined by winding and additional losses (loss of proximity effect and eddy currents), as well as thermal loads on various EM elements;
- overall dimensions and mass of the slotless EM: mass and size parameters of the slotless EM depend on the end winding axial protrusion length and its mass);
- working capacity of the slotless EM under load and overload: the structural scheme of the winding determines the scattering resistance (xs) and accordingly the magnitude of the armature reaction demagnetizing field;
- a structural diagram of the power electronics used in conjunction with the slotless EM; the harmonic composition of the voltage determines the size of the filters and depends on the winding type).

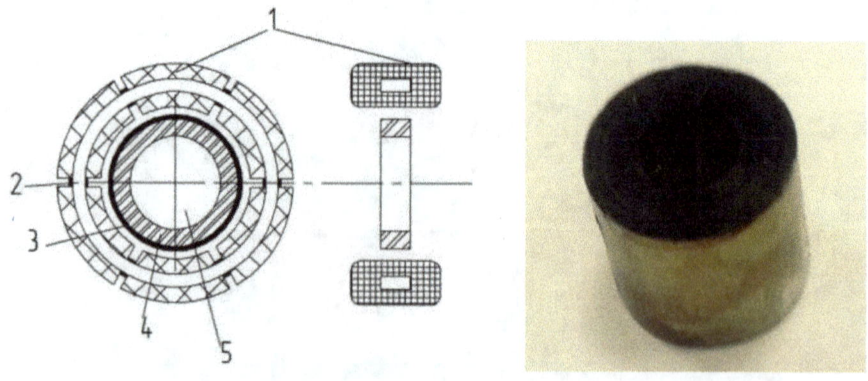

Figure 13. The slotless EM with the AMM stator magnetic core: 1 is a winding; 2 is a stator magnetic core; 3 is a rotor sleeve; 4 is the HCPM; and 5 is a shaft.

The manufacturing technology of the stator magnetic core determines:

- stator magnetic core losses; and
- the technological complexity of its manufacture, as well as the mass and size parameters of the slotless EM, which are determined by the fill factor of the stator magnetic core.

Thus, all the main parameters, which characterize the efficiency of the slotless EM, depend on the winding type, number of poles and stator magnetic core design. Therefore, the selection of these elements should be made with particular accuracy.

Typically, either a distributed winding or a toroidal winding is used in the slotless EM. Each of these windings has its disadvantages. The distributed winding is characterized by technological complexity of manufacturing, and the toroidal winding provides higher losses in copper due to the inability to use half of the winding. The possibility of using a tooth-coil winding in the slotless EM, and also a comparison of the effectiveness of a tooth-coil winding and a toroidal one in a two-pole and 10-poles slotless EM is considered in this work.

Two-pole slotless EM with a tooth-coil winding and its connection scheme is presented in **Figure 14a** and **b** accordingly. Ten-pole slotless EM with a toroidal winding and its connection scheme is presented in **Figure 14c** and **d** accordingly.

The tooth-coil winding allows achieving a minimum end winding axial protrusion length, and its manufacturing technology is simpler than the toroidal winding. At the same time, due to the MMF spatial harmonic, the tooth-coil winding should have a distorted voltage shape, which will lead to additional losses in the HCPM.

To evaluate the effectiveness of the design, four experimental slotless EMs were designed and manufactured: a 2-pole and a 10-pole slotless EMs with a HCPM and a toroidal winding, and a 2-pole and a 10-pole slotless EMs with a HCPM and a tooth-coil winding (**Figure 15**). To simplify the laboratory studies, the experimental models were performed with a low power for a reduced rotational speed of 3000 rpm, which also simplifies the requirements for bearings. The stator magnetic core of prototypes was low saturated. Thus, four stators and two rotors were made.

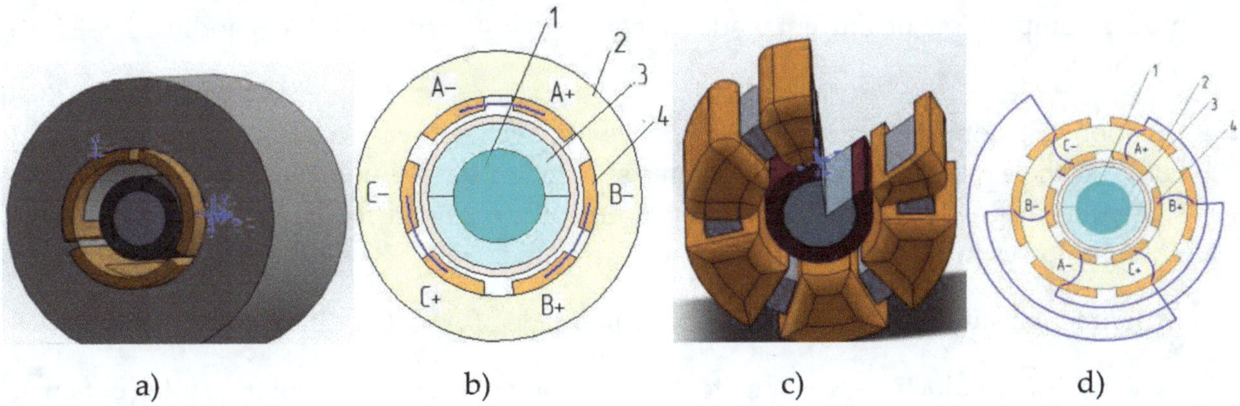

Figure 14. Slotless EM with a concentrated winding (a) and its connection scheme (b) and slotless EM with a toroidal winding (c) and its connection scheme (d): 1 is a rotor core; 2 is a slotless stator magnetic core; 3 is the HCPM; and 4 is a winding.

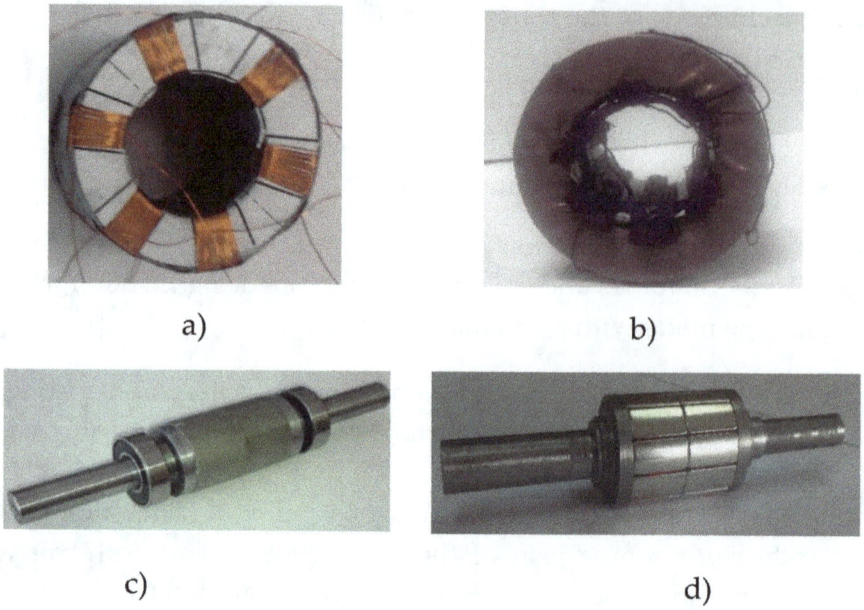

Figure 15. (a) The stator core with toroidal winding; (b) the stator core with concentrated winding; (c) 2-pole rotor; and (d) multi-pole rotor.

Analysis of various winding types of slotless EMs with HCPM can be made using only computer simulation methods, but at the same time, it does not allow to assess the technological manufacturing complexity. From the known works, it can be concluded that the main operating mode of the slotless EM with HCPM is the generator mode. One of the main characteristics of the generator mode for aerospace EMs is the voltage dependence on the load current, which is determined by the inductive resistance of the winding. According to the experimental results, four models and four computer models were developed in the Ansys Maxwell software package, and the simulation and experimental results are compared.

Therefore, the procedure for the design selection was as follows:

1. Design calculation and creation of experimental prototypes and its research under load in the generator mode. An external characteristic, oscillograms of current and voltage, harmonic composition of current and voltage, and estimation of the winding losses are obtained.

2. Creation of a computer model in the Ansys Maxwell software package. The discrepancy between the experimental and computer simulation data should not exceed 5%. At this stage, the magnetic field distribution patterns are obtained for each considered slotless EM, and losses in permanent magnets are estimated.

3. Experimental studies of losses in a stator core made of AMM at different frequencies.

4. Selection of topology. According to the experimental and computer results obtained, different topologies of slotless EMs with HCPM are evaluated.

As a research result, it was established that the two-pole slotless EM with a tooth-coil winding has the less steep external characteristic and the voltage decreasing is below

10–15% at a load. The steepest external characteristic is in 10-pole slotless EM with a tooth-coil winding. The open circuit voltage is different from the voltage at one-and-a half overload by 210%. This is explained by the fact that with increasing number of poles, the inductive scattering resistance increases significantly; thus, the multi-pole slotless EM with HCPM is essentially a high-reactance permanent magnet EM. This EM type is most applicable in aerospace engineering, since it provides the most reliable short-circuit protection of the slotless EM and the system in which it is installed. Thus, from the analysis of the experimental external characteristics, it can be concluded that the 10-pole slotless EM with a HCPM and a tooth-coil is most effective.

4.3. Soft magnetic composite materials (Somaloy)

Somaloy is a soft magnetic composites produced by Höganäs in the powder form for sintering and pressing. The advantage of these materials is the nonwaste-producing technology. The technological process of producing magnetic cores has a certain complexity. **Figure 16** shows the magnetization curves for 1P Somaloy 500P and 1P Somaloy 700P. **Table 3** presents characteristics of different soft magnetic composite materials. It can be seen that soft magnetic composite materials have significant specific losses, which limits their use in high-speed EMs. Due to the high adaptability of this material, it can find wide application in the EM with low rotational speeds.

4.4. Ironless stator magnetic cores

The ironless stator magnetic cores are magnetic cores made of nonmagnetic nonconductive materials. The ironless stator magnetic cores are used in superconducting EMs when significant magnetic flux density in the air gap leads to a supersaturation of the magnetic core. In addition, the ironless stator magnetic cores can find application in ultra-high-speed EMs with rotational speed above 1,000,000 rpm.

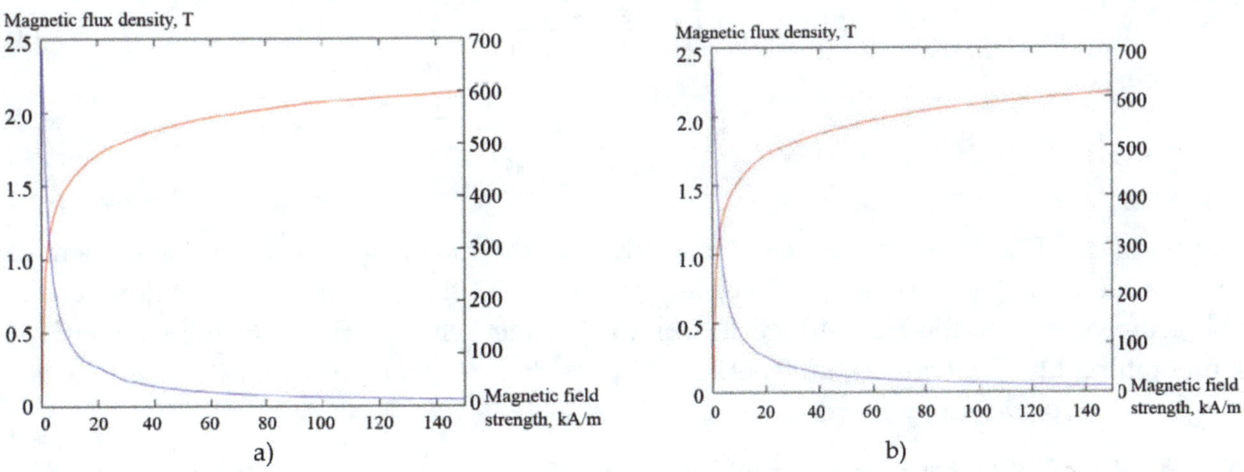

Figure 16. The magnetization curves for 1P Somaloy 500P (a) and 1P Somaloy 700P (b).

Material type	Magnetic flux density at a magnetic field strength of 10,000 A/m (T)	Specific losses (W/kg)			
		5 × 5 mm			15 × 15 mm
		100 Hz	400 Hz	1000 Hz	1000 Hz
1PSomaloy 130i	1.4	12	54	145	147
1PSomaloy 700	1.56	10	44	131	158
1PSomaloy 700 HR	1.53	10	46	134	145
3PSomaloy 700	1.61	10	46	137	189
3PSomaloy 700 HR	1.57	11	48	137	157
3P Somaloy 1000	1.63	10	46	144	287
5PSomaloy 700 HR	1.57	6	32	104	115

Table 3. Characteristics of different soft magnetic composite materials.

5. Permanent magnet types

To select permanent magnets of high-speed EMs, the demagnetization curve of permanent magnets at different temperatures (**Figures 17** and **18**), residual magnetic flux density (Br), coercive force by magnetic flux density (HcB) and magnetization (Hcj), maximum operating temperature, Curie temperature, temperature coefficients by magnetic flux density and coercive force were used.

5.1. Temperature coefficients by magnetic flux density and coercive force

Under the temperature influence, the coercive force and the residual magnetic flux density of the permanent magnets decrease (**Figure 18**). It is analytically determined as follows:

$$B_r(\Theta) = B_r\left(1 - \frac{k_{Br}(\Theta_{ВПМ} - 20)}{100}\right), \tag{14}$$

$$H_c(\Theta) = H_c\left(1 - \frac{k_{Hc}(\Theta_{ВПМ} - 20)}{100}\right), \tag{15}$$

where $B_r(\Theta)$, $H_C(\Theta)$ are, respectively, the values of the magnetic flux density and coercive force of the HCPM; B_r, H_C are, respectively, the values of the residual magnetic flux density and coercive force of the HCPM, indicated in the technical characteristics; $\Theta_{ВПМ}$ is the HCPM temperature; kBr is a temperature coefficient by residual magnetic flux density; and kHc is a temperature coefficient by coercive force.

In Eqs. (20) and (21), temperature coefficients are used constant, while the temperature coefficients of the HCPM are constant only at temperatures below 100–120°C. At higher temperatures,

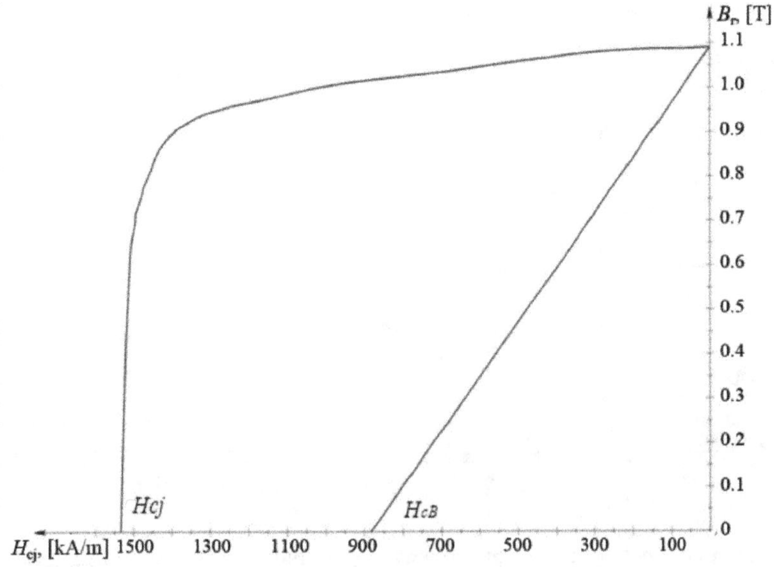

Figure 17. The demagnetization curve of Sm_2Co_{17} permanent magnets.

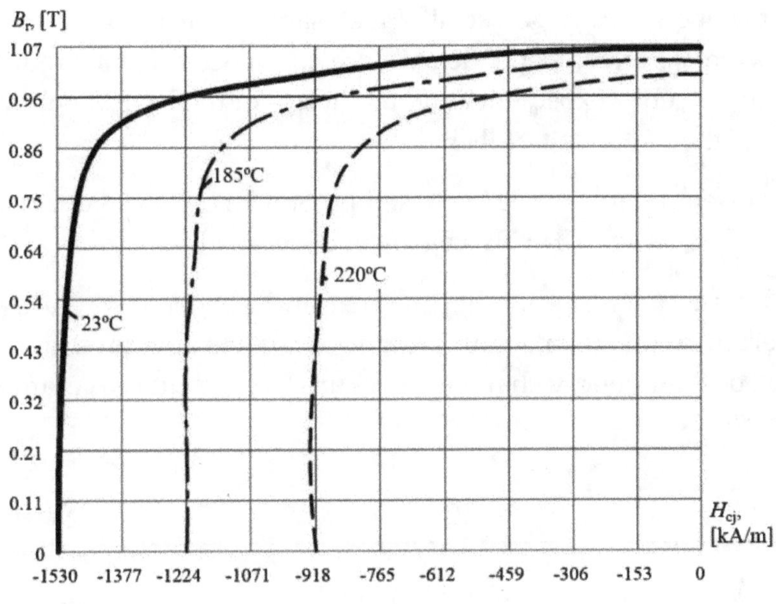

Figure 18. The demagnetization curve of Sm_2Co_{17} at different temperatures.

coefficients vary with temperature. Thus, the temperature characteristic of the HCPM demagnetization is nonlinear at temperatures above 100–120°C.

In modern high-speed EMs, several HCPM types are used: SmCo, SmCoFeCuZr, and NdFeB alloys. The main advantage of the HCPM based on SmCo and SmCoFeCuZr is their high temperature stability as well as corrosion resistance compared to NDFeB. The SmCo Curie temperature is above 800°C; the temperature coefficients by induction and coercive force are −0.04 and −0.3, respectively. The Curie temperature of NdFeB is above 300°C, and temperature coefficients by induction and coercive force are −0.11 and −0.6, respectively. NdFeBs have higher energy characteristics than SmCo by 20–30%. Nevertheless, their low temperature stability does not allow the efficient use of these high-energy characteristics. **Figure 19** shows

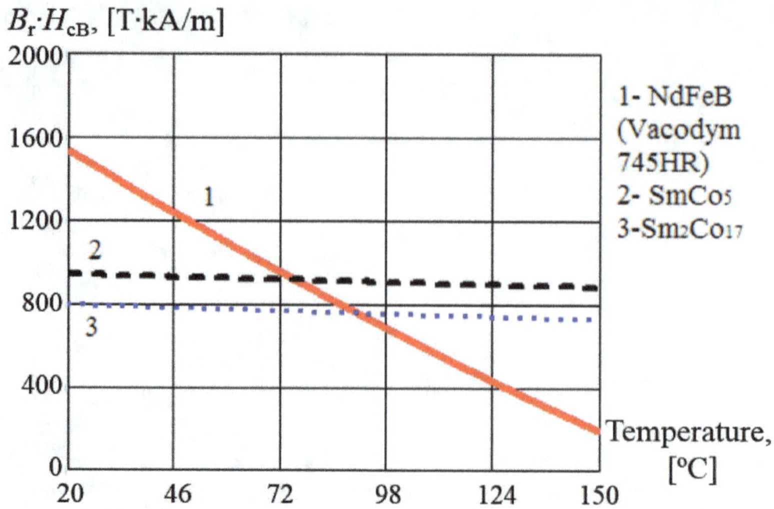

Figure 19. Characteristic dependences of SmCo, SmCoFeCuZr, and NdFeB at various temperatures.

the characteristic dependences of SmCo, SmCoFeCuZr, and NdFeB at various temperatures. For comparison, the most energetically powerful NdFeB sample Vacodym 745 HR with Br = 1.44 T, HcB = 1065 kA/m were selected. Dependencies are constructed for a product $B_r \cdot H_{cB}$. It was found that the use of NdFeB is expedient at temperatures of the HCPM below 60–70°C. **Table 4** presents properties of the SmCo and NdFeB.

In addition, the high-temperature HCPMs based on SmCoFeCuZr alloys with operation temperatures up to 700°C (T 500 and T 550) have been developed.

In the HCPM manufacture, resin-bonded magnet and rubber ferrite are often used. They have higher temperature coefficients than ordinary magnets and have mechanical properties similar to plastics, but their energy characteristics are lower than the ceramic or metal HCPM obtained by sintering.

Property	NdFeB	$SmCo_5$	Sm_2Co_{17}
Density (g/cm^3)	7.4	8.3	8.4
Young's modulus (GPa)	150	—	—
Cross-breaking strength (MPa)	210–290	180	150
Rupture strength (MPa)	70–130	40	35
Vickers hardness (HV)	600	450–500	500–600
Electrical resistance (Ohm/m)	$1.6 \cdot 10^{-6}$	$5 \cdot 10^{-7}$	$8.6 \cdot 10^{-7}$
Specific heat (J/kg · K)	440	370	340–390
Thermal conductivity (W/m · K)	9	12	13
The temperature coefficient of linear expansion along the magnetization direction (10^{-6}/K)	5	6	8
Temperature coefficient of linear expansion across the magnetization direction (10^{-6}/K)	−1	13	11

Table 4. Mechanical, electrical, and thermal of the SmCo and NdFeB.

5.2. Climate testing and long-term storage of various HCPM types

Results of corrosion and climate tests of Sm_2Co_{17}, $SmCo_5$, and NdFeB are presented in [9]. It is shown that Sm-Co alloys are susceptible to corrosion both during storage and in a heated storage room, but their magnetic properties change insignificantly (~ 1–2%). The greatest changes both in magnetic properties and in corrosion occur in the first storage year, and SmCo5 alloys are more resistant to long-term storage than Sm_2Co_{17} alloys. In addition, NdFeB is significantly susceptible to corrosion and practically has not been tested for long-term storage.

5.3. Radiation resistance of HCPM

For EM with HCPM for special areas of industry, including the space and nuclear industries, an important criterion in the HCPM selection is their resistance to radiation exposure. In [10], studies of the radiation resistance of Sm_2Co_{17}, $SmCo_5$, and NdFeB are presented. It was established that the radiation resistance of the SmCo is above 1018 neutrons/cm^2 and the radiation resistance of the NdFeB is lower. The magnetic property loss of NdFeB in radiation irradiation is in most cases caused not by the weak stability of NdFeB to neutron radiation but by their radiation heating. Therefore, it is not recommended to use NdFeB in EMs, which can be exposed to significant radiation exposure.

6. Conclusion

In this chapter, materials used in high-speed EMs with HCPM are presented. The application areas of high-speed EMs are shown. Practical recommendations for the selection of active and structural components of high-speed EMs are given. The mechanical strength calculation of the rotor sleeve is shown. The obtained results can be used in practice in the EM design and in the further research.

Author details

Flyur R. Ismagilov, Viacheslav Ye. Vavilov* and Valentina V. Ayguzina

*Address all correspondence to: s2_88@mail.ru

Ufa State Aviation Technical University, Ufa, Russia

References

[1] Dr. Ralph Funck Composite Materials in High Efficient Sleeve Applications of Electric Machines [Internet]. Available from: http://www.circomp.de/de/downloads/CirComp-high-efficient-sleeveapplications.pdf [Accessed: March 28, 2018]

[2] Kruchinina IY, Antipov VI. Problematic issues of creation high-speed mini-turbogenerators and ways of their decision (in Russian). Management Information Systems. 2012;**4**:25-34

[3] Ledovsky AN. Electrical Machines with High-coercive Permanent Magnets (in Russian). Vol. 169. Moscow: Energoatomizdat; 1985

[4] Golovanov DV, Kovarsky ME, Magin VV, Trunov IG. Methods of calculation of high-speed generators for gas-turbines. Questions of Electromechanics, 2012;**126**:3-8

[5] Tao Z, Ye X, Huiping Z, Hongyun J. Strength design on permanent magnet rotor in high speed motor using finite element method. Telkomnika Indonesian Journal of Electrical Engineering. 2014;**12**(3):1758-1763

[6] Zwyssig C, Kolar JW, Round SD. Mega-speed drive systems: Pushing beyond 1 million RPM. IEEE/ASME Transactions on Mechatronics. 2009;**14**(5):564-574

[7] Borisavljevic A, Polinder H, Ferreira JA. Enclosure design for a high-speed permanent magnet rotor. Power Electronics, Machines and Drives (PEMD);**2**:817-822

[8] McPherson MW, Hirzel AD. Stator used in an electrical motor or generator with low loss magnetic material and method of manufacturing a stator. Patent US2013278103

[9] Tyutnev AP, Sergeyev VV, Semyonov VT, Stanolevich GP. To a question of radiation firmness of permanent magnets on the basis of rare-earth elements (in Russian). Electromechanics Questions. 2013;**133**(2):18-27. VNIIEM

[10] Stanolevich GP, Fedoseev NV, Timakov SA, Borisov SL. Stability of rare-earth SmCo magnets to long storage (in Russian). Electromechanics Questions. 2014;**139**(2):29-34. VNIIEM

2

Designing a DC Motor Simulator Based on Virtual Instrumentation

Nicolae Patrascoiu

Abstract

This chapter proposes a state-space model for the DC motor built for separately excited voltage and considering two inputs: supply voltage and load torque. The three states of the resulted model are represented by angular speed, angular displacement, and current supply, and either of these states can be an output variable for the simulation model. Consequently, the system's model has two inputs and three outputs. Using this model, LabVIEW functions and programming structure of a simulator based on the virtual instrument is built, through which it is possible to observe the dynamic characteristics of the DC motor in different operating conditions. In this way, students can verify, by simulation, the operation of the DC motor as a dynamic system observing and measuring its reaction in different operating conditions.

Keywords: modeling, simulation, human-computer interface, front panel, diagram block virtual instrument

1. Introduction

One of the most used actuators in control systems is direct current (DC) motor. It is the means by which electrical energy is converted to mechanical energy. DC motors have a high ratio of starting torque to inertia and therefore they have a faster dynamic response. DC motors are constructed using rare earth permanent magnets, which have high residual flux density. As no field winding is used, the field copper losses are zero and hence, the overall efficiency of the motor is high. The speed-torque characteristic of this motor is flat over a wide range, as the armature reaction is negligible. Moreover, speed is directly proportional to the armature voltage for a given torque. The armature of a DC motor is specially designed to have low inertia [1].

The speed of DC motors can be controlled by applying variable armature voltage. These are called armature voltage controlled DC motors. Wound field DC motors can be controlled by either controlling the armature voltage or by controlling the field current [2]. In this chapter, we consider modeling and simulation of an armature controlled DC motor. There are many types of DC motor and their detailed construction is quite complex, but it is possible to derive the equations for a satisfactory dynamic model from basic electromagnetic relationships [3].

The general output variable of this actuator can be angular speed or angular displacement motion, but coupling the motor axle with wheels or drums and cables, translational motion can be obtained.

2. Building the simulation model

The physical model of an armature controlled DC motor [4] is given in **Figure 1**, where:

e_a, i_a—armature supply voltage and current; e_f, i_f—field voltage and current; R_a, L_a—armature winding resistance and inductance; e—back electromotive force (e.m.f.); $\omega(t)$—angular speed; $T_m(t)$—electromagnetic torque; $T_L(t)$—load torque.

When the armature is supplied with a DC voltage of e_a volts, the armature rotates and produces a back e.m.f e_b. The armature current i_a depends on the difference of e_a and e_b, so applying Kirchhoff's law it can obtain:

$$e_a(t) - e_b(t) = R_a \cdot i_a(t) + L_a \cdot \frac{di_a(t)}{dt} \tag{1}$$

The mathematical model of the mechanical system is:

$$T_m(t) = J \cdot \frac{d\omega(t)}{dt} + F \cdot \omega(t) + T_L(t) \tag{2}$$

where $T_m = k_m \cdot i_a$ is the torque T_m produced by the motor, J is the moment of inertia, F is the frictional coefficient and T_L is the load torque applied to the shaft of the motor considered to be time dependent. Based on these relationships considering that the back e.m.f. is proportional to

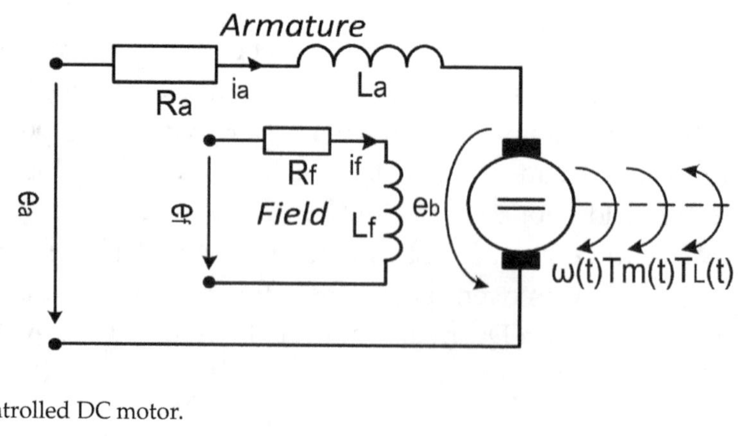

Figure 1. Armature controlled DC motor.

the speed of the motor, i.e., $e_b(t) = k_e \cdot \omega(t)$ and also considering that in SI (commonly used) $k_e \approx k_m = k$, is obtained:

$$\begin{aligned} \frac{di_a(t)}{dt} &= \frac{1}{L_a} \cdot e_a(t) - \frac{R_a}{L_a} \cdot i_a(t) - \frac{k}{L} \cdot \omega(t) \\ \frac{d\omega(t)}{dt} &= -\frac{F}{J} \cdot \omega(t) + \frac{k}{J} \cdot i_a(t) - \frac{1}{J} \cdot T_L(t) \end{aligned} \tag{3}$$

Considering the angular displacement $\alpha(t)$ instead of angular speed $\omega(t)$ as output variable, it is necessary to include the relationship between these:

$$\omega(t) = \frac{d\alpha(t)}{dt} \tag{4}$$

Generally, the mathematical model used is the transfer function [4, 5], but in this chapter it is proposed a Multiple Input Multiple Output (MIMO) mathematical model in a state-space form. Based on Eqs. (3) and (4), this model can build for use into a program simulation.

To build the state-space MIMO model, the input, state, and output vectors are defined [6]:

1. input vector $u(t)$, whose two components are represented by armature voltage $e_a(t)$ and load torque $T_L(t)$, that is,

$$u(t) = \begin{bmatrix} e_a(t) \\ T_L(t) \end{bmatrix} \tag{5}$$

2. state vector $x(t)$, whose three components are represented by armature current $i_a(t)$, angular displacement $\alpha(t)$ and angular speed $\omega(t)$, that is,

$$u(t) = \begin{bmatrix} i_a(t) \\ \alpha(t) \\ \omega(t) \end{bmatrix} \tag{6}$$

3. output vector $y(t)$, whose three components we consider to be the same with state vector components (Relation (8)) so it is possible to simulate these three physical variables.

Having these vectors, it is possible to write Eqs. (3) and (4) in the matrix form:

$$\begin{vmatrix} \dot{i}_S(t) \\ \dot{\alpha}(t) \\ \dot{\omega}(t) \end{vmatrix} = \begin{vmatrix} -\frac{R_a}{L_a} & 0 & -\frac{k}{L_a} \\ 0 & 0 & 1 \\ \frac{k}{J} & 0 & -\frac{F}{J} \end{vmatrix} \cdot \begin{vmatrix} i_S(t) \\ \alpha(t) \\ \omega(t) \end{vmatrix} + \begin{vmatrix} \frac{1}{L_a} & 0 \\ 0 & 0 \\ 0 & -\frac{1}{J} \end{vmatrix} \cdot \begin{vmatrix} e_a(t) \\ T_L(t) \end{vmatrix} \tag{7}$$

Act on systems general equations, this matrix equation can be written in the compact form:

$$\dot{x}(t) = A \cdot x(t) + B \cdot u(t) \tag{8}$$

Bring on output vector definition, the input-output equation can be written in the matrix form:

$$y(t) = \begin{vmatrix} 1 & 0 & 0 \\ 0 & 1 & 0 \\ 0 & 0 & 1 \end{vmatrix} \cdot \begin{vmatrix} i_S(t) \\ \alpha(t) \\ \omega(t) \end{vmatrix} + \begin{vmatrix} 0 & 0 \\ 0 & 0 \\ 0 & 0 \end{vmatrix} \cdot \begin{vmatrix} e_a(t) \\ T_L(t) \end{vmatrix} \tag{9}$$

or in the compact form:

$$y(t) = C \cdot x(t) + D \cdot u(t) \tag{10}$$

Having now the mathematical model, described by matrices A, B, C, D:

$$A = \begin{vmatrix} -\frac{R_a}{L_a} & 0 & -\frac{k}{L_a} \\ 0 & 0 & 1 \\ \frac{k}{J} & 0 & -\frac{F}{J} \end{vmatrix}; \; B = \begin{vmatrix} \frac{1}{L_a} & 0 \\ 0 & 0 \\ 0 & -\frac{1}{J} \end{vmatrix}; \; C = \begin{vmatrix} 1 & 0 & 0 \\ 0 & 1 & 0 \\ 0 & 0 & 1 \end{vmatrix}; \; D = \begin{vmatrix} 0 & 0 \\ 0 & 0 \\ 0 & 0 \end{vmatrix} \tag{11}$$

It is possible to simulate the DC motor. Several studies, such as [7] [8], show that computer simulation enhances learning effectiveness especially in technical disciplines such as studying the behavior of dynamic systems.

3. Designing of virtual simulation tool

A program developed in LabVIEW is called a virtual instrument (VI), and it has two components: the block diagram that represents program itself and the front panel that is the user interface [9].

3.1. Front panel of the virtual instrument

The front panel of the virtual instrument and is built with controls and indicators, which are the interactive input and output terminals. The Controls simulate the instrument input devices and supply data to the block diagram of the VI and the Indicators simulate the instrument output devices and display data that are acquired or generated [10, 11].

The front panel has two components, one is used to set the parameters of the simulation and another is used to collect the results obtained from simulation. The user via the Tab Control button accomplishes switching between the two components.

The first component, called Simulation Process, contains elements (Control type) through which the user can:

- set the motor parameters (DC Motor Parameters): R_a [Ω], L_a [H], k [$N \cdot m/A$], J [$kg \cdot m^2$], F [$(N \cdot m)/(rad \cdot s)$];

- set simulation time:
- choose, from the list, the shapes for the input variables and set their parameters (amplitude and duration);
- choose simulation previews for current and/or displacement and/or speed

An overview of the entire first component of the front panel is shown in **Figure 2**.

This part of the front panel also contains elements of graph indicator type through which a preview of the input and output variables can be displayed.

The second component, called Simulation Results, allows the user to the acquisition of information about the dynamic behavior of DC motor.

This contains Controls through which the user:

- can select the display state-space equations form and select the transfer function, both models are represented by coefficients calculated based on the DC motor parameters sets on the front panel, **Figure 3**; and
- can select and display the graphical results of the simulation, that is, the graphical responses for the imposed outputs shown in **Figure 4**.

For each selected output variable, a window opens, which displays parameters that are relevant to this dynamic system, such as the peak value (maximum value) and time of its occurrence, the steady-state value and its determining time, rise time, falling time, overshoot. For displacement, in case the load torque is greater than the active torque occurs motor reversing, a flashing message (OVER TORQUE) and a proper light indicator, also displaying the time of its occurrence, signal this situation like as shown in **Figure 4**.

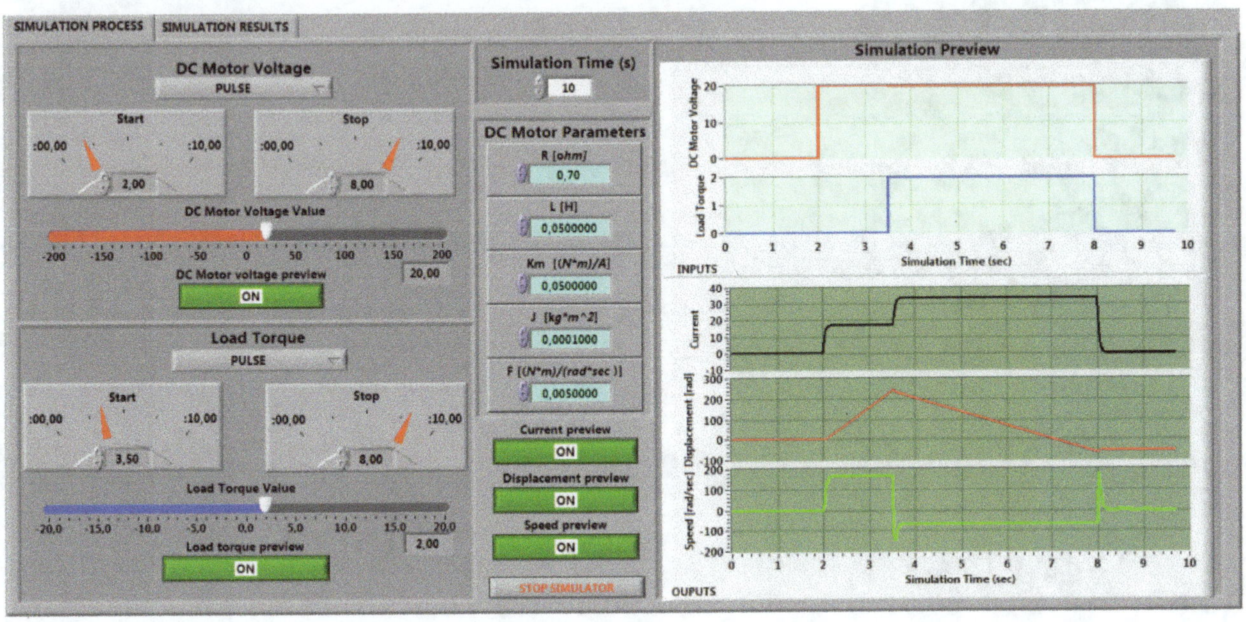

Figure 2. (a) Controls used to set the simulation parameters; (b) indicators used to preview of the simulation.

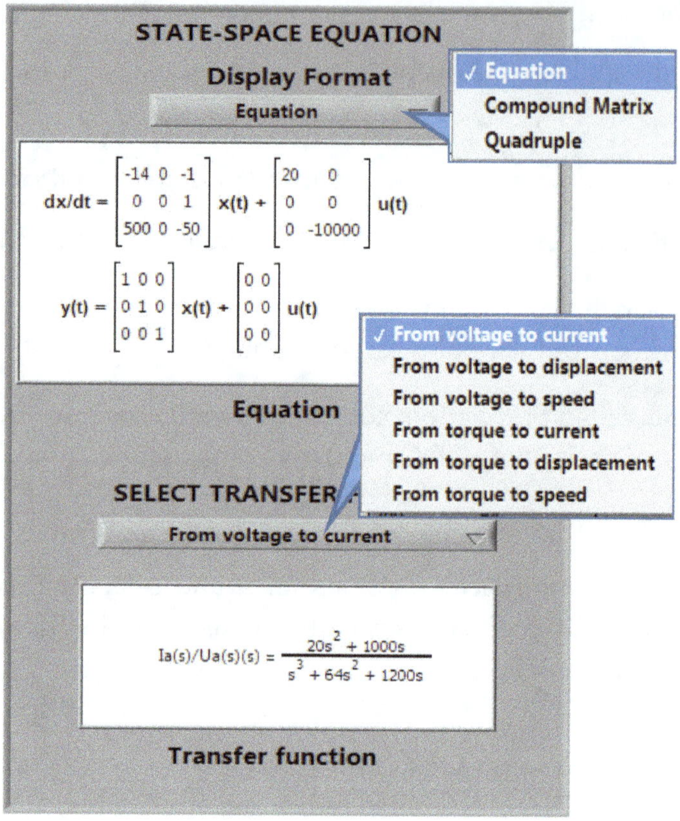

Figure 3. DC motor models.

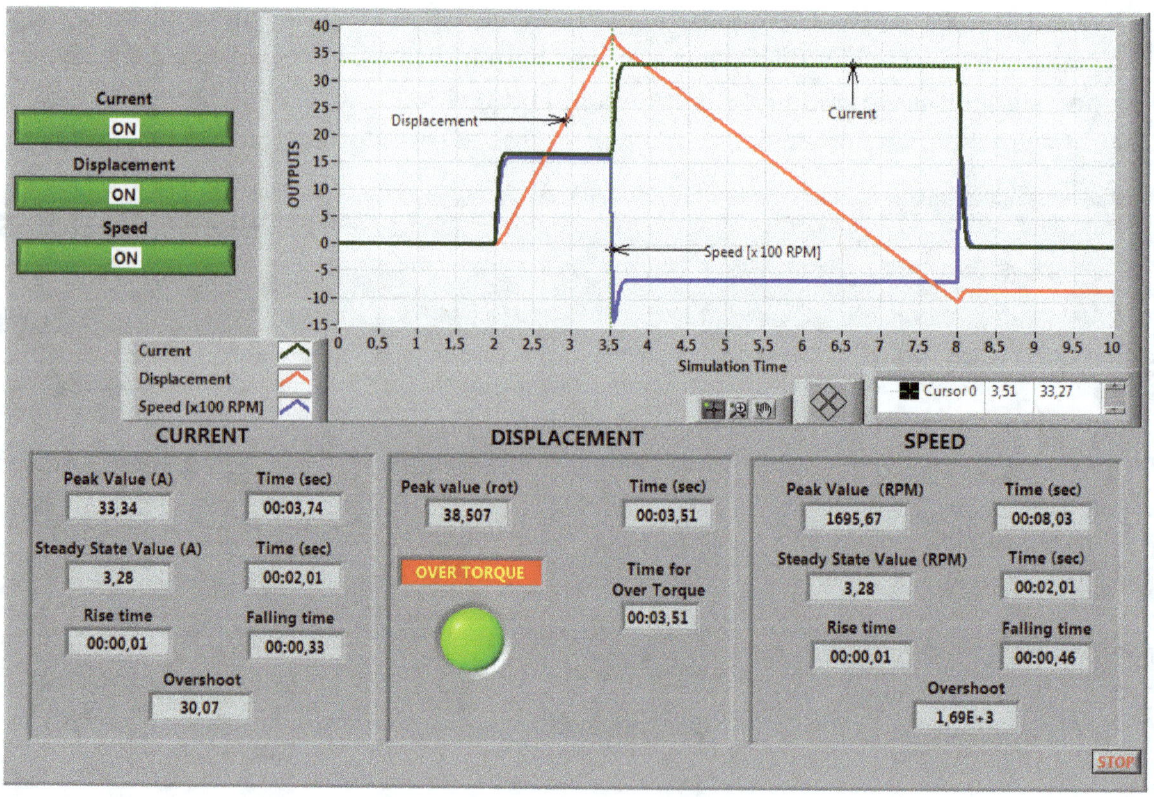

Figure 4. DC motor simulation results.

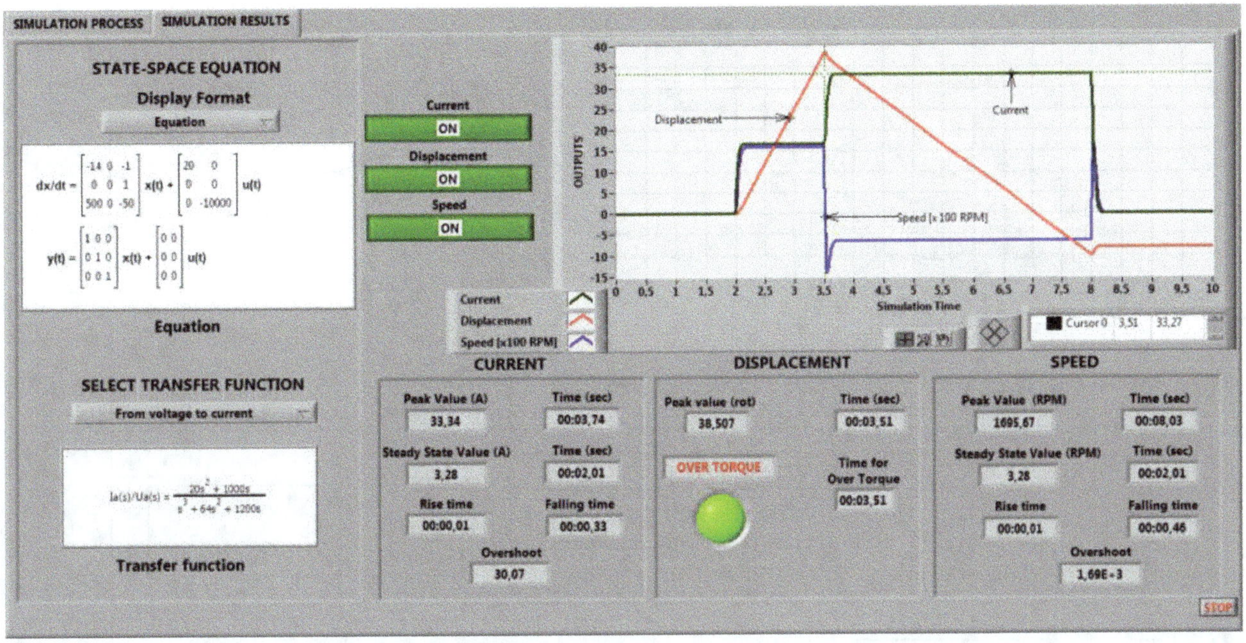

Figure 5. Second component of the front panel of the simulator.

The entire component SIMULATION RESULTS of the front panel is shown in **Figure 5**.

3.2. Block diagram of the virtual instrument

After the front panel is built, the code using graphical representations of functions to control the front panel objects is added. The block diagram represents the program under which the simulator runs and contains the graphical source code. Front panel objects appear as terminals on the block diagram. Block diagram objects include terminals, subVIs, functions, constants, structures, and wires, which transfer data among other block diagram objects.

The basic structure in constructing the virtual instrument is a While Loop that continues to run the program until the user presses the STOP SIMULATOR button on the front panel.

The program is running in two sequences according to the main algorithm shown in **Figure 6**. The user selects one of the shapes (step, pulse or ramp) and the values for the two variables input u_a and T_L and depending on constructive parameters of the DC motor, whose all values are set also by the user on the front panel, the state-space model is generated. By numerical simulation, based on this model, output variables are determined, and also, state-space and transfer functions models of the DC motor are displayed. The chosen output variables that represent the dynamic parameters of the system (maximum value, steady-state value, rise time, falling time, overshoot) are calculated.

The first sequence of the program and the MathScript Node are used to introduce the DC motor model into the program, and it is shown in **Figure 7**. The main element of the first sequence is the Control and Simulation Loop in conjunction with MathScript Node. Control and Simulation Loop is one of the components of the LabVIEW Control Design and Simulation Module. With this module, the plant and control models using transfer function, state-space

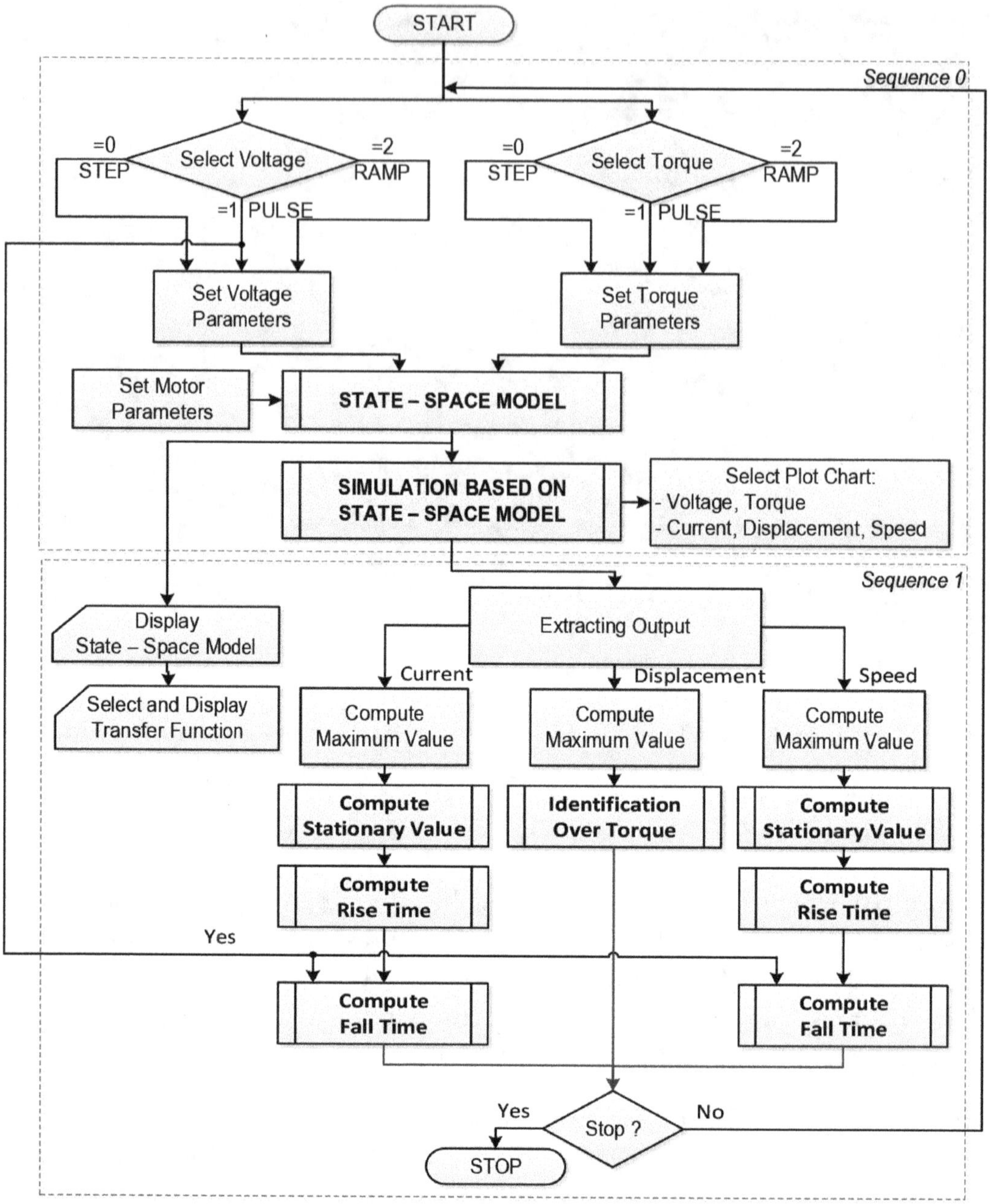

Figure 6. Main program algorithm.

model, or zero-pole-gain model can be constructed, and also system performance with tools such as step response, pole-zero maps, and Bode plots can be analyzed through simulation techniques. The simulation time is set from corresponding control on the front panel, and this is the only parameter of the loop.

The state-space model of the DC motor is placed into Control and Simulation Loop through a MathScript. MathScript is a high-level, text-based programming language. MathScript includes more than 800 built-in functions, and the syntax is similar to MATLAB and allows creating custom-made m-file, like in MATLAB. To process scripts in LabVIEW, LabVIEW MathScript Window or a MathScript Node can be used. If it is necessary to integrate MathScript functions (built-in or custom-made m-files) as part of a LabVIEW application and combine graphical and textual programming, as found in this application, MathScript Node is used [12, 13].

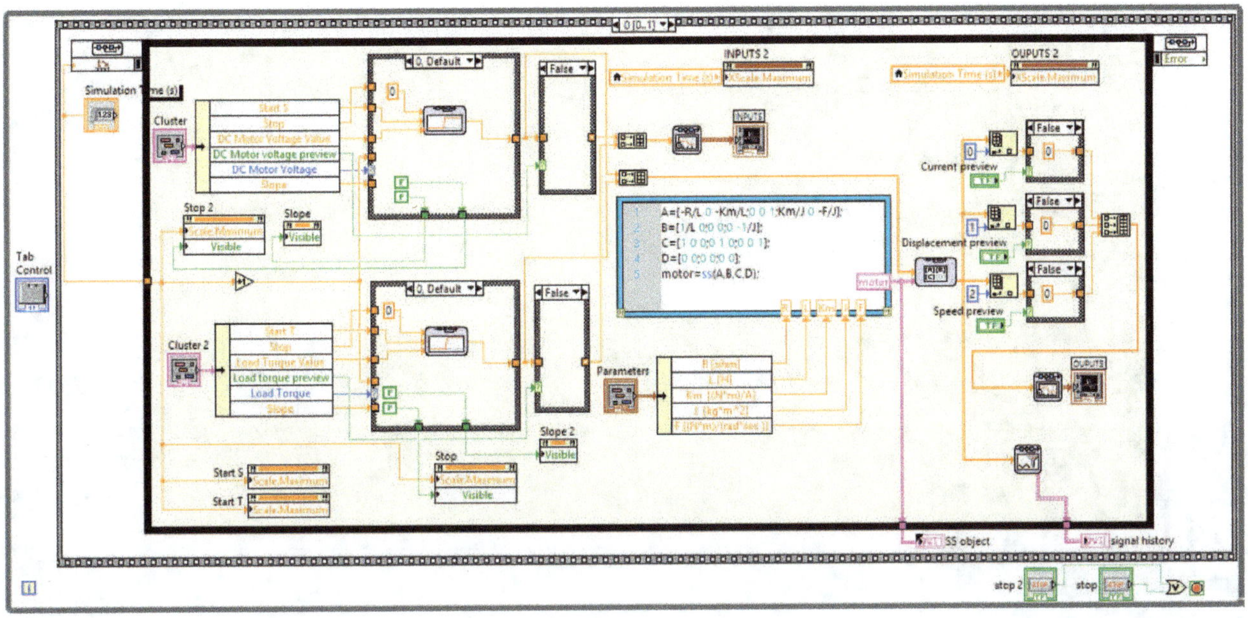

Figure 7. The first sequence of the simulator.

DC motor parameters are introduced into the model, and these can be changed interactively via the corresponding buttons on the front panel. The output from the MathScript Node is SS (state-space) object called motor and is a 2D matrix, which is a representation of the state-space system dynamic equations corresponding to the general forms (8) and (10). From this output, the matrices A, B, C, and D can be extracted, but also other information, such as properties, state names, transport delay. This object, together with the input vector, is used to generate the output vector $y^T = [ia, \alpha, \omega]$ through the LabVIEW simulation function State-Space. Each of the three components of the output vector can be identified using the Index Array function that returns the element or subarray of the n-dimensional array corresponding to the index value. After identifying any of these three components, these can be visualized on the front panel, through a graphic indicator. The obtained vector is also processed through the Collector function. This function collects a signal at each time step of the simulation and returns a history of the signal value and the time at which this function recorded each value in the history. Based on the signal history, the values of the dynamic regime, in the next sequence, can be identified.

For each of the two components of the input vector u_a respectively T_L standard signals are established, used to time domain analysis, namely: step, pulse, and ramp. The user can choose at any time and for each of the two components either of these signals through corresponding controls on the front panel. This choice is made by using a case structure and for the input selection, is used a control placed on the front panel button like is shown in **Figure 8**.

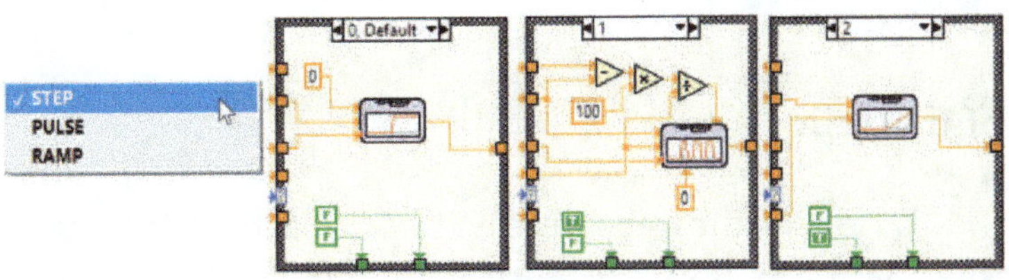

Figure 8. Selection of the input signals.

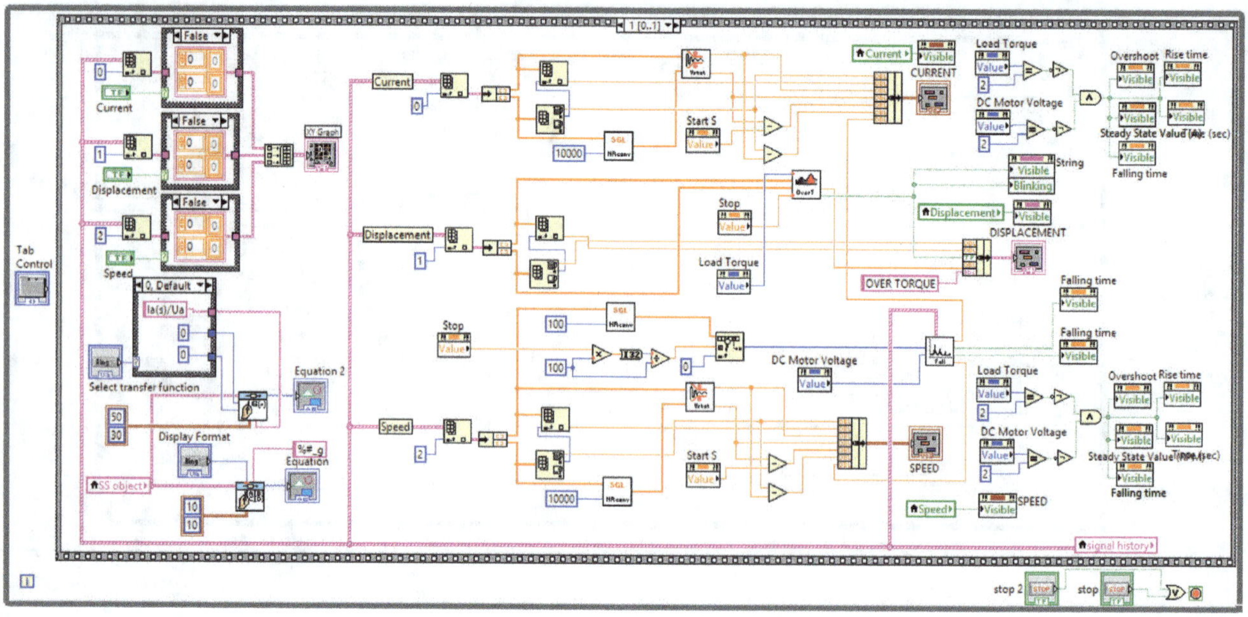

Figure 9. The second sequence of the program.

In the second sequence, shown in **Figure 9**, an identification of values is performed that characterize the operation in dynamic regime of the DC motor.

As mentioned earlier, the second sequence is used for extracting the information, obtained through simulation. Display of the equations of these models is done through *CD* Draw State-Space Function respectively CD Draw Transfer Function Equation, in Control Design and Simulation Toolkit.

By considering the DC motor like a MIMO system, having two inputs and three outputs are obtained six transfer functions that are displayed in a matrix form. To give to user the possibility to choose any of the six transfer functions, a control is provided on the front panel (Select transfer function) through which is generated the line and column index where the transfer function can be found (**Figure 10**).

As already mentioned, extraction of values that characterize the dynamic regime is made from the history signal. The history signal is a vector obtained as output from MathScript Node and represents the result of simulation. Just as in the first sequence, extraction of components from this vector is done through the Index Array function that returns the subarrays for the three outputs. Each of the three subarrays is composed of the values of output variables and values of the time variable on the simulation interval so that it is also possible to determine the corresponding time for each value that characterizes the dynamic regime.

3.3. Determination of dynamic regime values

The peak value that is the maximum absolute value of each of the output variables is computed using Array Max&Min function that returns the maximum and minimum values found in subarrays, along with the indexes for each value [7]. Since the peak value can be positive or

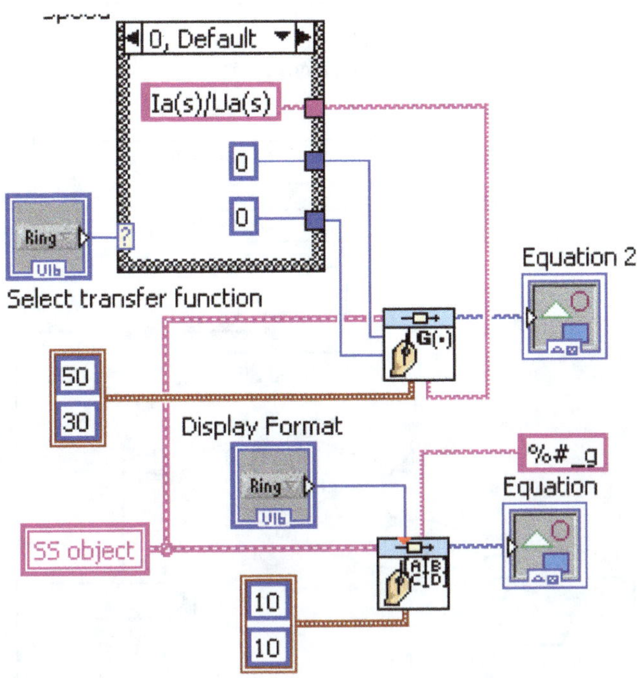

Figure 10. Functions used to display the equations of the model.

negative, depending on the direction of rotation, the maximum value is displayed for the maximum positive value and the minimum value for the maximum negative value. Selection is made through a CASE structure based on the comparison of absolute values corresponding to the maximum and the minimum values of displacement. Once determined the maximum value, based on its index, the moment of its appearance is searched.

The steady-state value *Sv* and the moment of its appearance *TSv* are computed using a SubVI named *Vstat* whose algorithm is shown in **Figure 11**.

This value is determined for each of the three components of the output vector. It considers the steady state to be reached if the difference between two consecutive absolute values for outputs variable is less than the constant $c = 10^{-6}$.

The algorithm shown above applies to current, to displacement and/or to speed, and each of them can be represented by the subarray *S(n)*. The moment of appearance of the steady-state value *TSv* is determined by reading the index *i* of the *Sv* in subarray *S(n)*, and with this index, the value with the same index *i* is searched in subarray *T(n)*, which is an array of the time values. Implementing of this algorithm in the virtual instrument structure is made by the SubVI *Vstat*, and this is shown in **Figure 12**.

For extracting the values *S(i)* of the subarray *S(n)* a while loop is used, whose condition terminal *i* is incremented until it fulfills the condition:

$$|S(i)|-|S(i-1)| \leq c \tag{12}$$

The rise time is computed through the difference between time *TSv*, when is identified the steady-state value, and the time when is generated the proper input variable. The time when is generated the input variable is taken from controls which determine the moment of start through a local variable or Property Node [9] StartS, as shown in **Figure 13**.

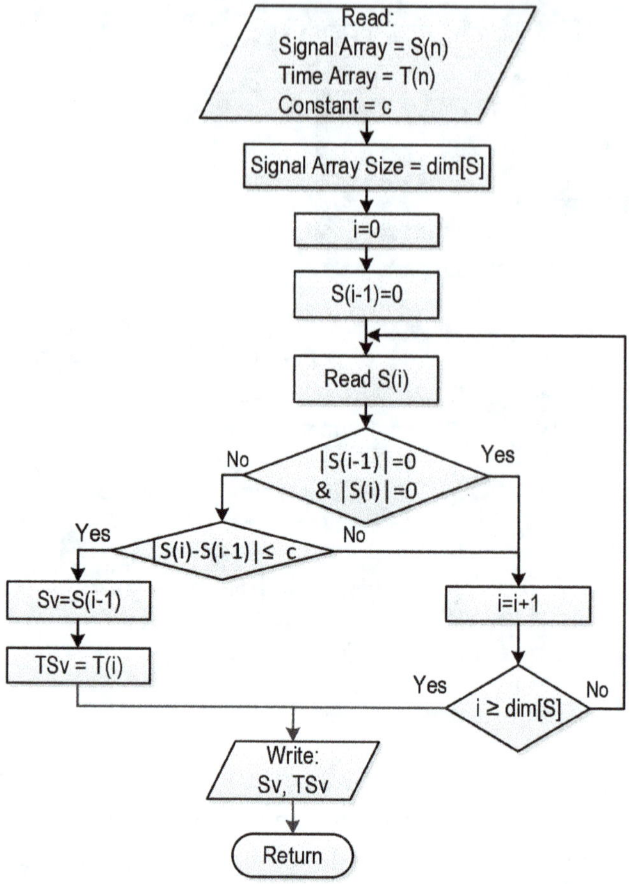

Figure 11. Algorithm used to compute steady-state value.

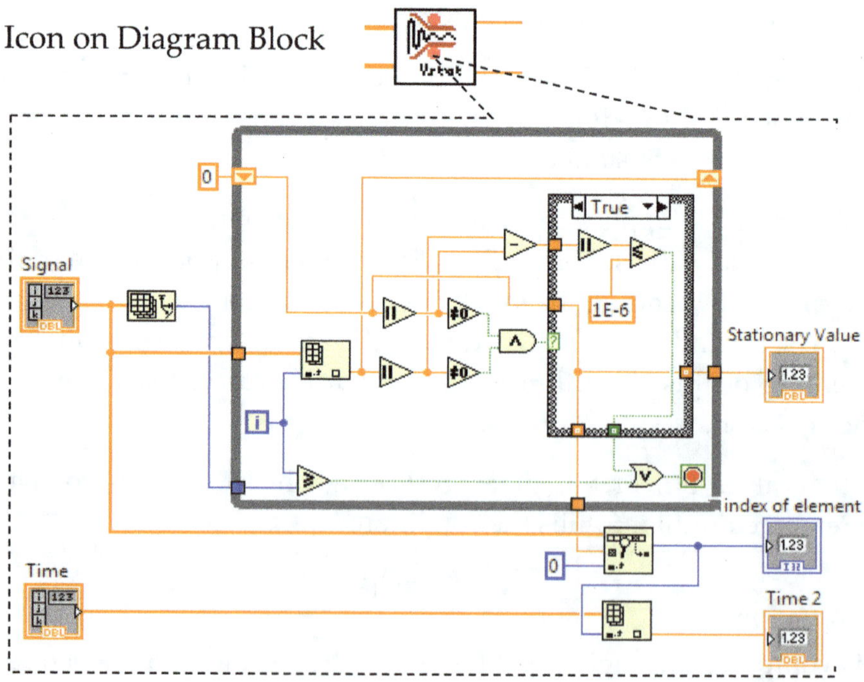

Figure 12. Implementation of the algorithm *Vstat*.

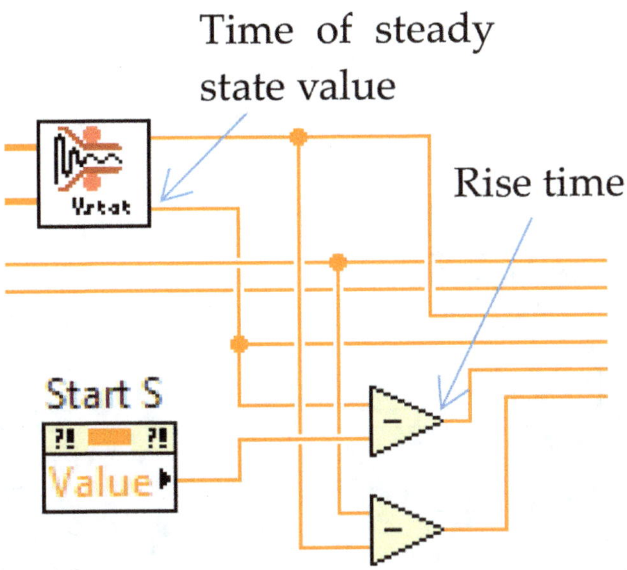

Figure 13. Computation of the rise time.

It is known that, if the load torque exceeds the active torque of the electrical machine, the "overturn" occurs. Therefore, it is necessary that the simulator be able to determine the overturning moment.

Identifying the phenomenon of "overturn" and the overturning moment is achieved using a SubVI named SENS, whose algorithm is shown in **Figure 14**.

Identifying the phenomenon of overturning is done by analyzing the sequence of values *Sd(n)* for angular displacement and the determination of the overturning moment is done by comparing the successive values *Sd(k-1)* and *Sd(k)* in the sequence considered. Thus, the phenomenon is considered to occur if:

$$\mathrm{Sd}(k)\text{-}\mathrm{Sd}(k\text{-}1) \leq c_1 \tag{13}$$

where $c_1 = -0.01$ is a negative constant used to make the comparison.

If the torque shape is a pulse, the stop moment *TTs* and its index are identified in the sequence of time values and based on this index are removed all values with the index higher than this, from sequence *Sd(n)*. Obviously, the comparison is done only if two successive values from the sequence *Sd(n)* are different from zero.

By fulfilling the condition (13), the value *Sot* (=*Sd(k-1)*) is identified, corresponding to the overturn, in the sequence of values for angular displacement *Sd(n)*, starting with the appearance of the inversion of rotation. To signal the state of overturn, the value "true" is assigned to the boolean variable *Ovt*, otherwise the variable receives the value "false."

Indication of the overturning state is made on the front panel by a red color of a LED indicator (that otherwise is green) and by blinking message "OVERTURN," that becomes visible only when this phenomenon occurs.

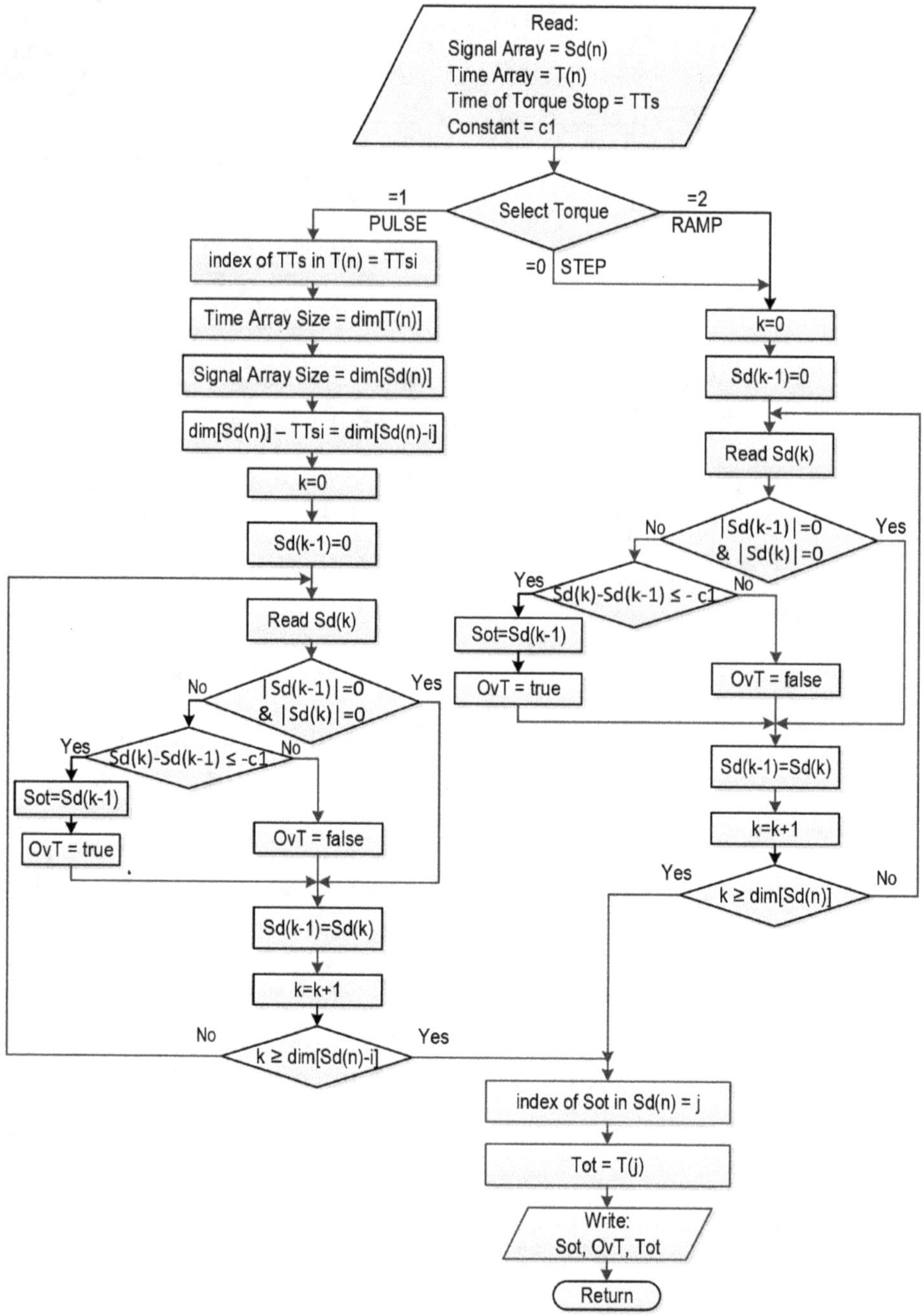

Figure 14. Algorithm used to identify the phenomenon of "overturn" and overturning moment.

In cases where the torque is step or ramp shape, considering the fact that these shapes did not have the stop time, it is not necessary to identify this moment in the times sequence values. In these cases, the overturn moment *Sot* is searched by identifying the condition given by (15) for successive values in sequence of values for angular displacement *Sd(n)*. Completion of the algorithm is carried on, as in the case of the torque having pulse shape.

After identifying the value *Sot*, corresponding to the overturning moment, its index is searched in the sequence of values for angular displacement *Sd(n)*. Based on the index *j* thus determined, the moment *Tot*, the moment when overturn occurs in the time array sequence *T(n)* is searched.

Implementation of the algorithm in the virtual instrument structure is made by the SubVI Over Torque, **Figure 15**, and this is realized through a While Loop that runs until the condition (13) is reached or until all values from the sequence of values for the angular displacement *Sd(n)* are tested. Depending on the shape of the input signal (Signal Type), one of the three cases is selected, noting that cases 0 and 2 (step and ramp) are the same.

In the execution of the search operation in the array sequences, 1D Array Search Function is used, that searches for the element *(Sd(k-1))* in a 1D array *(Sd(n))* starting at start index for that is set with 0 value. Because the search is linear, it is no need to sort the array before calling this function and LabVIEW stops searching as soon as the element is found. To make the index increments after each test for two consecutive values from the sequence array, the iteration (*i*) terminal of the While Loop is used that provides the current loop iteration count, which is zero for the first iteration.

Another value of the dynamic regime of DC motor that can be determined by the simulator is the fall time of the DC motor response. This value can be calculated, and it will be displayed as well, only for a pulse shape input. To display the value of this variable, proper indicators are used, which become visible on the front panel only if the input signal type is pulse shape.

To determine this value, the algorithm is used as shown in **Figure 16**.

Proper indicators became visible on the front panel only for a pulse shape input, when choosing the value true for the boolean variables *B1* and *B*, otherwise their values will be false.

In the sequence of the time values, the index *j* is identified, corresponding to the stop moment *TTs*, so from the sequence of the current and speed values, all values that have the index greater than *j* are removed. They are thus obtained for both current and speed variables, two sequences of values each with *S* size, and the values in these sequences are reordered starting at index 0.

Determination of fall time of the DC motor response is accomplished by identifying the moment when an absolute value for current or speed is zero (or less than a constant c_2, close to zero), according to the relations:

$$\begin{aligned} |\mathrm{Sc}(i)| &< c_2 \\ |\mathrm{Ss}(i)| &< c_2 \end{aligned} \tag{14}$$

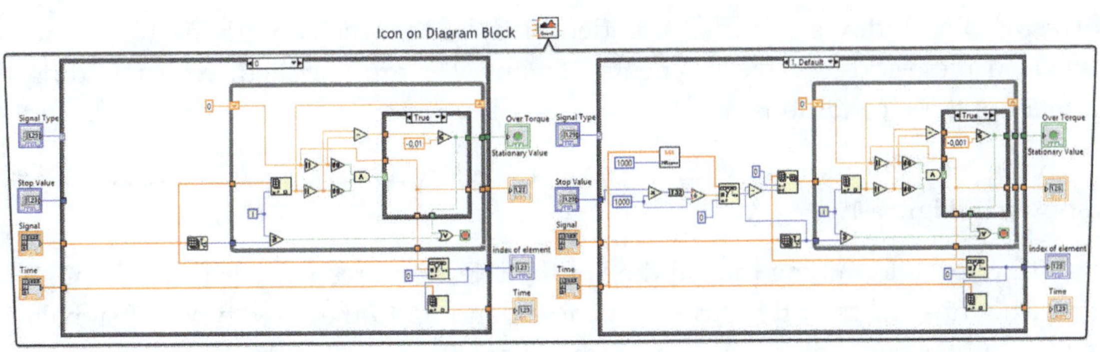

Figure 15. Implementation of the algorithm Over Torque.

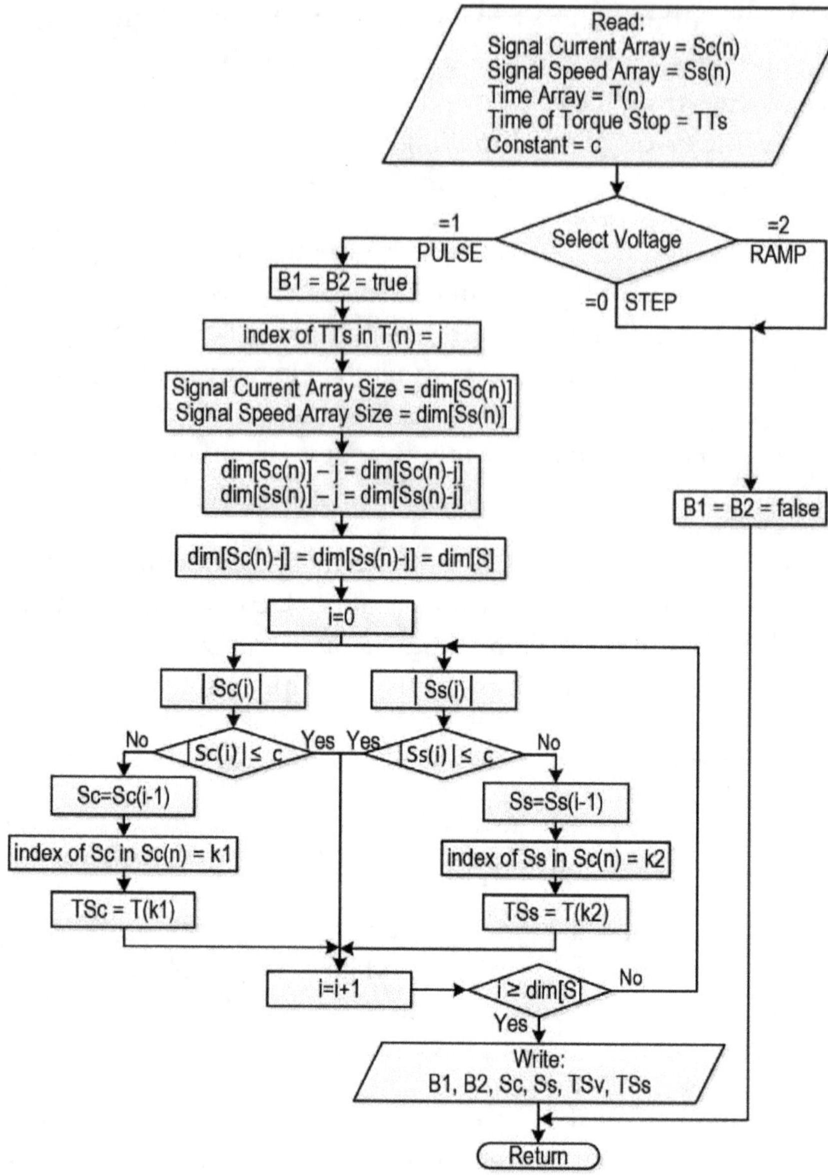

Figure 16. Algorithm used to identify the fall time of the DC motor response.

where |*Sc(i)*| and |*Ss(i)*| represent absolute values extracted from the sequences of values for current and speed, respectively.

The corresponding indexes k_1 and k_2 are identified for that the conditions (16) are achieved, and based on these indexes, the moments *TSc* and *TSs* are searched, corresponding to the achievement of these conditions.

Implementing the algorithm in the virtual instrument structure is made by the SubVI *Fall,* and this is shown in **Figure 17**.

The algorithm for determining the fall time of the DC motor response is made through a case structure. The three cases of this structure, corresponding to the three types of signals generated for DC motor command, are selected by controlling the DC Motor voltage from front panel through the value signal slide.

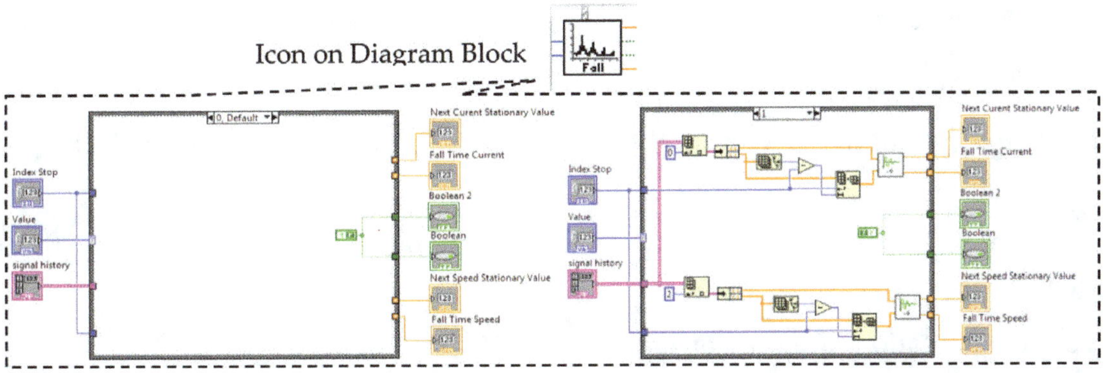

Figure 17. Implementation of the algorithm Fall.

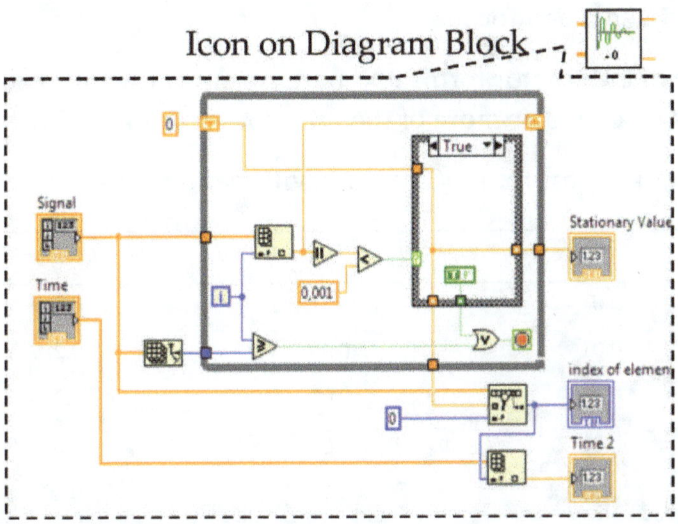

Figure 18. The SubVI Stationary Value Fall.

As seen in **Figure 17**, actual implementation of the algorithm takes place only in the sequence 1, which corresponds to a step shape command type for which is possible the compute this parameter of the dynamic regime. In the same sequence, boolean variables *B1* and *B2* are set to true. In the other two sequences, only the boolean variables B1 and B2 are set to false values.

In the structure of this SubVI, there is another SubVI (**Figure 18**) called Stationary Value Fall that determines just the moments in which the absolute values of current and displacement are 0 (or smaller than c_2, constant very close to zero). Its operation is similar to determine the stationary value in the SubVI Over Torque and consists in checking the differences between two successive values for displacement or current and to determine when this difference is zero (or less than constant c_2 whose value is set in this case at 10^{-3}).

4. Some results of simulation

To illustrate the operation of the simulator, several captures for the following elements of the front panel are presented:

- DC motor voltage and torque, in the graphical form, used as inputs variable shapes;
- DC motor parameters;
- simulation results in the graphical form for current, displacement and velocity used as outputs variables;
- numerical results and values for dynamical regime parameters, obtained from the simulation process.

Two types of simulations are used as follows:

- DC motors with different parameters are used and their simulation is done under the same conditions regarding the shape and parameters of input signals, that is, voltage DC motor and respectively, load torque;
- for the same type of DC, motor forms and parameters of the input signals can be modified or not, but in this case, one or more of the DC motor parameters are changed.

The simulation is performed for four DC motors with parameters presented in below table.

	Motor A	Motor B	Motor C	Motor D
R [Ω]	2.0	1	0.699	2.3
L [*H*]	0.0169	0.5	0.279	0.00845
Km [*N.m/A*]	0.283	0.01	0.215	0.66
J [$kg.m^2$]	0.0112	0.01	0.00279	0.052
F [*(N.m)/(rad.s)*]	0.058	0.1	0.00415	0.002

Considering the first type of simulation, to the four DC motors, a voltage as the control signal and a load torque having the shape and parameters, such as shown in **Figure 19** are applied. The voltage applied to the DC motors has a pulse shape, with an amplitude of 24 volts, from 2 to 8 s and the load torque increases linearly up to 1 N.m also from 2 s on an interval of 3.6 s.

In **Figure 20a–d**, the results of simulations corresponding to the four DC motors are presented.

As observed on the displacement graph, in the cases (a), (b), and (d), due to high load torque, overturning phenomenon occurs, and this phenomenon is indicated by turning on a LED, a blinking message, and identifying its moment of appearance. The peak values for current,

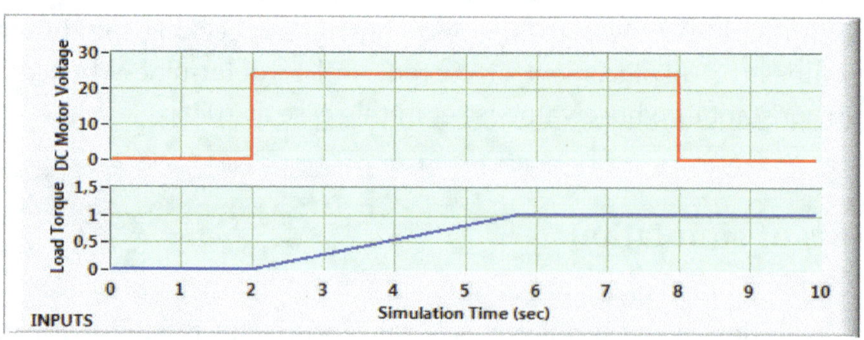

Figure 19. The inputs shapes used for simulation.

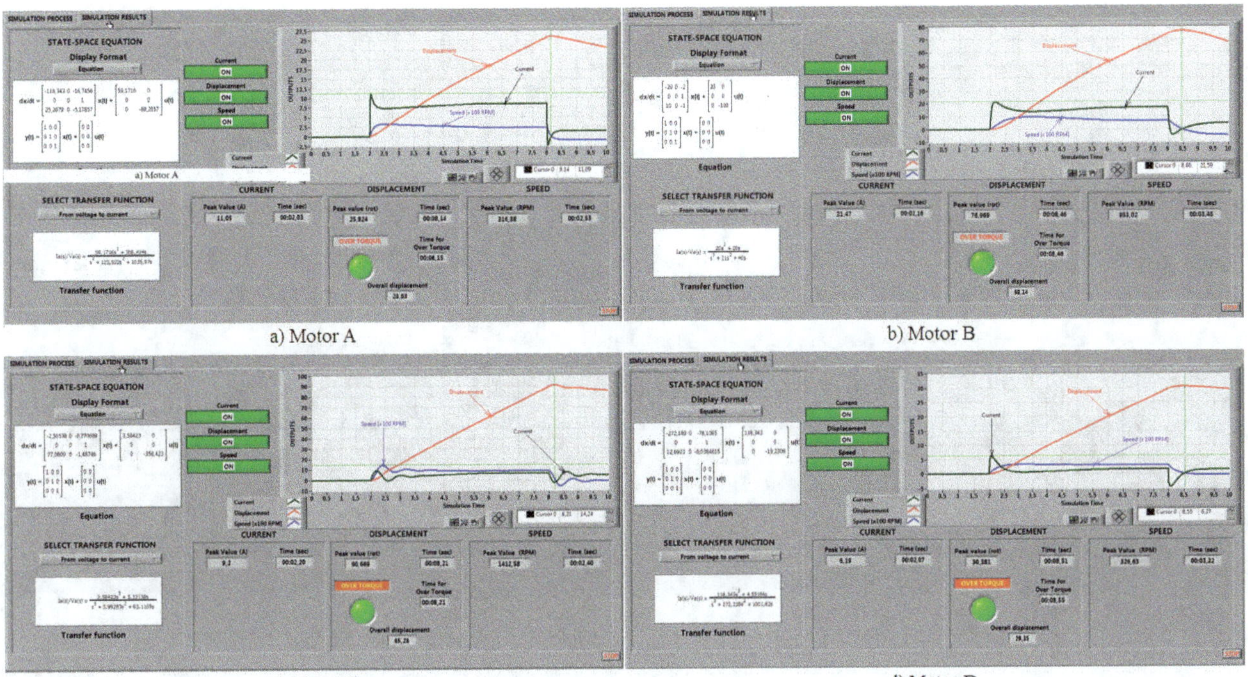

a) Motor A b) Motor B c) Motor C d) Motor D

Figure 20. The results of the motors simulation. (a) Motor A. (b) Motor B. (c) Motor C. (d) Motor D.

displacement and speed with its moments of appearance and also, the overall displacement are displayed. Evidently, the user can select, for each DC motor, any of the three output variables or combinations thereof and the type of the transfer function.

For the second type of simulation, the DC motor D is chosen, and that it will be supplied with voltage under the shape of a step, impulse and ramp and also, that will be under the influence of a load torque that has a step shape with a front increased linearly.

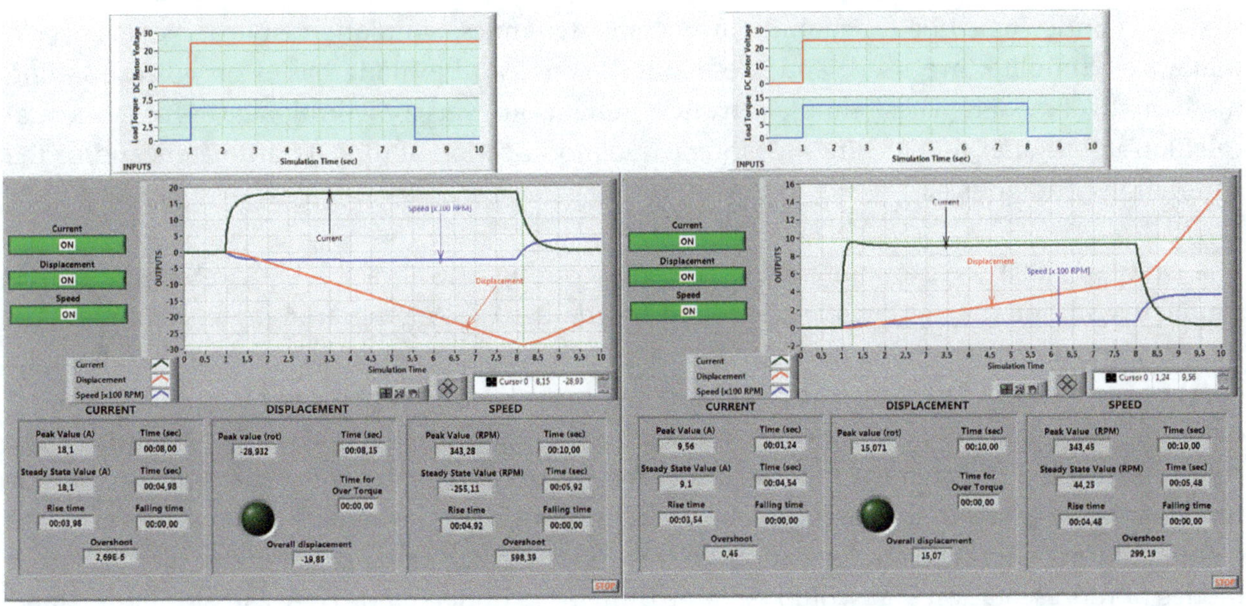

Figure 21. Increasing the load torque amplitude of 6–12 N.m.

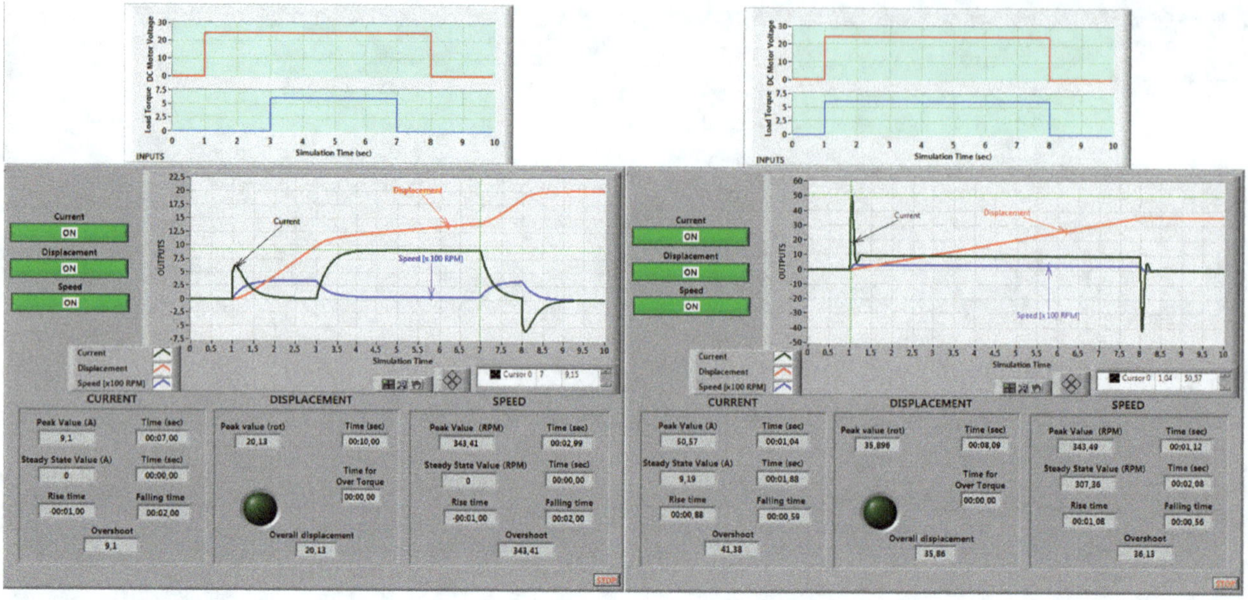

Figure 22. Decreasing the armature resistance of 2.3–0.3 Ω.

In **Figures 21** and **22**, the simulation results are compared for the DC motor powered by 24 V for a load torque increasing from 6 to 12 N.m, respectively for a load torque at 6 N.m and armature resistance decreasing from 2.3 to 0.3 Ω.

5. Conclusions

It is known that computer simulation is an experimental method based on a set of techniques by which operating with virtual things, the functioning of a real system or process is understood. To understand the functioning of systems dynamics, simulation environments such as Matlab or Simulink are used, and these simulations can be made based on corresponding mathematical models. These simulation environments are very powerful in terms of numerical calculation but offer few facilities for interactive modification of model parameters and/or the simulation conditions.

The simulator that we propose combines the computing power of the Matlab-Simulink environment with interfacing facilities provided by the LabVIEW programming environment. In this way, the users can understand the dynamics of a system such as a DC motor, following its chart of response to different stimuli or different operating conditions. On the other hand, by obtaining the mathematical model expressed by a transfer function in various operating conditions of the DC motor, students in modeling complex systems in which the DC motor represents a subsystem can use this one. Getting results in a graphical form allows users to compact large amounts of information and easily recognize these phenomena then by other modes of representation that would be more difficult to understand. They can also understand the effects of changes in the dynamic system parameters of its operation.

In addition, because LabVIEW is mainly a programming environment for data acquisition, it is possible to make a comparison between data acquired from a real system and data obtained by simulation from a mathematical model of this system by creating a hardware-in-the-loop structure.

The simulator can be extended to study other systems with minimal changes, using corresponding mathematical models.

Acknowledgements

An important part of this chapter represents the papers of the author which have been presented at the 5th International Multidisciplinary Scientific Symposium "Universitaria" (SIMPRO 2012) conference, at the 13th International Conference on Engineering of Modern Electric System (EMES 2105) and at the 14th International Conference on Engineering of Modern Electric System (EMES 2107).

Some of these papers have been written by a team of authors, and the author of this chapter has the acceptance of his co-authors for their use in the realization of this chapter.

Author details

Nicolae Patrascoiu

Address all correspondence to: nicolaepatrascoiu@upet.ro

System Control, Computer Engineering and Electrical Engineering Department, University of Petrosani, Romania

References

[1] Jagan NC. Control systems. 2th ed. Hyderabad: BS Publications; 2008

[2] Alexandru F. Machines and Electric Drives (in Romanian). Bucharest: Ed. Tehnica; 1986

[3] Golnaraghi F, Kuo BC. Automatic Control Systems. 9th ed. USA: John Wiley & Sons, Inc; 2010

[4] Dorf RC, Bishop RH. Modern Control Systems. 12th ed. USA: Prentice Hall; 2011

[5] Matko D, Karha R. Simulation and Modelling of Continuous System. New York: Prentice-Hall; 1992

[6] Patrascoiu N. Modeling and simulation of the DC motor using Matlab and LabVIEW. International Journal of Engineering Education. 2005;**21**(1):49-54

[7] Trundle KC, Bell RL. The use of a computer simulation to promote conceptual change: A quasi-experimental study. Computers & Education. 2010;**54**(4):1078-1088

[8] White R. Computational Mathematics. Models, Methods, and Analysis with MATLAB. USA: Chapman & Hall/CRC; 2004

[9] Beyon JY. LabVIEW Programming, Data Acquisition, and Analysis. New York: Prentice Hall; 2001

[10] Faraco G, Gabriele L. Using LabVIEW for applying mathematical models in representing phenomena. Computers & Education. 2007;**49**(3):856-872

[11] Bitter R, Mohiuddin T, Nawrocki M. LabVIEW. Advanced Programming Technique. 2nd ed. New York: CRC Press; 2007

[12] Halvorsen HP. LabVIEW MathScript. Telemark: Tutorial, Telemark University College, Department of Electrical Engineering, Information Technology and Cybernetics; 2011

[13] Marcu MD, Popescu FG, Pana L. Modeling and simulation of power active filter for reducing harmonic pollution using the instantaneous reactive power theory. Environmental Engineering and Management Journal. June 2014;**13**(6):1377-1382

3

Battery Electric Propulsion Systems for Competition Racing Applications

Sanath Alahakoon and Mark Burridge

Abstract

Electric propulsion systems applied in racing karts is a matured technology. This is confirmed by the fact that the Fédération Internationale de l'Automobile (FIA) has established a recognised technical formula for sanctioned competitive electric kart racing. Electric kart racing was introduced into Europe in 2017 with a set of technical regulations that are quite open at this early stage. This allows for experimentation of different electric powertrain configurations to enhance their development. The recent introduction of electric kart racing has provided the incentive to further develop electric powertrains to maximise their performance potential. The design and construction procedure necessary that was undertaken in the process of engineering a high-performance battery electric powertrain for a racing kart application for competitions is presented in this chapter. The design ensures that the electric propulsion system is capable of matching the performance characteristics of a comparable racing kart powered by an internal combustion engine (ICE). The implementation process of the design on a recognised, competition racing kart rolling chassis together with its performance evaluation data is also presented.

Keywords: battery, electric propulsion, racing kart, drive, controller

1. Introduction

Electric propulsion systems applied in competition racing karts are at a point of their development such that the technology is ready to be introduced into electric kart racing for sanctioned competitions. The following discussion in this chapter provides a basic approach to the design and construction process that is required in engineering an electric powertrain dedicated to a racing kart application for competitions. The procedure is demonstrated through a working example of an electric propulsion system orientated for high performance. The objective is to make this a stimulus for encouraging further development of more advanced electric powertrains suited to competition applications.

The final objective of the design is to end up with an electric powertrain that is capable of at least matching the competition racing cart performance characteristics powered by an internal combustion engine (ICE) with similar capacity. The chapter presents the documentation of the design procedure required, construction, and testing carried out to determine the actual performance. The electric powertrain will be installed into an existing rolling kart chassis, which has been designed specifically for the sport of kart racing.

In order to ensure functionality, reliability and minimised project costs at this introductory level, only proven electric vehicle (EV) technologies, which are readily commercially available have been integrated in the design and construction process.

The following main outcomes will be delivered by the overall research project:

- Specifications of an EV propulsion system applicable on an electric kart platform suitable for the lower levels of competition kart racing
- Produce a ***working prototype*** of an electric vehicle powertrain from an engineering design to meet a specific vehicle performance specification suitable for racetrack competition
- Results obtained through experimental verifications to confirm the dynamic performance capabilities achieved by the complete electric powered vehicle system.

2. Methodology

This research project must work within the following parameters to assist in achieving the working prototype outcome:

- Proceed with a simple design with proven reliable technology to handle the continuous high speed operation inherent to a competition environment
- Designed and constructed from 'off the shelf' readily available commercial products and components to facilitate easy construction, reliable operation and regular maintenance requirements
- Robust and reliable components from reputable manufacturers and suppliers to survive the rigours of an unforgiving and sometimes punishing operating environment
- Minimising the cost

The above constraints lead into the inevitable comparison between a battery electric powered kart and an ICE powered machine. The latter is indeed a very simple yet powerful mechanical vehicle for its compact size, with a high power to weight ratio indicative of the sport. It is these performance attributes that make the ICE powered racing kart so attractive as an introduction to the sport of competitive motor racing. In order to compete with the status quo, and present an attractive viable alternative, the proposed electric kart must exhibit similar characteristics.

This research project requires achieving three main work functions:

1. The design that includes modelling of the components and verification through simulations
2. The fabrication followed by assembly of the powertrain into the kart chassis
3. Testing and evaluation from on track performance data

The flow chart in **Figure 1** summarises the method of how the design and development was carried out to achieve the final outcomes.

The process commences with the establishment of the preferred EV platform and architecture [1] followed by an initial performance specification to set the performance goals for the electric vehicle as a complete unit with the powertrain installed.

This is followed by an initial determination of all resistive forces acting on the kart to establish the tractive effort [2] required to propel the vehicle forward to comply with the intended

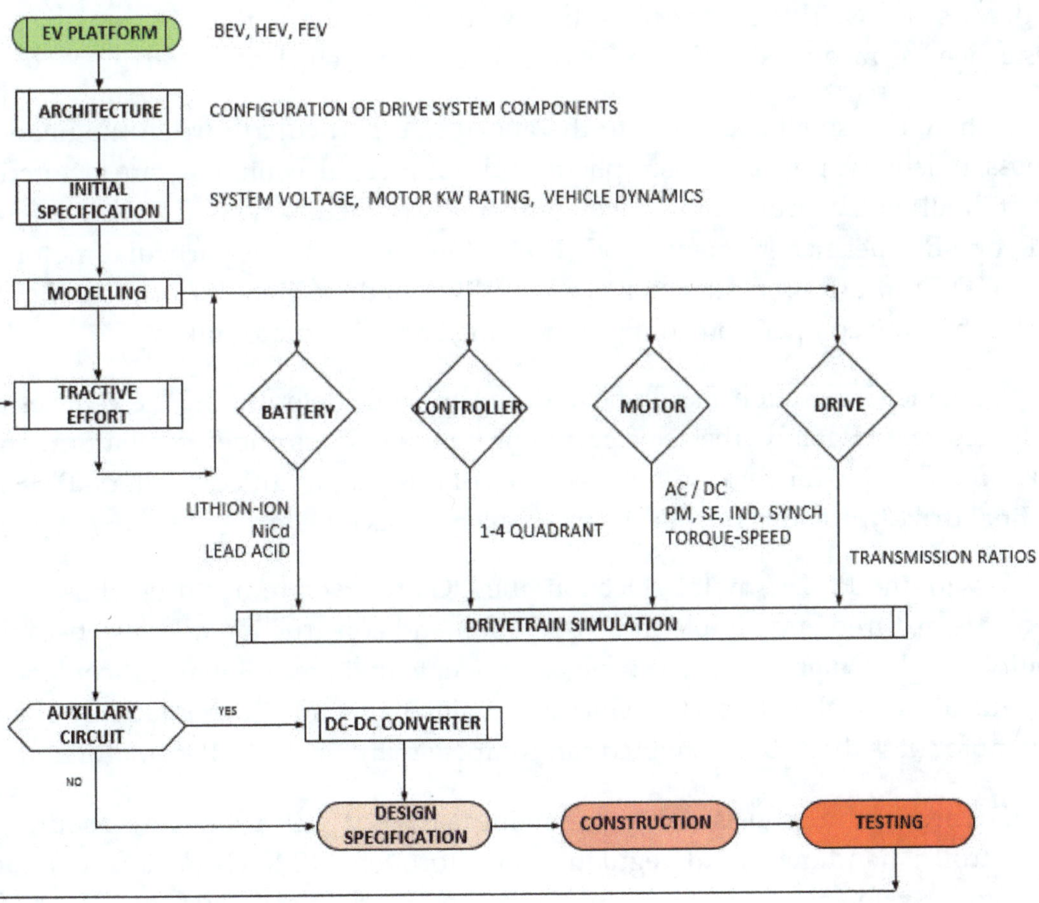

Figure 1. Methodology flow chart.

performance specification. Otherwise known as force modelling, this is a well-known and extensively documented design method employed by electric vehicle designers.

The main components identified in the architectural arrangement are either modelled or simulated individually with alternate options from different manufacturers to select the most appropriate items to achieve the vehicle's performance criteria. All chosen components are then brought together into the powertrain simulation to confirm their performance potential as a complete system.

The selected components represent the final design and are subsequently sourced from both national and international suppliers for assembly into the electric powertrain mounted within the rolling kart chassis. The complete vehicle is then bench tested and finally road tested to confirm the actual performance data with the initial specifications set down at the beginning of the research project.

3. Platform and architecture

In current practice, pure DC motor-powered drivetrains are in decline [3] and are limited to light recreational type electric vehicles only. AC motor-driven powertrains are without doubt the preferred method for all new battery electric vehicle (BEV) and hybrid electric vehicle (HEV) drive systems. This is especially true when designing both full-size passenger road vehicles as well as racetrack performance orientated-type vehicles.

However, there is a small exception to the above rule particularly when the financial cost, robustness, reliability, powertrain simplicity and commercial availability are factored into the argument. It all centres around the development of a particular type of DC motor—the axial flux (AF) brushed permanent magnet DC (PMDC) motor. With this particular motor installed as part of a DC powertrain platform, it is especially suited for high-performance light electric vehicles such as the competition-configured racing karts and motor bikes.

Therefore, the motor selection should be undertaken while determining the vehicles platform and architecture. Inevitably, the selected motor type will determine the platform and hence the make-up of the powertrain system. Section 6 of this chapter titled "Battery, BMS, controller and final drive" provides further detail on the approach taken.

Combined with the PMDC motor selection, the DC architecture has been chosen since this technology is matured and simple in its operation and construction and also because of the availability of light weight electric vehicle components that are suitable and compact. An AC system, with all of its inherent performance and maintenance advantages lacks a suitable small capacity motor (less than 20 kW), which can compete with the AF PMDC motor at this stage.

The BEV DC architecture consists of the following: a battery pack as the power source, DC-DC motor controller handling speed regulation and torque, a PMDC traction motor generating the angular velocity and torque, and a coupling mechanism to transfer the mechanical power output of the motor to the rear axle for the purpose of driving the vehicle forward. An

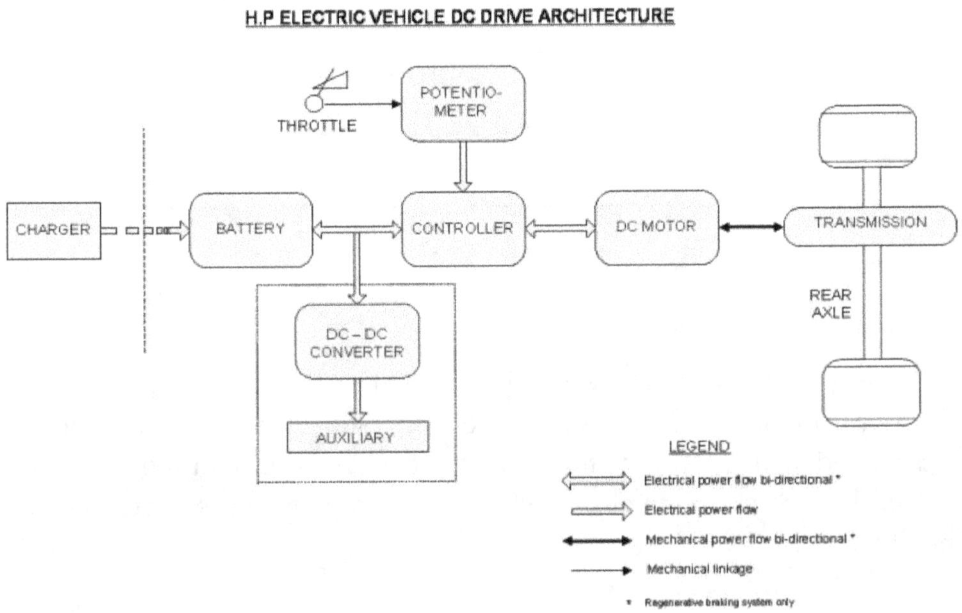

Figure 2. BEV DC drivetrain architecture.

optional inclusion could be an auxiliary circuit that controls any other vehicle accessories not directly associated with the operation of the electric drivetrain. **Figure 2** diagrammatically summarises this arrangement.

4. Initial specification

Since the objective of this research project is targeted at the introductory stage of sport karting, it is assumed that the only necessary requirement is to produce an electric kart system that has similar performance characteristics that are comparable with an equivalent ICE-powered kart competing in the senior entry-level classes of the sport at present. Performance characteristics of this class are the following:

Power plant: Yamaha or IAME air-cooled 100 cc ICE, restricted to 11 hp. (8.2 kW), or unrestricted to 16 hp (11.9 kW) [4].

Top speed: 77–104 km/h depending on port restrictions and track layout [5].

Acceleration: ICE acceleration figures unavailable. The recently released Bosch® e-kart rated at 20 kW 48 V has a claimed acceleration figure of 5.5 ms^{-2} for 0–100 km/h [6].

The above speed and acceleration figures are very much dependent on mass, engine horsepower, gear ratio, and the combined forces retarding forward motion. Bearing in mind these limitations, they do however provide a realistic base line from which this research project can aim for in its attempt to at least replicate comparable performance data. With the unavailability of any acceleration figures for an ICE powered kart, the Bosch® e-kart acceleration figure has been utilised instead and scaled according to the rated output power of the motor.

Criteria	Objective
Velocity	75–85 km/h
Acceleration	2.7–4 ms^{-2}
Range	10 km min

Table 1. Vehicle performance specification.

Vehicle range, or the total distance travelled before battery recharge is required is not a major concern with this particular EV application, due to the short track distance and limited number of laps associated with each race for such vehicles. A nominal range of 10 km has been nominated based on the short driving cycle experienced by these machines. Performance specifications of the vehicle are listed in **Table 1**.

The electric powertrain parameters are specified according to the desired platform operating voltage level and the main architecture component requirements, that is, the battery pack, battery management system (BMS), controller, traction motor and final drive. The desired voltage level is a trade-off between maximum power requirements and project cost. The higher voltage level results in more expensive powertrain components to handle the increased load currents. Commercially available components for light electric vehicles are readily available up to 72 and 96 V in some cases depending on the platform adopted. At the other end of the voltage scale, system levels at 36 V and below are more suited to the slower-paced recreational leisure type vehicles, which are definitely well short of performance objectives of the scope of this research project.

Consequently, a voltage range of between 48 and 72 V appears to be the level which is sustainable for vehicle type and DC platform of this research project. The remainder of the powertrain components are specified from a review of the available technology and their suitability to the DC platform. **Table 2** summarises the powertrain parameters.

Component	Type
Platform and architecture	BEV with DC powertrain
System voltage	48–72 V
Rated motor power and type	10–15 kW PMDC brushed
Controller	'Chopper' PWM 1–4 quadrant
Battery pack	Lithium-ion
Battery management system (BMS)	Cell balancing with pack voltage limits
Final drive	Single gear chain/belt drive

Table 2. Powertrain specification.

5. Modelling and simulation

Intended operating environment must first be evaluated for the design of the EV powertrain to satisfy the given design specifications. There are a number of opposing forces resisting the forward motion of any electric vehicle by virtue of its application. There is a direct impact of these same forces on the capability of the powertrain to overcome them, while producing high enough momentum in the forward direction to drive the vehicle forward, meeting a predetermined cruising velocity and rate of acceleration [6]. Hence, it is crucial to determine the feasible final velocity and acceleration performance characteristics associated with a particular powertrain configuration in relation to the vehicle, to which the power train is going to provide traction drive capability.

The force modelling procedure in its various formats is well known to vehicle designers and has been extensively documented elsewhere. The procedure is summarised in the main equations described below:

The main Equations [2] associated with the force modelling procedure and applied to the design of the powertrain system include the following:

$$F_{rr} = \mu_{rr} mg \text{ (rolling resistance force)} \tag{1}$$

$$F_{ad} = \frac{1}{2}\rho A\, C_d v^2 \text{ (aerodynamic drag)} \tag{2}$$

$$F_{hc} = mg \sin(\psi) \text{ (hill climbing force)} \tag{3}$$

$$F_{la} = ma \left(\text{linear acceleration force}\right) \tag{4}$$

$$F_{\omega\alpha} = I\frac{G^2}{\eta_g r^2} a \text{ (motor angular acceleration force)} \tag{5}$$

Combining Eqs. (1)–(5) results in the total tractive effort described in Eq. (6) to propel the kart forward

$$F_{te} = F_{rr} + F_{ad} + F_{hc} + F_{la} + F_{\omega a} \tag{6}$$

The motor power requirements at a given velocity is then determined from Eq. (7) taking into account efficiency losses in the drivetrain and can be described as

$$P_{te} = \frac{F_{te} \times v}{\eta_g} \tag{7}$$

Motor torque and speed relationship for a DC motor is given by Eq. (8), which takes into account the back EMF generated by the motor.

$$T = \frac{K_m \phi E_s}{R_a} - \frac{(K_m \phi)^2}{R_a} \omega \tag{8}$$

The equation describes the linear correlation between a decreasing speed with increasing torque depending on the rotor or armature resistance and the resultant back EMF. This type of torque-speed relationship creates an ideal condition for the regulation of motor speed via a DC motor controller.

Eq. (8) can subsequently be modified into the characteristic torque Eq. (9) below with the inclusion of the rated current specified by the motor manufacturer to calculate the rated torque value:

$$T = K_m \phi I \tag{9}$$

The all-important acceleration potential of the vehicle can then be modelled using Eq. (10)

$$\frac{G}{r} T_{max} n_g = \mu_{rr} mg + \frac{1}{2} \rho A C_d v^2 + \left(m + 1 \frac{G^2}{\eta_g r^2} \right) \frac{d_v}{d_t} \tag{10}$$

The force modelling and simulation were performed using a combination of calculations in MS Excel®, which were then applied into the Matlab/Simulink® applications for evaluation of the dynamic behaviour of the vehicle.

Simulated output velocity results for the acceleration process are displayed below in **Figure 3**.

The modelling and simulation from **Figure 3** confirmed that a minimum motor power rating of 10 kW will be required for this application and hence three commercially available AFPM

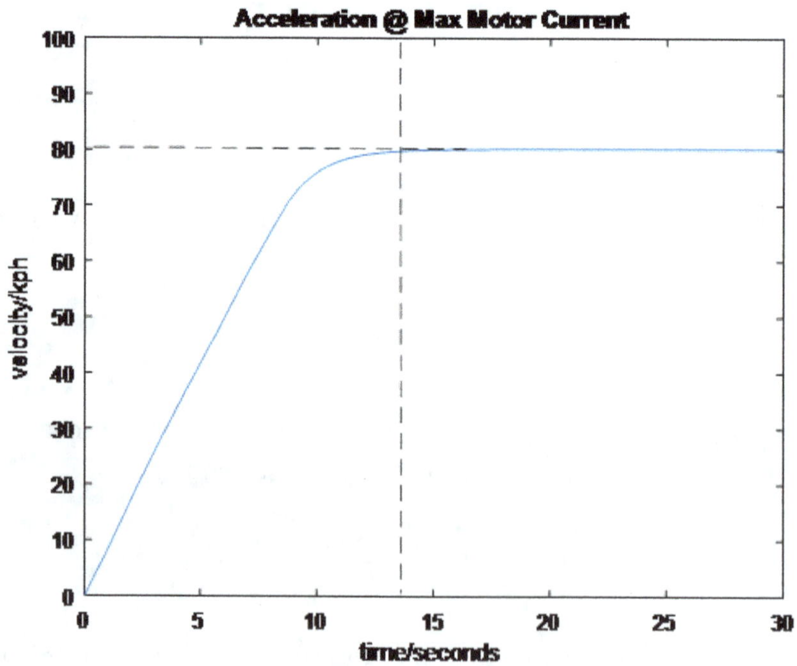

Figure 3. Vehicle acceleration under maximum load.

motors were selected for comparison. **Table 3** presents the ratings of one example of the three motor types, which were evaluated for the project kart. The motors listed in **Table 3** are by no means limited to these three examples. A limited number of model variants are available for selection from each manufacturer, with varying levels of power, torque and voltage ratings depending on the design requirement.

The selection of a traction motor for this project required the following points in the order of priority to be taken into consideration:

- Cost and commercial availability
- Performance characteristics, such as power to weight ratio, power density, efficiency, torque, speed.
- Compact, robust, proven and reliable design
- Motor control technology and simplicity
- Motor cooling requirements, air or liquid cooled.
- Motor to driven wheel final drive mechanical coupling arrangement.

Motor type	PM DC	PM SM	Yokeless DD
Manufacturer	LMC®	Motenergy®	YASA Motors®
Model No.	LEM-200 95	ME-1114	YASA-400
Voltage rated (V)	48	48	400
Voltage peak (V)	60	72	
Power rated (kW)	10	10	50–70
Power peak (kW)	18	24	90
Current cont. (A)	250	180 DC/125 AC	
Current max. (A)	400	600 DC/420 AC	
Efficiency peak (%)	92	92	95
Power to weight rated/peak (kW/kg)	0.90/1.636	0.63/1.512	2.1/3.75
Torque rated (Nm)	28	65	250
Commutation	Brushed	Sine/cosine encoder	
Control technology	PWM voltage/speed current/ torque	PWM frequency inverter	
Country of origin	UK	USA/China	UK
Application examples (published)	Sodikart®, Zerokart®	EV power®	

An equivalent motor is also manufactured by Saietta Group Limited in the UK.

Table 3. Motor comparison.

5.1. A motor with similar characteristics is also manufactured by Magnax in Belgium. Yokeless DD motor

The Yokeless DD motor has been a recent innovation designed for full-size electric vehicles in mind with a higher voltage level and with the preferred mechanical coupling via 'in wheel' direct drive. This drive method eliminates the requirement for a geared coupling, should that be the desired outcome. Clearly, this method is not currently practical for an electric kart situation, due to the small diameter wheels fitted onto the rolling chassis. For an electric kart to take advantage of a DD-type motor, it will require a complete redesign of the rear axle assembly. This will involve mounting the motor directly onto the rear axle through a splined hub, thus eliminating the need for the pinion drive and chain driven sprocket assembly currently in existence. This arrangement will have the added benefit of increasing drivetrain efficiency with the elimination of the geared reduction drive.

The DD motor typically operates between 200 and 400 V. Installed into a kart chassis, the motor will need to run at a lower system voltage level of around 100 V or less. This is necessary to maintain an acceptable battery mass within the limitations of the physical dimensions of the vehicle. This motor requires oil cooling to maintain a desirable operating temperature. The addition of an oil reservoir, pump, power supply and associated hosing may be problematic with available space limitations and increased weight penalty.

Though the topology of the motor is quite simple, the manufacturing process required to produce such a motor can be quite complex and expensive depending on the design adopted by the particular manufacturer. This motor has been included because it represents the future direction in electric vehicle powertrains with its high performance capabilities and compact design [7, 8].

5.2. Permanent magnet synchronous motor

On the other hand, the axial flux permanent magnet synchronous motor (AF PMSM) is very adaptable for an electric kart application, albeit from only a single manufacturer. Readily available from commercial retailers in a number of different power ratings, it provides a very cost-effective alternative. It also displays superior torque speed characteristics compared to its DC counterpart. Previous models of the motor listed in **Table 3** employed a single stator and a single rotor topology with an inbuilt cooling fan using Hall sensor control feedback on rotor position. Later versions of the same motor have been upgraded with a single stator dual rotor with the Hall sensors replaced with a sine/cosine encoder for higher resolution rotor position feedback.

The disadvantages of the PMSM include a lower power to weight ratio and a requirement for an internal fan for cooling purposes. The control of these types of electrical machines requires feedback algorithms incorporating famous synchronous frame (dq) control strategies with variable speed drives. The AC platform, which the PMSM finds integral to its operation, could be somewhat more susceptible to the effects of the harsh operating environment, typified for racing karts. The sensitive feedback electronics in the motor and controller will need to withstand a high impact, high vibration environment of risk and early failure.

5.3. Brushed PMDC motor

The obvious disadvantage with this type of motor is the brushed commutation, which hindered the efficiency in the older traditional radial flux (RF) motors. With advances in brush metallurgy and an inclined brushed mounting arrangement, this problem has been somewhat reduced [10]. Regular brush maintenance is still required to maintain the motor at its peak efficiency [11].

The advantages of this motor outweigh the brush maintenance issue at least at the beginning of its serviceable life. With a high power to weight ratio, good torque speed characteristics, unforced air cooling, compact aspect ratio and robust high quality construction, it is an ideal motor for electric kart applications. In addition, with the motor operating at the lower voltage levels of 48 and 60 V, it is still possible to obtain respectable overall vehicle performance. The lower voltages require less battery mass, thereby assisting in reducing the total powertrain weight. In conjunction with a light weight AF PMDC motor, the end result is a kart with a gross vehicle mass of 82 kg, which is of similar weight to an equivalent ICE powered machine and meets the FIA minimum weight restriction of 80 kg [12].

Combined with simple open-loop torque/speed control from a matched controller, the brushed PMDC motor is a straight forward installation with minimal complexity involved. This simplicity is in line with the overall philosophy of competitive kart racing, which is noted for its minimalist yet robust engineering.

6. Battery, BMS, controller and final drive

6.1. Battery pack

The lithium-ion based chemistry is the most viable chemistry with high power density [9], which can be applied to the battery pack requirements for high performance vehicle applications [13].

There were two common types of lithium-ion chemistries available to choose from when determining the battery pack configuration; Lithium-ion phosphate ($LiFePO_4$) and the lithium-nickel-cobalt-manganese-oxide (LNCM) [14, 15]. The latter has a higher specific energy and performance ratings but can be considerably more expensive to obtain. The $LiFePO_4$ was selected due to lower cost and a recent upgrade in the manufacturers production techniques resulted in an increase in specific energy.

The $LiFePO_4$ battery pack consists of 16 cells with 20 Ah rating at a nominal voltage of 3.2 V for a total nominal pack voltage of 48 V. Various constant current discharge (C) rates have been modelled in Matlab® to provide an estimation of time to 80% Depth of Discharge (DoD) for the battery pack. **Figure 4** provides the details. It must be noted here that it is possible to obtain similar information readily from the battery manufacturers.

It is important to choose the battery supplier/manufacturer wisely by doing some preliminary research on failure rates experienced in the field on any particular brand of battery. Other factors to consider are

- The chemical stability of lithium based batteries is limited over a narrow temperature range
- Maintain each cell voltage within manufacturers upper and lower limits during both charging and discharging
- Maintain voltage and charge equilibrium between individual battery cells

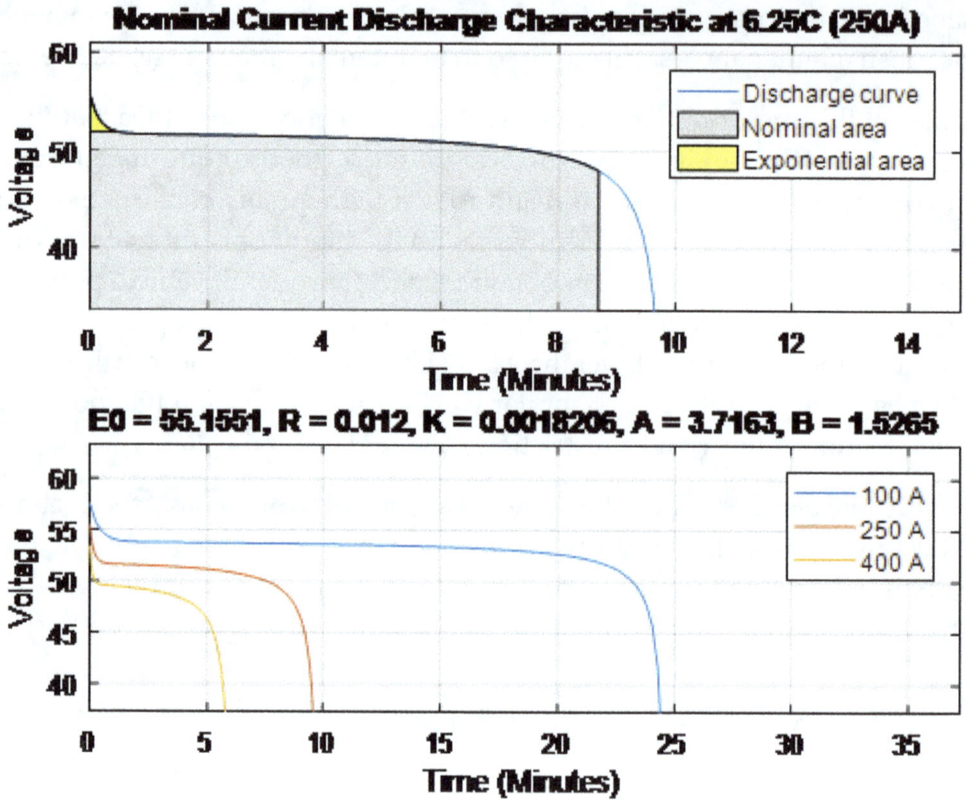

Figure 4. Battery discharge rates.

- Installation of a battery management system (BMS) is mandatory (in order to successfully maintain the conditions mentioned in the two previous points) to avoid premature battery cell/pack failure.
- The battery pack must be housed in a robust IP54 rated housing to prevent mechanical damage

6.2. Battery management system

The BMS consists of 2 × (120 A 51.2 V) modules connected in parallel with each module having the following specification:

- Overvoltage protection 3.75 V per cell
- Under voltage protection 2.1 V per cell
- Full battery pack charge voltage 3.6 V × 16 = 57.6 V
- Balance voltage 3.6 V
- Continuous rated current 120 A
- Over current protection 240 A
- Short circuit protection 320 μs
- Over temperature protection 70°C

6.3. Controller

The PMDC motor requires the simplest of electronic semi-conductor control circuits of all the motor variants previously listed. In a PMDC motor, the speed is governed by output voltage from the controller, which in turn determines armature current for motor torque. The voltage output from controller to the motor is a square wave pulse width modulated (PWM) signal, which produces the voltage variation and hence motor speed regulation, otherwise known as a 'chopper', where the switching frequency of a MOSFET controls the average voltage output with a segmented square waveform.

A DC 450 A controller with single quadrant (Q1) forward motoring capability only was initially selected with the following representative circuit in **Figure 5**.

The controller circuit depicted in **Figure 5** does not provide regeneration capability, whereby when the motor is producing negative torque it will act as a generator and recharge the battery pack via the DC controller. To achieve this function and enhance the charge capability of the battery in the order of 15% or more, a controller with the full bridge circuit depicted in **Figure 6** is required.

Otherwise known as a four quadrant (Q4) controller, it will provide both forward and reverse motoring as well as forward and reverse regeneration capability.

The controller is the one critical device in the complete powertrain which is susceptible to early failure, or underperformance, from poor workmanship, abuse or incorrect setup. Therefore when selecting a suitable controller, the following points were taken into consideration:

- Preference given to controllers which are built for Q4 operation.
- Controllers and their respective circuitry are manufactured to operate with a specific type of electric motor. Each motor type listed above in Section 6 Traction Motor requires its own individually matched controller for successful operation.
- The high frequency switching operation from the MOSFETs can generate considerable heat during high vehicle speed, and high amperage loading situations so inherent in a competition setting. The controller must be mounted in a location with good airflow for heat dissipation.

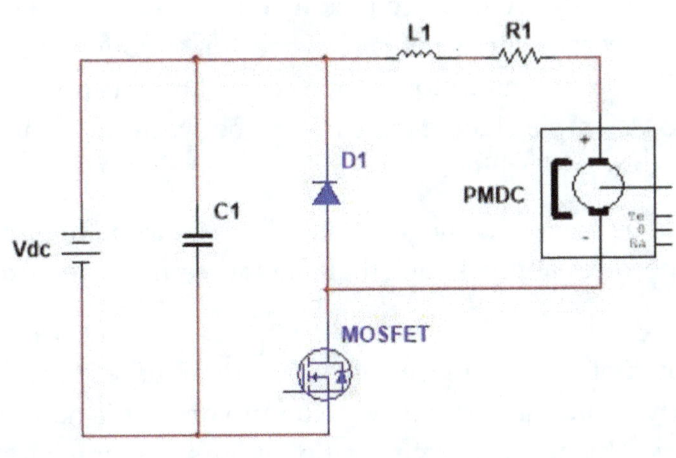

Figure 5. DC Q1 drive circuit.

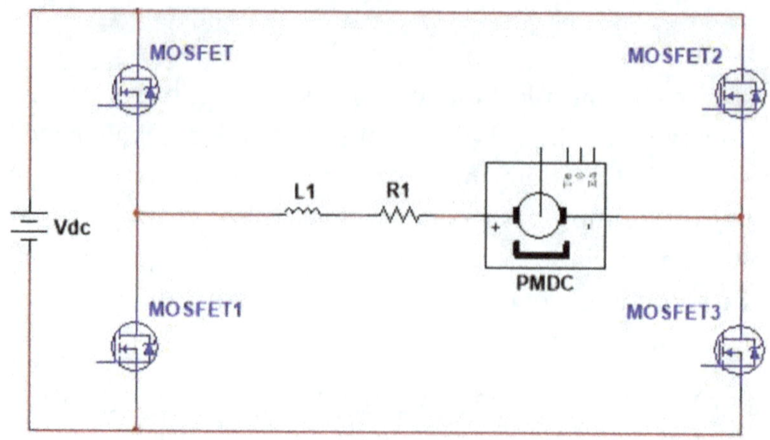

Figure 6. DC Q4 drive circuit.

- The controller must be rated above the maximum current rating of the motor.
- Most controllers are programmable for various functions relating to the operation of the motor. The programming software can either be freely available and user friendly with minimum hardware requirements or quite expensive with specific hardware and software requiring complicated programming techniques. Support from the manufacturer/supplier may not be forthcoming in some instances.
- The controller for the project kart was selected from a well-known brand name manufacturer with an established reputation in the electric vehicle industry. The additional cost paid dividends with ease of setup, technical support and reliability.

6.4. Final drive

The final drive fitted to the kart consists of a pinion gear on the output shaft of the motor driving a chain driven sprocket on the rear axle—a very simple arrangement though subject to drivetrain losses particularly with the chain drive component.

The final drive arrangement can be utilised to modify the speed and torque characteristics of the electric motor to values more in tune to the kart's performance potential to suit a particular racetrack setting. That is to say that the kart can be provided with additional top speed at the expense of torque on a fast-flowing racetrack where maximum velocity is more important than acceleration. And the opposite is the case when the situation is reversed. A slow track with many corners will demand more torque for acceleration as opposed to a high terminal velocity for the vehicle. A simple change in gear ratio between the motor output shaft pinion gear and rear axle sprocket will achieve this result.

This requirement could be somewhat negated should the kart be fitted with a motor exhibiting sufficient torque speed capabilities such that all that would be required is a direct 1:1 ratio final drive.

This ideal situation does not exist in this instance, so a final drive consisting of a single reduction gear and chain drive onto the rear axle with a ratio of 2.3:1 is necessary to accommodate the above requirements. This particular gear ratio favours maximum velocity at the expense of torque (i.e., fast track setup) when taken from the rear axle.

7. Final design

The final design of the electric kart powertrain is a result of direct performance comparisons through modelling and simulation and the evaluation of particular products marketed for light electric vehicles. The technical and production expertise of each manufacturer was an important consideration, when evaluating competing products to ensure a reliable and robust powertrain. The final result is depicted in the schematic of **Figure 7**.

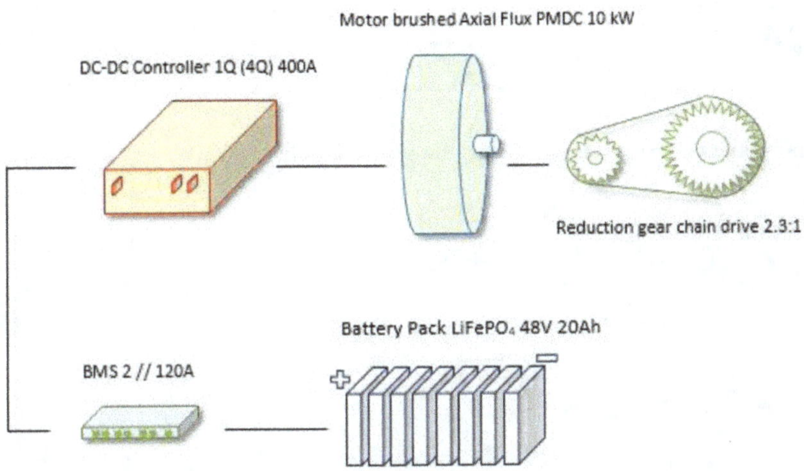

Figure 7. Final design summary.

8. Construction

The complete vehicle with the main powertrain components are depicted in **Figures 8** and **9**.

Figure 8. PMDC 10 kW motor and Q1 450 A controller.

9. On-track testing

The electric kart has been road tested to determine its performance drive potential. **Figure 10** provides a plot of velocity over distance for one such test session. A maximum speed of just over 90 km/h was obtained on a short road section with the vehicle still under acceleration.

Figure 9. 48 V Lithium-ion battery pack with BMS and the prototype electric kart.

Figure 10. Vehicle velocity plot.

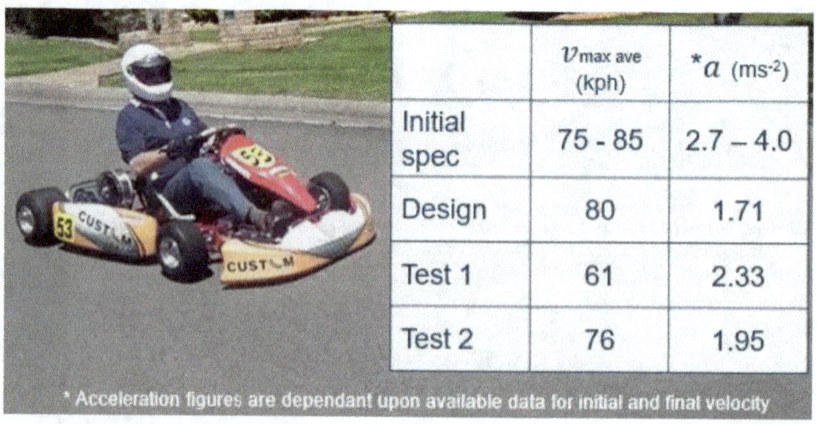

	$v_{max\ ave}$ (kph)	$^*a$ (ms^{-2})
Initial spec	75 - 85	2.7 – 4.0
Design	80	1.71
Test 1	61	2.33
Test 2	76	1.95

* Acceleration figures are dependant upon available data for initial and final velocity

Figure 11. Vehicle performance test data.

A summary of all road test data is presented in **Figure 11** with a comparison against initial design data.

10. Conclusion

The DC 48–72 V platform with the brushed axial flux PMDC motor with a high power to weight ratio, simple torque speed control and compact construction is ideal at the introductory level for the sport of kart racing. However with the exception of the lithium-ion battery pack, the remaining DC components have probably limited scope for further development to improve their performance levels.

For the more advanced levels of the sport, preference should be given to a higher system voltage of 96 V using a PMSM with an axial flux design and closed-loop motor control. This will ensure the maximum performance potential is extracted from the electric vehicle platform utilising readily available EV technology.

The development of the DC electric powertrain detailed above for the application of competition karting provides an insight into what can be achieved with what is essentially well established, proven and reliable technology. Consequently, this particular DC powertrain should be ahead in the reliability stakes whenever it is required to operate at its maximum capability out on the racetrack. The harsh operating environment will quickly destroy any components, which are the least bit fragile in both their construction and operation. A light vehicle AC high-performance powertrain will also need to demonstrate the same qualities as its DC counterpart for it to succeed in this environment.

Author details

Sanath Alahakoon* and Mark Burridge*

*Address all correspondence to: s.alahakoon@cqu.edu.au and mark_burridge@hotmail.com

School of Engineering and Technology, Central Queensland University, Rockhampton, Australia

References

[1] Ehsani M, Gao Y, Gay SE, Emadi A. Modern Electric, Hybrid Electric, and Fuel Cell Vehicles—Fundamentals, Theory, and Design. Boca Raton: CRC Press; 2005

[2] Larminie J, Lowry J. Electric Vehicle Technology Explained. Chichester, West Sussex, England: John Wiley & Sons, Ltd; 2003

[3] Chan KT. Electric Vehicle Machines and Drives Design, Analysis and Application. 1st ed. Singapore: John Wiley & Sons; 2015. p. 38

[4] Karting Australia. Karting Classes [Internet]. 2017 [Updated: 2018]. Available from: http://www.karting.net.au/how-to-get-started-in-karting [Accessed: January 13, 2018]

[5] Miller MotorSports. The Yamaha KT 100 Kart Engine [Internet]. 2014. Available from: https://www.youtube.com/watch?v=Y_lhrfPQ-pg [Accessed: January 13, 2018]

[6] Robert Bosch. Bosch Racing Karts [Internet]. 2016. Available from: http://www.bosch-presse.de/pressportal/de/en/bosch-makes-racing-karts-clean-and-quiet-44889.html [Accessed: January 13, 2018]

[7] Woolmer TJ, McCulloch MD. Analysis of the yokeless and segmented armature machine. IEEE International Electric Machines & Drives Conference, Antalya. 2007. pp. 704-708. DOI: 10.1109/IEMDC.2007.382753

[8] Moreels D, Leijnen P. High efficiency axial flux machines. 2018;**1**(2):1-20

[9] Wikipedia. Power Density [Internet]. 2010 [Updated: November 17, 2017]. Available from: https://en.wikipedia.org/wiki/Power_density [Accessed: January 20, 2018]

[10] Sharkh SMA, Mohammad MTN. Axial field permanent magnet DC motor with powder iron armature. IEEE Transactions of Energy Conversion. 2007;**22**(3):6

[11] LMC. Instructions for Servicing LMC Motors. Available from: http://www.ellert.info/files/LSM.pdf [Accessed: January 20, 2018]

[12] FIA Sport Technical Department. Technical regulations for electric karts (E-karting). 2018:**15**. Available from: file:///C:/Users/alahakos/Downloads/fia_aec_-_technical_regulations_for_electric_karts_e-karting_-_2018.pdf [Accessed: January 20, 2018]

[13] Alahakoon S, Leksell M. Emerging energy storage solutions for transportation—A review: An insight into road, rail, sea and air transportation applications. In: 2015 International Conference on Electrical Systems for Aircraft, Railway, Ship Propulsion and Road Vehicles (ESARS); Aachen; 2015. pp. 1-6

[14] Burridge M, Alahakoon S. The design and construction of a battery electric vehicle propulsion system–high performance electric kart application. In: IOP Conference Series: Earth and Environmental Science; Vol. 73; 2017. DOI: 10.1088/1755-1315/73/1/012016

[15] Miller P. Automotive lithium-ion batteries. Johnson Matthey Technology Review. 2015; **59**:8-9

4

Single-Phase Motors for Household Applications

Damiano D'Aguanno, Fabrizio Marignetti and Francesco Faginoli

Abstract

Single-phase motors are widely used in household applications. Shaded-pole and split-phase capacitor-start single-phase induction motors are very popular for their ruggedness and their comparatively low cost. Recently, line-start single-phase motors are gaining market shares. However, their superior efficiency and torque density are counterbalanced by the higher cost of the rotor construction due to the magnets. This chapter compares the main structures of single-phase line-start motors, presenting their lumped parameter models and the finite element analysis. The equivalent circuits of the single-phase induction motor and of the line-start permanent magnet are derived. Different rotor structures for single-phase line-start permanent magnet (PM) motors are compared. The finite element method (FEM) is used to compare the characteristics of the motors. Motors with the same stator have been tested. No-load and load tests have been performed and compared to the FEM simulations and to the analytical model. Finally, the performances of line-start PM motors are compared to the shaded-pole induction motors in terms of torque density and efficiency.

Keywords: single-phase line-start permanent magnet motors, shaded-pole induction motor, performance analysis, permanent magnets, energy efficiency

1. Introduction

Energy saving is an important aspect of sustainable development in modern society. In this field, electrical machines play a fundamental role in industrial, commercial and residential applications. It is well known that the energy consumed by electrical machines represents the largest part of the total consumption of electricity in the industrial sector. Higher efficiency can lead to the significant reduction of fossil fuel consumption and also of the environmental impact of human activities. For this reason, nowadays and worldwide, all products for industry or residential applications are classified on the basis of their energy efficiency.

Single-phase induction motors are used in household applications due to their robust and simple construction and to their capability for being attached directly to the single-phase grid without using power converters [1, 2]. Split-phase and shaded-pole single-phase induction motors (SPIM) represent today the most common single-phase general purpose motors. As SPIM are inherently not self-starting when directly connected to the grid, they use an auxiliary winding to improve the starting capability. The most significant characteristics of the SPIM are as follows: robust and relatively cheap construction and capability to withstand large overloads. In comparison with three-phase induction motors [3] and to other machine types, SPIMs have a much lower efficiency due to their higher copper and core losses [4].

The motors used in domestic appliances often have a small-rated power, less than 2 kW and run at a constant speed [5].

In this power range, the smaller the motor, the lower the efficiency of the machine. This is mainly due to the fact that both the iron loss and the copper loss are significant in comparison with the rated power. In fact, the stator core of small motors is generally not annealed; the air-gap length is relatively large and the resistance of the stator windings is comparatively large.

Permanent magnet synchronous motors (PMSMs) provide higher efficiency and high-torque density, although they need an inverter for normal operation [6]. Due to dramatic improvements in the magnetic and thermal properties of permanent magnet (PM), materials over the past 40 years, alongside with considerable cost reduction, PM synchronous motors have gained popularity both in the inverter-fed and in the line-start categories [7, 8].

Due to their simple structure and direct connection to the grid, single-phase line-start PM motors (SPLSPMM) represent a good alternative to the induction motor as they produce significant energy savings in the long term. SPLSPMMs are structurally similar to single-phase induction motors with the addition of permanent magnets glued or embedded in the rotor. Line-start PM motors have higher efficiency than SPIMs and operate at near unity power factor [9]. Also, they can be supplied by a three-phase supply source and may be provided with a rotor cage [10, 11]. This motor type is suitable for use in devices such as drain pumps and electric fans [12]. Line-start PM motors start like induction motors and run synchronously like any other type of synchronous motor.

Compared to the widespread induction motors, PM motors with direct online starting ability have higher efficiency, high power factor, low sensitivity towards voltage variations and compact size. They also have the additional advantage of achieving higher power density, besides the capacity to operate at synchronous speed [13, 14].

In particular, the SPLSPMM can be used instead of conventional induction motors for applications like pumps, air conditioners and fans [15]. However, the PM synchronous motor working at line frequency has a major drawback during the starting transient as the stator iron bore must be accurately profiled to increase the starting torque and improve the ability to synchronise with a load attached to its shaft.

During motor start-up, the acceleration torque of the SPLSPMM motor is the average cage torque (if the cage is present) minus the load torque. The permanent magnets on the rotor also generate a braking torque which decreases the starting torque and reduces the ability of the rotor to synchronise. The optimisation of the design of these motors improves the output torque as well as their overall efficiency.

Figure 1. Line-start prototype rotor (right), stator (centre), shaded-pole rotor (left).

The technical literature has dealt with this issue, and different works have investigated how to improve the efficiency of SPLSPMM [16, 17].

Even if the superiority of SPLSPMMs with regard to SPIM is well known, SPIMs are still widely used in many different home appliances.

This chapter analyses the main structures of single-phase motors and compares different rotor structures suitable for SPLSPMM, characterised by different magnet arrangements.

One shaded-pole induction motor (as in **Figure 1**) used in home appliances is compared to different structures of low-cost single-phase line-start PM motor (SPLSPMM) [18, 19] to numerically assess its performance improvement.

The lumped parameter models of the single-phase induction motor and of the single-phase line-start PM motor are presented [20, 21] alongside with their equivalent circuits.

The equivalent circuit of the SPLSPMM is very similar to that of the SPIM. Thanks to this, the SPLSPMM can be easily considered as a particular case of a SPIM with the presence of permanent magnets in the rotor; this leads to an easier comparison of their overall performances.

A comparison of the rotor structures is therefore made through FEM analysis. The optimal solution is experimentally tested.

The study aims at numerically assessing the performances of SPLSPMM and its efficiency in comparison with the shaded-pole SPIM with the same volume and weight [22]. The motors compared basically share the same structure.

The comparison is made by using the finite element method (FEM), the analytical model and the experiments.

2. Efficiency of single-phase motors

Single-phase shaded-pole induction motors are widely used, but their efficiency is low. Their low performances are due to their intrinsic characteristics. The need for sustainability has led to international regulations on energy efficiency. The IEC 60034-30-1 standard, which was published in March 2014 classifies the motors into four levels of energy efficiency (IE1–IE4).

The IEC 60034-30-1 standard applies both to single- and three-phase motors. The classification is different for motors with different number of pole pairs. The European Union has transposed the IEC 60034-30-1 standard, introducing a timeline for energy efficiency of all motors produced in the power range 0.75–375 kW. Since 2017, all the manufactured electric motors must have efficiency at least equal to the IE3 class (or IE2 if the motor is supplied by an inverter). It is expected that the range is extended in the future. Optimising the motors of everyday use could save huge quantities of energy and keep the competitiveness of industries.

In the literature, many studies are aimed at improving the output characteristics of induction motors [23, 24]. Single-phase line-start induction motors are of two main types:

- split-phase induction motors;
- shaded-pole induction motors.

The first use an auxiliary wining provided with an external impedance. The value of the impedance is chosen by taking different aspects into consideration:

- *Reverse field cancellation*: A capacitor is chosen in order to cancel, as far as possible, the inverse torque present in the machine, to obtain a circular field at the air gap and eliminate the alternative torques with double pulsation with respect to the supply. This technique, however, allows to optimise the performances at one operating speed;
- *Minimisation of the reverse torque/direct torque ratio*: The external impedance is chosen to maximise the performance in a wide speed range;
- *Maximisation of electromagnetic torque.*
- *Maximisation of the torque/current consumption ratio;*

To achieve the same objectives, the shaded-pole induction motors use an auxiliary winding which is generally short-circuited and spatially lagging from the main winding. The impedance of the auxiliary winding introduces the necessary time-lag.

3. Line-start permanent magnet motors

Line-start permanent magnet synchronous motors are structurally similar to single-phase induction motors, except for the permanent magnets located on the rotor. The efficiency of SPLSPMM is higher in comparison to conventional SPIM, and furthermore, they can operate with a close to unity power factor. SPLSPMM are mainly used for household applications, such as refrigerators, compressors or extractor hoods/fans. However, these advantages lead to an increase in production costs. As the cost of high-energy permanent magnets is decreasing, it is possible to expect SPLSPMM gained a wider share of the market, to meet the regulations on energy efficiency.

SPLSPMM combine the advantages of permanent magnet motors to those of the cage rotor. The squirrel cage provides the asynchronous start capability, while the action of the magnets disturbs the transient phase. Another advantage of cage-rotor SPLSPMM is that they can be attached directly to the grid.

At steady state, the motor operates at synchronous speed. At synchronous speed, the currents can be reduced. In fact, in an induction machine, the torque is obtained at the price of the speed difference between the main flux and the rotor. This phenomenon generates Joule losses in the rotor bars and to a lesser extent rotor core loss. Furthermore, the magnetising current required to generate the magnetic field, determines additional losses in the stator. In synchronous machines, the magnetic field is produced by the armature winding and by the excitation, with the majority of the magnetic flux produced by the excitation, while a smaller amount of the reactive power is absorbed by the grid during operation. Moreover, in the case, the excitation field is obtained using permanent magnets; there are no copper losses in the rotor and virtually no core loss.

The main limitations of line-start machines (similar to those of SPIM) are that the air-gap field is elliptic and the starting capabilities are limited. Fortunately, the most limitations can be solved by suitably shaping the magnetic circuit.

The mathematical models of the SPIM and of the SPLSPMM are based on the decomposition of the main fluxes into *direct* and *quadrature* components [25–28].

In this chapter, the mathematical model of the shaded-pole induction machine and of the line-start permanent magnet single-phase machine is presented. The performances of both motor types are compared. The model uses the space vectors [23] to describe the distribution of the induction in the air gap.

3.1. The equivalent circuit of the single-phase induction motor

Single-phase induction motors are widely used in low-power applications (up to a few kW). The construction of these machines is similar to the three-phase version, with one single-phase stator winding and one rotor cage. However, they achieve a lower power density. The stator winding, which generally occupies the two-thirds of the stator periphery is supplied with a sinusoidal voltage, which causes a sinusoidal MMF, too. The magnetic field distribution in the air gap has a fixed position while its amplitude varies sinusoidally as the current.

3.1.1. MMF and torque generated by the main winding

The air-gap magnetic field generated by a single-phase winding is:

$$B(\alpha, t) = B_M(t)\, cos(p\alpha) \tag{1}$$

where p is the number of pole pairs, and α is the angular coordinate in the stator reference.

If in Eq. 1 is set:

$$B_M(t) = N\xi \frac{4}{\pi} \frac{1}{2p\delta} \mu_o\, i(t) \tag{2}$$

with N number of turns per pole pair; ξ the winding factor; δ the width of the air gap; μo the permeability of vacuum.

Considering the case of a sinusoidal supply:

$$i(t) = I_M\, cos(\omega t) \tag{3}$$

Using Eq. (1), one has:

$$B(\alpha, t) = k\, I_M\, cos(\omega t)\, cos(p\alpha) \tag{4}$$

which represents an electromagnetic wave varying its amplitude with time. Eq. (4) can be easily rewritten by using trigonometric assumptions:

$$B(\alpha, t) = \frac{k\, I_M}{2}\, [cos(\omega t + p\alpha) + cos(\omega t - p\alpha)] \tag{5}$$

This means that the magnetic field of a single-phase winding can be achieved as the sum of two different fields with the same amplitude and with different verse of rotation (**Figure 2**).

These two fields produce the same effect on the rotor. The field rotating in the same direction as the rotor is called *direct field* while the other *reverse field*. Similarly, the electromechanical torque (T) can be considered as the sum of the direct torque T_d, caused by the direct field and of the reverse torque T_i, caused by the reverse field. Obviously, the values of these torques depend on the speed of the rotor. T_d and T_i are equal if the speed is zero, that is, the slip is equal to one (because at zero slip the magnetic fields having equal amplitudes, rotate with same speed but opposite directions). In all other operating points the values of torque are different.

Hence, two slips can be defined, one direct and one reverse:

$$s_d = \frac{n_o - n}{n_o} \tag{6}$$

$$s_i = \frac{n_o + n}{n_o} \tag{7}$$

In order to study the SPIM, a simplification can be introduced by considering the motor as the union of two three-phase machines. The system can be studied with the technique of the superposition of the effects.

The main drawback of the pure single-phase induction machine is that is not self-starting, because at the starting point the resulting torque is null. When the rotor rotates, a non-zero net torque arises.

3.1.2. *Reverse field cancellation*

In order to solve the problem of cancelling the reverse field produced by the primary winding of the single-phase induction motor, different techniques can be applied. Generally, this is done by adding

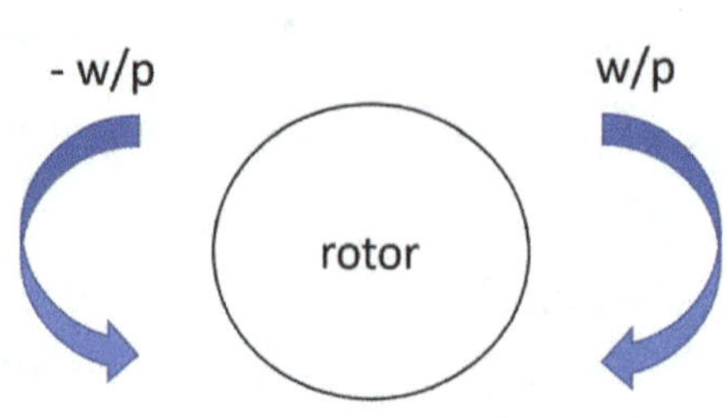

Figure 2. Field rotation directions.

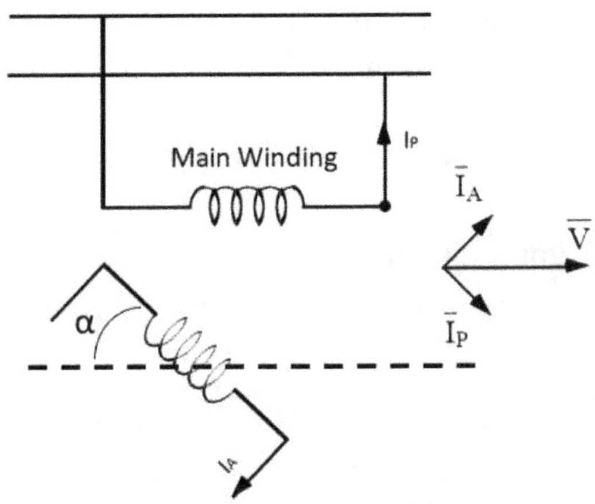

Figure 3. Simplified circuit diagram of a split-phase induction motor.

an auxiliary winding with a magnetic axis displaced γ electric radians from the main winding and supplying the two windings with currents mutually displaced in time of an angle φ (**Figure 3**).

The auxiliary winding produces the additional field *Ba*:

$$B_a(\alpha,t) = B_{aM}(t)\, cos(p\alpha + \gamma) \tag{8}$$

If γ is positive, then the auxiliary winding is lagging. If the auxiliary winding is supplied with a current lagging φ behind the current in the main winding in Eq. 5, the flux density distribution becomes:

$$\begin{aligned} B_a(\alpha,t) &= k_a I_{aM} cos(\omega t - \varphi) cos(p\alpha + \gamma) = \\ &= \frac{k_a I_{aM}}{2} \cdot [cos(\omega t - p\alpha - \gamma - \varphi) + cos(\omega t + p\alpha + \gamma - \varphi)] \end{aligned} \tag{9}$$

The resultant flux density in the air gap is achieved summing Eq. (9) to Eq. (5) and consists of two rotating fields, one in the *direct direction*:

$$B_d = \frac{k\, I_M}{2} cos(\omega t - p\alpha) + \frac{k_a\, I_{aM}}{2} cos(\omega t - p\alpha - \gamma - \varphi) \tag{10}$$

and the other one in the *reverse direction*:

$$B_r = \frac{k\, I_M}{2} cos(\omega t + p\alpha) + \frac{k_a\, I_{aM}}{2} cos(\omega t + p\alpha + \gamma - \varphi) \tag{11}$$

To ideally cancel the reverse field, two conditions must be considered:

1. The MMF amplitudes of the primary and auxiliary fields must be equal:

$$k_a \cdot I_{aM} = k \cdot I_M,$$

2. The following relationship between the geometrical and the time phase lags must hold:

$$\gamma - \varphi = \pi,$$

Finally, the maximum direct field amplitude is achieved if:

$$\gamma = -\varphi. \tag{12}$$

In the case, the two conditions for reverse flux cancellation are achieved, the resultant flux density in the air gap is achieved by summing Eq. (9) to Eq. (5):

$$B(\alpha,t) = k\,I_M\,[cos(\omega t - p\alpha) + cos(\omega t - p\alpha - \gamma - \varphi)] \tag{13}$$

Eq. (13) represents one field rotating in one direction, like in a three-phase machine. Obviously, the perfect cancellation may be achieved only at one working point, for example, at start-up or at nominal load.

In split-phase machines, condition 2 is achieved by displacing the auxiliary winding of $\gamma = \pi/2$ that means $\varphi = -\pi/2$. The minus sign means that the auxiliary current *leads* the main current.

Therefore, the split-phase SPIM includes two stator windings: one *main winding* and one *auxiliary winding*, displaced of 90°. If the auxiliary winding is used for starting, it can be excluded when the machine reaches a fixed operating condition.

The current lag of $\pi/2$ is produced by introducing a capacitor in series to the auxiliary winding. This is necessary to provide the self-starting capability and to improve its performance, so that the phase displacement between the currents circulating in the two stator windings, creates an imbalance between the direct torque and the reverse torque. This phase shift is possible because the ohmic-capacitive nature of one of the winding (due to the presence of the capacitor).

Since the auxiliary winding can be disconnected over a fixed speed, capacitor-start induction machine can be divided into permanent capacitor topology (**Figure 4**) and starter capacitor topology (**Figure 5**).

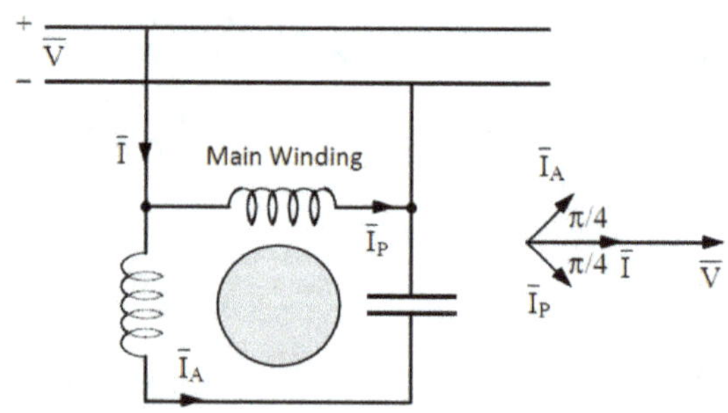

Figure 4. Permanent capacitor induction motor.

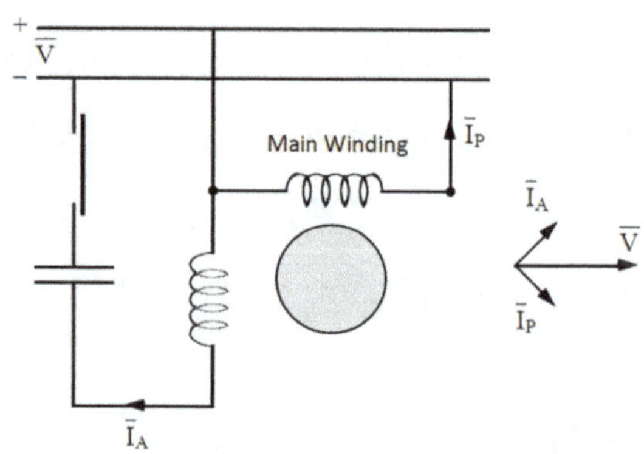

Figure 5. Capacitor-start single-phase induction motor.

For the capacitor-start motor, the value of the capacity is selected to achieve the desired starting performance, while for the permanent capacitor motors, it is generally the result of a trade-off between performances under different load conditions.

In shaded-pole SPIMS, the auxiliary winding is, instead, composed of two short-circuited windings wound around the pole shoes. The angle ψ is varied by varying the number of the short-circuited turns until the desired performance is reached.

3.2. Analysis of the single-phase induction motor

Based on the air-gap flux density distribution in Section 3.1.2, it is possible to compute the EMF induced in the stator and rotor windings. Afterwards, the electromagnetic torque is obtained and finally the equivalent circuit is derived.

3.2.1. Resultant air-gap flux density

The air-gap flux density is the sum of the flux density distribution of the stator windings and of the cage:

$$B(\alpha, t) = B_{stator} + B_{cage} \tag{14}$$

where B_{stator} and B_{cage} can be derived from Eqs. (10) and (11):

$$\begin{aligned} B_{stator} &= \sqrt{2}\,\frac{\mu_0}{\pi p}\,N_i\xi_i\,Re[I_{fs}e^{j(\omega t - p\alpha)} + I_{bs}e^{j(\omega t + p\alpha)}] \\ B_{cage} &= \sqrt{2}\,\frac{\mu_0}{2\pi p\delta}\,Re[I_{fr}e^{j(\omega t - p\alpha)} + I_{br}e^{j(\omega t + p\alpha)}] \end{aligned} \tag{15}$$

where I_{fs} and I_{fr} are the direct and the reverse component of the stator currents system, respectively. If the cancellation of the reverse flux is not perfect, the amplitudes of these current components are different. The same is for the cage.

At sinusoidal steady state, the main harmonic of the air-gap flux density is:

$$\begin{aligned} &B(\alpha, t) = \sqrt{2}\,\frac{\mu_0}{\pi\delta}\,N_i\xi_i\,Re[(I_{fs} + \kappa I_{fr})e^{j(\omega t - p\alpha)} + (I_{bs} + \kappa I_{br})e^{j(\omega t + p\alpha)}] \\ &\text{with} \quad \kappa = \frac{1}{2pN_i\xi_i} \end{aligned} \tag{16}$$

3.2.2. Stator back EMFs

The EMF of one stator winding can be obtained by summing the EMF of each coil.

$$e_{is} = -\frac{1}{2}\,pDl \sum_{\nu=0}^{q_i - 1} z_{\nu i}\,\frac{d}{dt} \int\limits_{-\frac{\pi}{2p} - \frac{\pi(q_1 - 1)}{Q} + \frac{2\pi\nu}{Q} + \frac{\alpha_i}{p}}^{\frac{\pi}{2p} + \frac{\pi(q_1 - 1)}{Q} + \frac{2\pi\nu}{Q} + \frac{\alpha_i}{p}} B(\alpha, t)\,d\alpha \tag{17}$$

where D is the diameter of the machine; $z_{\nu i}$ is the number of conductors in slot.

The EMF induced in the stator windings can be evaluated as:

$$e_{is} = -\sqrt{2}X_m\beta_i\, Re[j(I_f e^{-j(\alpha_i-\alpha_i^*)} + I_b e^{-j(\alpha_i-\alpha_i^*)})\, e^{j\omega t}] \tag{18}$$

where the following equation was assumed:

$$X_m = \frac{\mu_0}{\pi\delta}\,\omega D l (N_i \xi_i)^2$$

3.3. Rotor back EMFs

The EMFs in the rotor bars are obtained in the same way as before.

$$e_{kr} = -\frac{1}{4}\, Dl\, \frac{d}{dt} \int_{-\frac{\pi}{2p}+\omega_r t+\frac{2\pi(k-1)}{m_r}}^{\frac{\pi}{2p}+\omega_r t+\frac{2\pi(k-1)}{m_r}} B(\alpha,t)\, d\alpha \tag{19}$$

The EMF induced in each rotor bar is the sum of two sinusoidal components with pulsation $(\omega + p_{\omega r})$ e $(\omega - p_{\omega r})$:

$$e_{kr} = -\sqrt{2}\,\frac{\mu_0}{2\pi p\delta}\, DlN_i\xi_i\, Re[j(I_f\,(\omega - p\omega_r)\, e^{-j(\omega-p\omega_r)t}\, e^{j\frac{2\pi p(k-1)}{m_r}} + \\ + I_b\,(\omega + p\omega_r)\, e^{j(\omega+p\omega_r)t}\, e^{j\frac{2\pi p(k-1)}{m_r}}] \tag{20}$$

3.4. Electromagnetic torque expression

The electromechanical torque can be evaluated as:

$$T = -\frac{D^2 l}{4} \int_0^{2\pi} B(\alpha,t)\, \Theta_r(\alpha,t)\, d\alpha \tag{21}$$

with $B(\alpha, t)$ the distribution of the induction in the air gap and $\Theta_r\,(\alpha, t)$ the rotor current density:

$$\Theta_r(\alpha,t) = \frac{2}{D}\frac{d}{d\alpha}[\Delta F_r(\alpha,t)] \tag{22}$$

where $\Delta F_r\,(\alpha, t)$ is the MMF caused only by the rotor currents:

$$\Delta F_r(\alpha,t) = \frac{\delta}{\mu_0}\, B(\alpha,t) = \frac{\delta}{\mu_0}(\sqrt{2}\,\frac{\mu_0}{2\pi p\delta}\, Re[\frac{I_{fr}}{\rho}\, e^{j(\omega t-p\alpha)} + \frac{I_{br}}{\rho}\, e^{j(\omega t+p\alpha)}]) \tag{23}$$

Eq. 22 becomes:

$$\Theta_r(\alpha,t) = \frac{2\sqrt{2}N_i\xi_i p}{D\pi}\, Re[-jI'_{fr}\, e^{j(\omega t-p\alpha)} + jI'_{br}\, e^{j(\omega t+p\alpha)}] \tag{24}$$

By substituting Eq. 24 and Eq. 16 in Eq. 21 and finally solving, the expression of the electromagnetic torque is obtained:

$$T = -X_m\frac{p}{\omega}\, Re[jI_f I'_{fr} - jI_b I'_{br} + j(I_f I'_{br} - I_b I'_{fr})\, e^{j2\omega t}] \tag{25}$$

where the following equation was assumed: $I_f = I_{fs} + \rho I_{fr}$ and $I_b = I_{bs} + \rho I_{br}$.

Therefore, the electromagnetic torque in a single-phase induction motor is the sum of three terms:

1. Direct electromagnetic torque:

$$T_d = -X_m \frac{p}{\omega} Re[jI_f I_{fs}]$$

2. Reverse electromagnetic torque:

$$T_i = -X_m \frac{p}{\omega} Re[-jI_b I_{bs}]$$

3. Electromagnetic torque with pulsation 2ω:

$$T_a = -X_m \frac{p}{\omega} Re[j(I_f I_{bs} - I_b I_{fs})\, e^{j2\omega t}]$$

3.5. Mathematical model

The mathematical model of the SPIM is obtained by considering the fundamental harmonic of the air-gap flux density distribution.

The impedances of the two circuits (**Figure 6**), including the external impedance are: $Z_1 = R_1 + jX_1$ and $Z_{t2} = R_2 + R_c + j(X_2 + X_c)$.

The currents I_1 and I_2 can be expressed as a function the direct and reverse components of the stator currents:

$$\underline{I}_1 = \frac{\underline{I}_{fs} + \underline{I}_{bs}}{2} \tag{26}$$

$$\underline{I}_2 = \frac{\underline{I}_{fs} + \underline{I}_{bs}}{2j\beta} \qquad \beta = \frac{N_2\xi_2}{N_1\xi_1} \tag{27}$$

The equations referred to the various nets of the equivalent circuits are:

where

$$\begin{aligned} +\underline{V} &= \dot{A}\underline{I}_{fs} + \dot{B}\underline{I}_{bs} \\ \pm\underline{V} &= \dot{C}\underline{I}_{fs} + \dot{D}\underline{I}_{bs} \end{aligned} \tag{28}$$

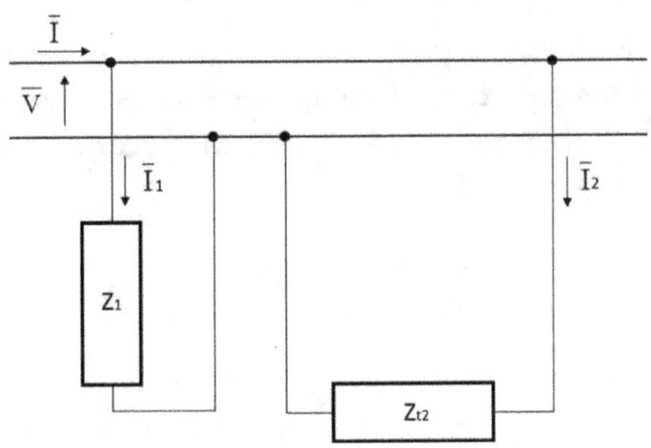

Figure 6. Grid connection of a capacitor-start single-phase induction motor.

where

$$\dot{A} = \frac{(R_1 + jX_1)}{2} + jX_m \cdot [1 - \frac{jX_m}{\frac{R_r'}{s} + jX_t}]$$

$$\dot{B} = \frac{(R_1 + jX_1)}{2} + jX_m \cdot [1 - \frac{jX_m}{\frac{R_r'}{2-s} + jX_t}]$$

$$\dot{C} = \frac{[R_2 + R_c + j(X_2 + X_c)]}{2j\beta} + \beta X_m [1 - \frac{jX_m}{\frac{R_r'}{s} + jX_t}]$$

$$\dot{D} = -\frac{[R_2 + R_c + j(X_2 + X_c)]}{2j\beta} + \beta X_m [1 - \frac{jX_m}{\frac{R_r'}{2-s} + jX_t}]$$

The expression of the electromagnetic torque is finally recalled.

$$T_m = -\frac{p}{\omega} X_m^2 [\frac{\frac{R_r'}{s}}{(\frac{R_r'}{s})^2 + j(X_m + X_r')^2} I_{fs}^2 - \frac{\frac{R_r'}{2-s}}{(\frac{R_r'}{2-s})^2 + j(X_m + X_r')^2} I_{bs}^2] \tag{29}$$

3.6. The equivalent circuit of the single-phase induction motor

The equations, which define the mathematical model of the single-phase induction motor with an auxiliary winding and external impedance, can be graphically translated into equivalent circuits.

Referring to the model Eqs. (28) after some manipulations, one can write:

$$\begin{aligned} \underline{V} \frac{\beta \pm j}{\beta} &= [\frac{\dot{Z}_2 + \dot{Z}_c}{2\beta^2} + jX_m (1 - \frac{jX_m}{\frac{R_r'}{s} + jX_t})] I_{fs} + (I_{fs} + I_{bs}) \frac{[\dot{Z}_1 - \frac{\dot{Z}_2 + \dot{Z}_c}{\beta^2}]}{4} \\ \underline{V} \frac{\beta + j}{\beta} &= [\frac{\dot{Z}_2 + \dot{Z}_c}{2\beta^2} + jX_m (1 - \frac{jX_m}{\frac{R_r'}{2-s} + jX_t})] I_{bs} + (I_{fs} + I_{bs}) \frac{[\dot{Z}_1 - \frac{\dot{Z}_2 + \dot{Z}_c}{\beta^2}]}{4} \end{aligned} \tag{30}$$

4. The line-start permanent magnet synchronous motor

The model of the line-start PM machine is derived hereafter on the basis of the model of the shaded-pole machine without any auxiliary winding.

4.1. Stator winding flux density distribution in the air gap

In a single-phase machine with p pole pairs, the magnetic field in the air gap is fixed in space and variable in time due to the structure of the motor and its winding. The armature winding is therefore unable to create a rotating field in time and space. The first harmonic of the stator magnetic field can be written as:

$$B_s(\alpha, t) = K\ i_s(t) \cdot cos(p\alpha) \tag{31}$$

where K is

$$\mu_0\ f\ \frac{N\ K_w}{2\delta}. \tag{32}$$

with f form factor, which is

$$f = \frac{4}{\pi} \tag{33}$$

in the case of rectangular magnetomotive force (MMF). The equation of the stator winding flux density is:

$$B_s = \sqrt{2}\frac{\mu_0}{\pi p} N K_w Re\left[I_{fs}e^{j(\omega t - p\alpha)} + I_{bs}e^{j(\omega t + p\alpha)}\right] \tag{34}$$

This field can be divided into two components, namely, the direct and the reverse field.

4.2. Total air-gap flux density of SPLSPMMs

The air-gap flux density of line-start single-phase PM synchronous motors is the sum of two contributions: the stator winding flux density and the permanent magnet flux density distribution B_{pm}. In some cases, an incomplete squirrel cage may be added to improve the starting performance and a further flux density distribution B_c is added:

$$B_p\,(\alpha, t) = B_s + B_c + B_{pm} \tag{35}$$

with:

$$B_s = \sqrt{2}\frac{\mu_0}{\pi p} N_i \xi_i Re[I_{fs}e^{j(\omega t - p\alpha)} + I_{bs}e^{j(\omega t + p\alpha)}] \tag{36}$$

$$B_c = \sqrt{2}\frac{\mu_0}{2\pi p\delta} Re[I_{fr}e^{j(\omega t - p\alpha)} + I_{br}e^{j(\omega t + p\alpha)}] \tag{37}$$

$$B_{pm} = B_M cos(p\alpha - \omega_r t + \beta) \tag{38}$$

At sinusoidal steady state and ω_r rotor speed, one has:

$$\begin{aligned} B_{\alpha,t} = \sqrt{2}\frac{\mu_0}{\pi\delta} N_i \xi_i Re[(I_{fs} + \rho I_{fr})e^{j(\omega t - p\alpha)} + (I_{bs} + \rho I_{br})e^{j(\omega t + p\alpha)}] \\ + B_M cos(p\alpha - \omega_r t + \beta) \end{aligned} \tag{39}$$

where I_{fs} and I_{bs} represent the forward and backward components of the stator current. Each flux density component gives rise to a contribution to the back EMF.

The more accurate model of a SPLSPMM is that of Ref. [14], which is based on [1]. The model analyses a SPLSPMM with an asymmetrical stator winding, an auxiliary winding and PMs on the rotor causing braking and pulsating torques. The model is based on a combination of symmetrical components and d-q axis theory. For a motor without an auxiliary winding, the V_q symmetrical component is zero, while $\underline{V_d} = 2V_m$. At synchronous steady state, the forward voltage is synchronised with the rotor, and the model can be simplified using the average of the apparent d-axis impedance, while the q-axis part is unnecessary. The resulting d-axis component model at steady state is represented as the equivalent circuit of **Figure 7** for a SPIM and **Figure 8** for a SPLSPMM.

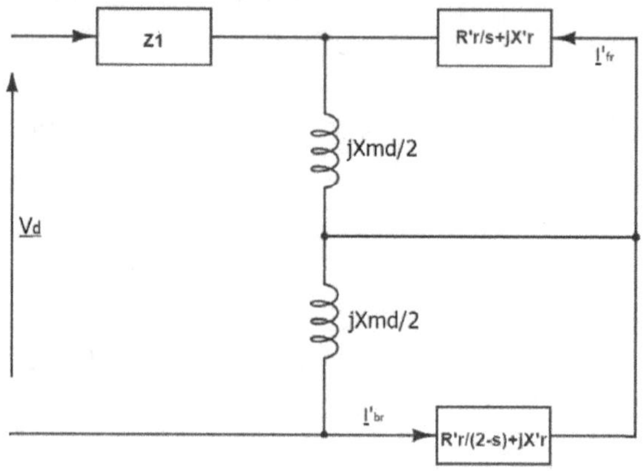

Figure 7. Equivalent circuit of a single-phase induction machine.

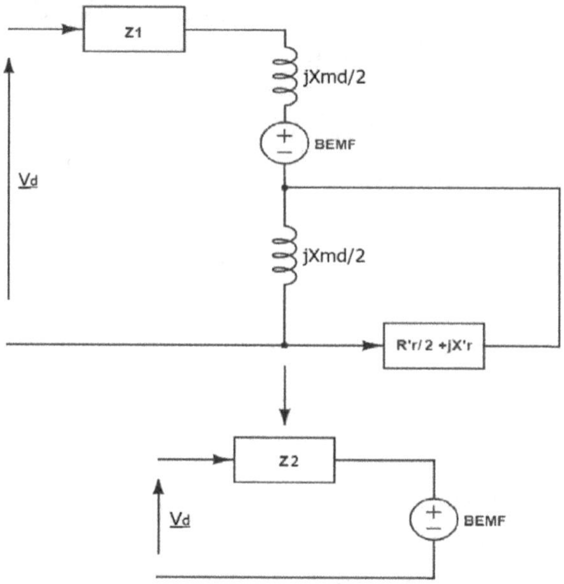

Figure 8. Equivalent circuits of a single-phase SPLSPMM machine, where Z1 is the stator impedance; X_{md} is the d-axis magnetisation reactance referred to the stator (main) winding; R_r^I is the rotor resistance referred to the stator winding; X_r^I is the rotor leakage reactance referred to the stator winding; *BEMF* is the equivalent back EMF caused by the presence of the magnets; Z2 is the average of the apparent d-axis impedance Z2 = [Z1 + jX_{md}] + [j*Xmd*//(Rr^1/2 + jXr^1)] [14].

5. Selection of the optimum configuration

5.1. Machine specifications and no-load field simulation

Different machine configurations have been numerically compared with the help of the finite element method (FEM). All configurations share the same stator core, which is the same as the core of the SPIM used as a benchmark, while the number, the type and the shape of rotor magnets are different. An incomplete cage is present on the rotor to improve the starting capability of the motor.

Table 1 shows the main dimensions and specifications of the considered motors. The winding is made of 1723 turns with a 0.2 mm copper wire, the material used for the cage is aluminium.

The first configuration is directly derived from the shaded-pole induction motor and includes two surface-mounted NdFeB magnets. The FEM motor model with the flux density contour is shown in **Figure 9(a)**. The volume of the magnets required by this structure is considerable and also the manufacturing cost is larger than that of the following solutions; therefore, it will be included here only for the sake of completeness. The incomplete cage is used to ensure starting in all configurations.

The second configuration is similar to the first, it is made of two NdFeB magnets but instead of being surface mounted they are buried inside the rotor. The FEM motor model with the flux density contour is in **Figure 9(b)**.

The third configuration is based on the second and includes two further inset NdFeB magnets to increase the field in the *d*-axis of the rotor. The FEM motor model with the flux density contour is in **Figure 9(c)**.

In order to further investigate the role of the middle magnets, a fourth configuration is proposed. It is the same as the third, although the two middle magnets are closer to the air gap. The FEM motor model with the flux density contour is in **Figure 9(d)**.

Finally, a low-cost configuration is presented. It is the same as the third configuration; it is equipped with two ceramic middle magnets instead of the inset NdFeB magnets. The FEM motor model with the flux density contour of this configuration is in **Figure 9(e)**.

5.2. Flux linkage

The flux linkage is numerically calculated for each of the above structures. The flux linkages of the different configurations are shown in **Figure 10**.

It can be seen that, while the first configuration produces the highest flux linkage, there is no significant difference among the other configurations. Based on cost considerations, configurations 2 and 4 must be preferred.

The configuration chosen is shown in **Figure 1** and in **Figure 9(b)**, with two Grade 32 NdFeB magnets embedded in the rotor. This configuration offers the maximum flux density with the lowest increase in the cost of manufacturing the machine. Furthermore, this configuration is the easiest to produce.

Element	Dimension
Stator height	67.6 mm
Stator width	84.6 mm
Rotor diameter	34.24 mm
Air gap	0.4 mm
Lamination length	19 mm
Poles	2
Turns	1723

Table 1. Prototypes main dimensions and specifications.

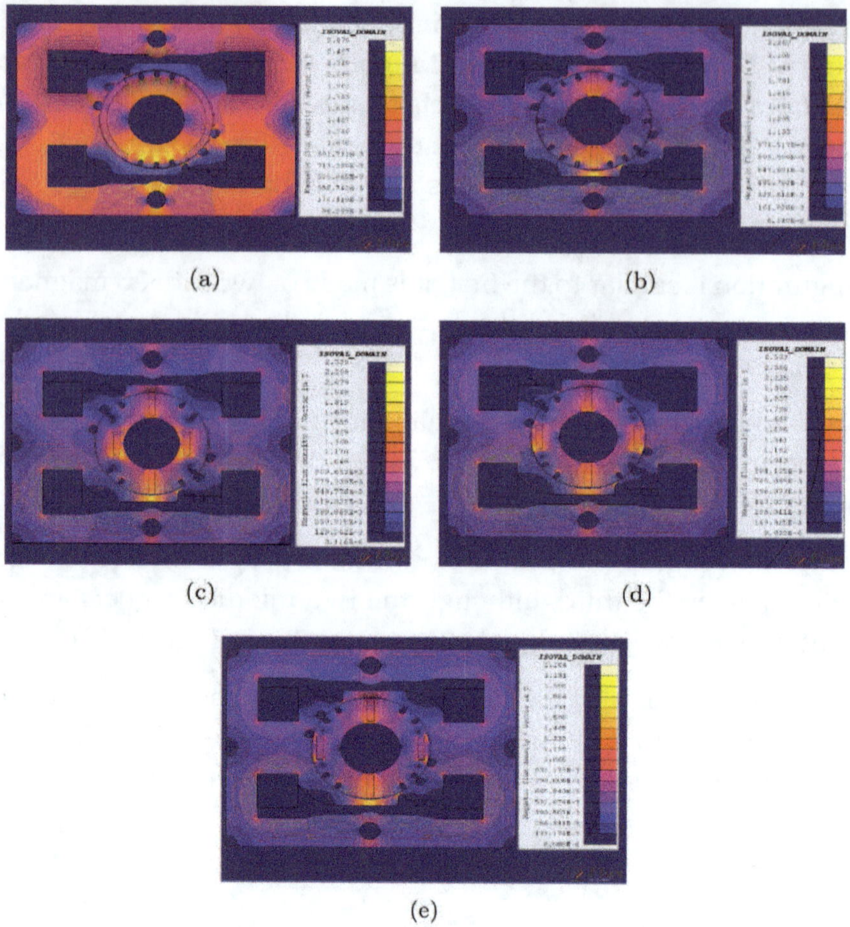

Figure 9. No load FEM simulations. (a) FEM simulation result of the first rotor configuration with two surface NdFeB magnets (model 1). (b) FEM simulation result of the second rotor configuration with two inset NdFeB magnets (model 2). (c) Third rotor configuration with with four NdFeB (position 1) simulation result (model 3). (d) Configuration with four NdFeB magnets (position 2) simulation result (model 4). (e) Configuration with two NdFeB magnets and two ceramic magnets (position 1) simulation result (model 5).

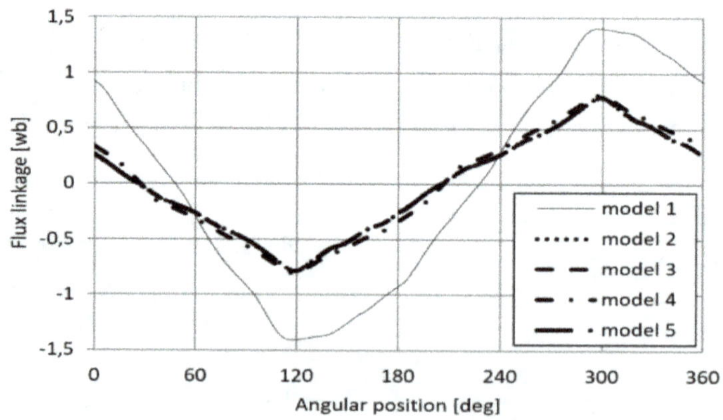

Figure 10. Flux linkages of the different structures for different rotor positions.

The back EMF can be calculated from the flux linkages as:

$$e = \frac{d\phi}{dt}\frac{d\theta}{dt} = \frac{d\phi}{dt}\omega_r \quad (40)$$

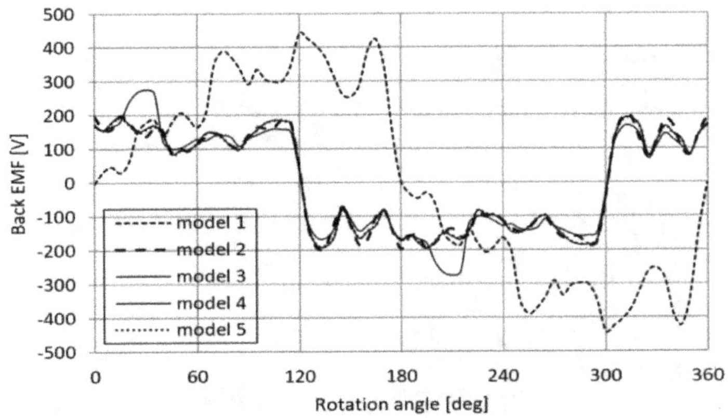

Figure 11. Predicted back EMFs versus rotor position model 1, 2, 3, 4, 5.

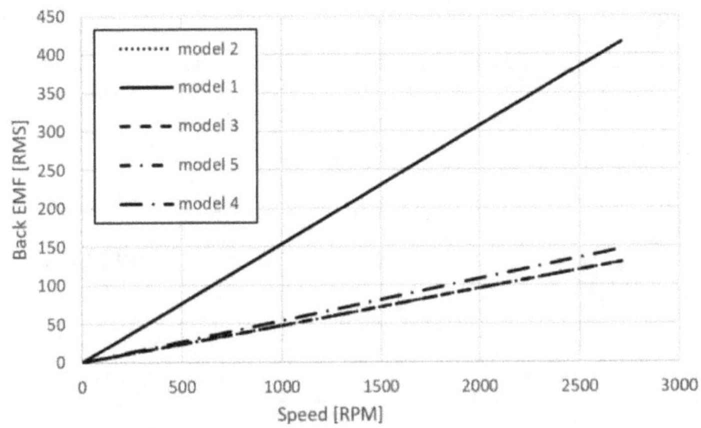

Figure 12. Back EMFs for different rotor speed different solutions.

Using Eq. (40), the back EMF is computed over a complete rotation of the rotor and its RMS value is extracted. On the other hand, the trends for the predicted back EMF are shown in **Figure 11**. The RMS value of the back EMF versus the speed (for the different solutions) is shown in **Figure 12**. One can notice that the trend is, as expected, approximately linear. These values are compared to the experimental results in Section 6 (for the chosen solution).

6. Experimental tests

To compare the SPLSPMM's performances with the SPIM's, the SPLSPMM has been supplied with a single-phase inverter implementing a V/Hz ramp.

6.1. Parameter identification of SPLSPMMs

In order to obtain the machine model parameters and study its performances, both no-load and load tests have been performed. The test bench is shown in **Figure 13**.

The equivalent circuit parameters, derived from tests, are shown in **Table 2**, while **Figure 14** shows the trend of the magnetisation reactance as a function of the voltage (which has been implemented in the mathematical model).

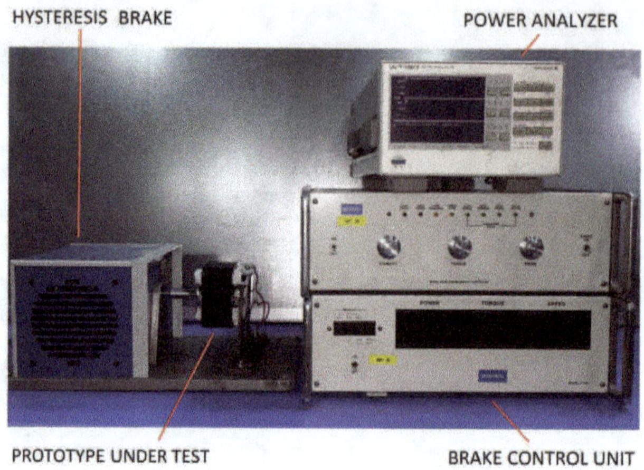

Figure 13. Test bench.

Parameter	Value
Rs	76.37 Ω
Xs	40.21 Ω
R_r^I	10.19 Ω
X_r^I	20.10 Ω
Xmd	113.34 Ω

Table 2. Equivalent circuit parameters.

For the open circuit test, the machine is coupled to an induction machine and the terminals of the motor under test (MUT) are open. The terminal voltages are measured at different values of the speed. The trend of the back EMFs RMS amplitudes is approximately linear as shown in **Figure 15**. From the same figure, it is possible to verify that the analytically predicted trend is very close to the experimental results. The slight differences are caused by the manufacturing process (which may have degraded the magnet superficially) and by the mathematical approximations.

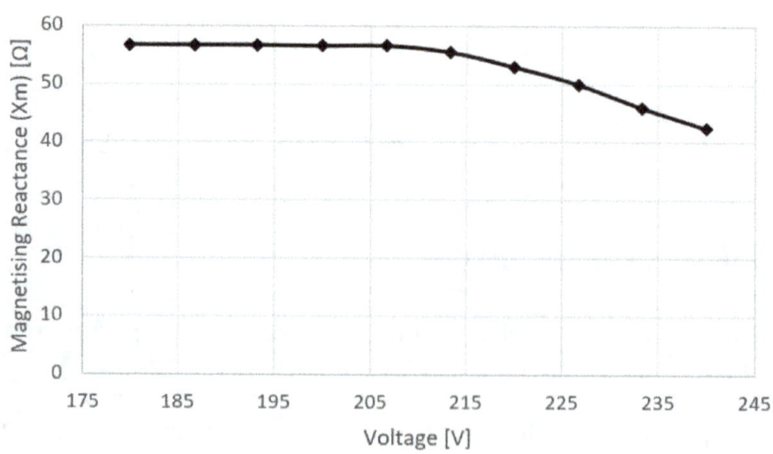

Figure 14. Magnetisation reactance as a function of voltage.

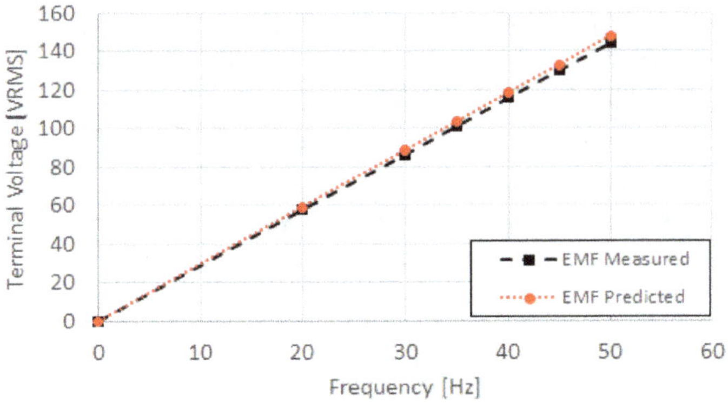

Figure 15. Back EMFs comparison.

The value and waveform shape of the back EMF at 2710 rpm (45.2 Hz) are shown in **Figure 16**. Considering the same operating point, a FEA simulation is performed, and the behaviour of the machine is shown in **Figure 11**.

The predicted and the experimental voltage RMS values are the same and the shapes of the back EMFs are approximately the same, except for the larger ripple presented by the experimental data, which is due to the rotor cage slotting.

In **Table 3**, the rated operating point of the shaded-pole machine is compared with two operating points of the prototype. Considering a fixed frequency of 50 Hz and a variable voltage, the maximum torque is delivered when the motor is supplied with 230 V, while the maximum efficiency is achieved when supplied with 190 V. More results are shown in Section 6.2.

The data shown in **Table 3** demonstrate that only with a slightly different rotor configuration of the machine, the performances improved considerably. This means that the motor can be used both in a completely different field and in similar applications, as in the case we are considering, with significantly better performances.

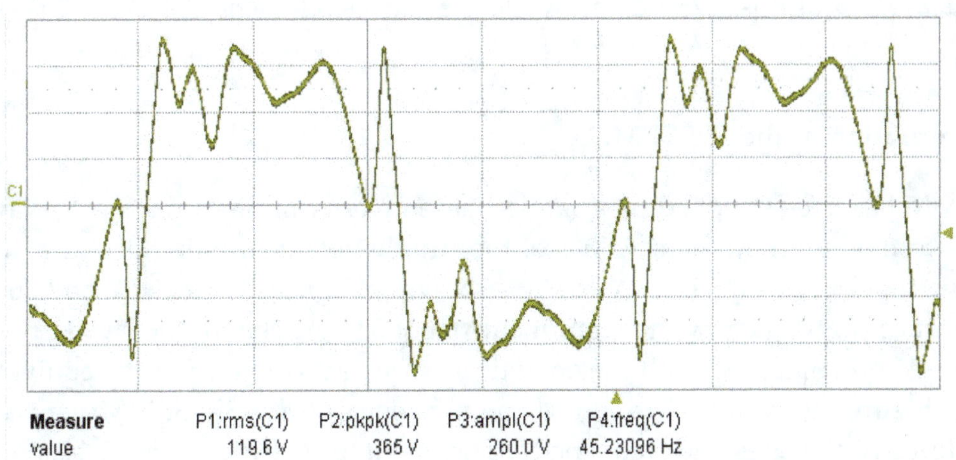

Figure 16. Measured back EMFs versus rotor position.

Quantity	Shaded-pole machine (rated values)	Prototype maximum torque O.P.	Prototype maximum efficiency O.P.
Current [A]	0.214	0.92	0.75
Voltage [V]	230	230	190
Electric power [W]	68	148	123
Rated speed [RPM]	2600	3000	3000
Power [W]	27	76.8	89.6
Rated frequency [Hz]	50	50	50
Efficiency [%]	12.00	62.44	73.59

Table 3. Motors characteristics.

6.2. Characterisation of the SPIM

The SPIM has one stator winding (main winding) and one short-circuited winding, while the rotor is of the squirrel-cage type. The single-phase induction motor without its auxiliary winding is not self-starting. When the motor is connected to a single-phase supply, the main winding carries the alternating current, which produces a pulsating magnetic field. To generate torque, an auxiliary winding is needed.

The SPIM used shares the same magnetic circuit as the SPLSPMM. The induction machine is supplied with its rated voltage and frequency, hence 230 V and 50 Hz, and the load torque is increased with steps between 0 and the maximum torque.

In order to have an idea of the performances of the shaded-pole motor here are some data. The maximum torque achievable by the motor in these conditions is 1.26 kg cm at 2076 rpm. The supply current of the motor is 0.82 A in this operating point. The input power is 107 W and the output power is 27 W, hence the efficiency in this point is 25.23%. The maximum SPIM efficiency of 28.16% is achieved at 2350 rpm, with an input current of 0.78 A and an output torque of 1.14 kg cm. The same test has been performed with different values of supply voltage between 180 V and 240 V. The highest value of torque, in these conditions, is delivered at 240 V and it is 1.23 kg cm, while the highest efficiency of 27.03% is obtained at 210 V.

6.3. Characterisation of the SPLSPMM

The SPLSPMM has been supplied by a single-phase inverter implementing a V/f law. The test has been performed with different values of the supply voltage ranging between 180 and 230 V. The torque trend versus the input current is shown in **Figure 17**. At 230 V and 3000 rpm the achieved torque is 2.91 kg cm with an input current of 0.92 A. The torque has been experimentally evaluated and compared with the analytical torque achieved using the equivalent circuit (**Figure 18**). **Figure 19** shows the trend of the efficiency of the SPLSPMM versus the input current at different voltages and frequencies. The maximum efficiency is 62.44% with 0.75 A, 2.49 kg cm at 3000 rpm. The highest torque value is delivered at 220 V and it is 2.6 kg cm, while the most efficient point is obtained at 190 V with 73.59%.

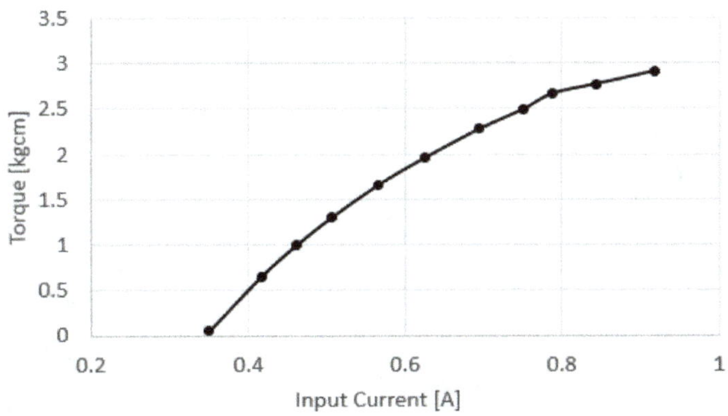

Figure 17. The SPLSPMM's trend of the torque with the input current by *V/f* inverter supply.

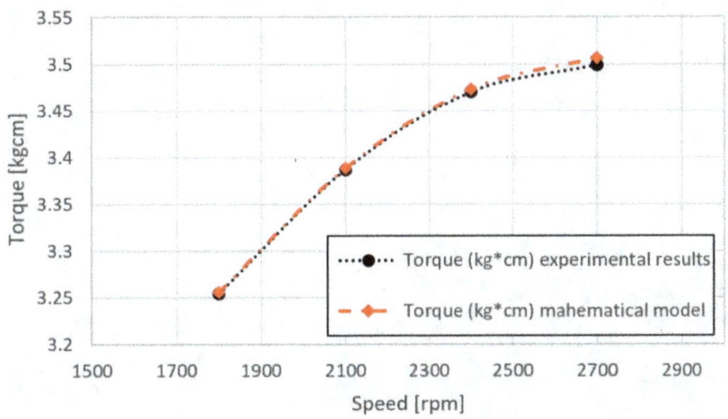

Figure 18. SPLSPMM torque trend comparison between experimental results and mathematical model.

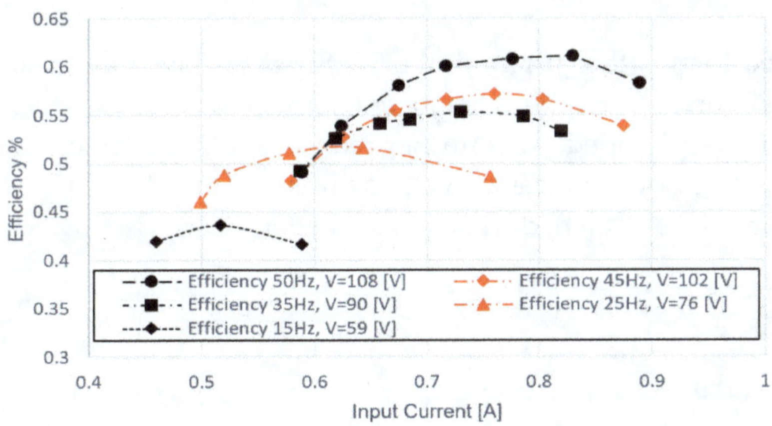

Figure 19. The SPLSPMM's efficiency at different frequencies.

6.4. Comparison between the SPLSPMM and the SPIM

To assess the performances of the two motor types, the SPLSPMM and the SPIM have been tested in the same operating conditions. The performance comparison is made by connecting the two motors to a sinusoidal power supply at 50 Hz frequency and variable voltage. The efficiency characteristic and the torques are compared in **Figures 20** and **21**.

The two figures highlight the supremacy of the SPLSPMM in terms of torque density and efficiency. The superiority is mainly due to the mechanism of torque production of the

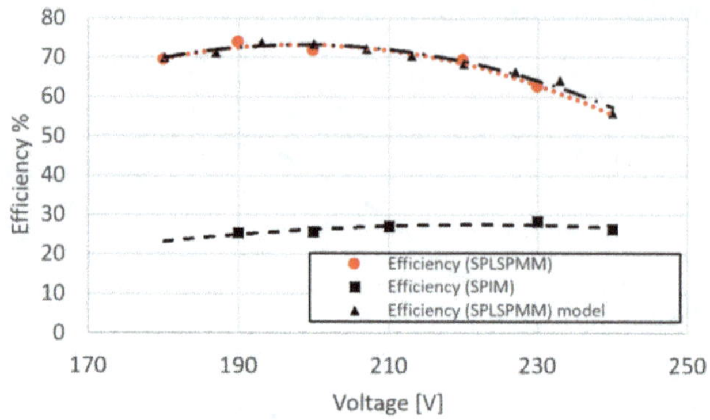

Figure 20. Efficiency comparison.

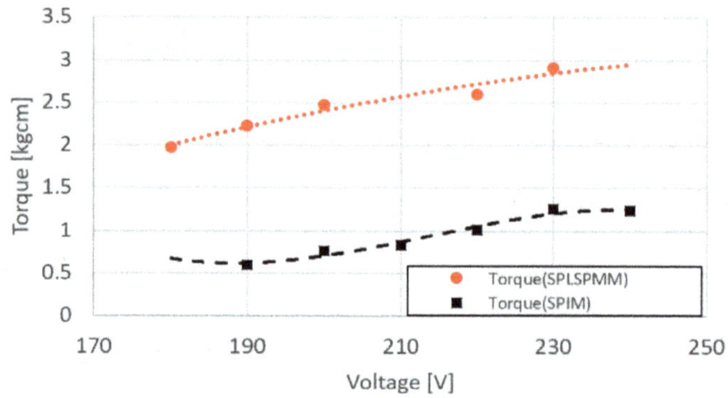

Figure 21. Torque comparison.

SPLSPMM. At fixed torque, the current of the SPLSPMM is lower, reducing the Joule loss in the stator winding. Moreover, the loss component in the rotor cage is reduced and the absence of the auxiliary winding (shaded pole) further reduces the Joule losses. The increase in iron losses due to magnets is not significant. Hence, as was imaginable, thanks to a simple change in the rotor structure, the overall performances get better.

7. Conclusion

This chapter has proposed a mathematical and experimental analysis of a SPLSPMM. Different structures of SPLSPMMs have been simulated with FEM and compared. A prototype has been built, analysed and tested.

Starting from the mathematical model of the single-phase shaded-pole motor, an equivalent circuit has been proposed for the SPLSPMM. The equivalent circuit has taken into account the presence of both magnets and rotor cage.

The experimental results include no-load and load tests carried out at different voltages and frequencies. The results show a significant performance improvement of the SPLSPMM in comparison to the classical SPIM. The maximum efficiency of the SPLSPMM is higher than 70%, while the efficiency of the SPIM is not higher than 30%.

Nomenclature

α	stator reference angular coordinate
N	number of turns
ω	grid pulsation
φ	angle between the stator and rotor reference
p	number of pole pairs
Kw	winding factor
μ_0	vacuum magnetic permeability
δ	air-gap width
Bpd	direct field
Bpi	reverse field
Ifs	direct component of the stator currents
Ifr	direct component of the rotor currents
Ibs	reverse component of the stator currents
Ibr	reverse component of the rotor currents
BM	residual magnet induction
β	initial angle in rotor reference
ωr	rated motor speed
ρ	stator-to-rotor winding turns ratio
φ	flux linkage
θ	rotor mechanical angle

Author details

Damiano D'Aguanno[1], Fabrizio Marignetti[1*] and Francesco Faginoli[2]

*Address all correspondence to: marignetti@unicas.it

1 Department of Electrical and Information Engineering, University of Cassino and South Lazio, Italy

2 Faber Spa, AN, Italy

References

[1] Miller TJE. Single-phase permanent-magnet motor analysis. IEEE Transaction On Industry Applications. May 1985;**IA-21**(4):651-658

[2] Zelik NG, Dogru UE, Ergene LT. Comparison study of drive motors for cooker hood applications. In: 2014 16th International Power Electronics and Motion Control Conference and Exposition; Sep. 2014. pp. 1252-1258

[3] Varga J, Basic D. Analysis of the characteristics of single phase shaded pole induction motor with two short-circuited auxiliary phases. IEEE Transactions on Energy Conversion. Dec. 1997;**12**(4):269-274

[4] Mansour Ojaghi SD. Analytic model for performance study and computer-aided design of single-phase shaded-pole induction motors. IEEE Transactions on Energy Conversion. Dec. 2016;**32**:649-657. DOI: 10.1109/TEC.2016.2645641

[5] Pessina G, Morra E. Optimization and design of the shaded pole single phase asynchronous motor. In: 2008 IEEE Power and Energy Society General Meeting—Conversion and Delivery of Electrical Energy in the 21st Century. July 2008. pp. 1-4

[6] Fang L, Lee BH, Hong J-P, Nam H. Rotor saliency improved structural design for cost reduction in single-phase line-start permanent magnet motor. In: Energy Conversion Congress and Exposition, 2009. ECCE 2009; IEEE. Sep. 2009. pp. 139-146

[7] Takegami T, Hasegawa M, Tsuboi K, Hirotsuka I, Nakamura M. Basic characteristics of a single-phase line-start permanent magnet synchronous motor. In: 2011 International Conference on Electrical Machines and Systems (ICEMS); Aug. 2011

[8] Shamlou S, Mirsalim M. Design, optimisation, analysis and experimental verification of a new line-start permanent magnet synchronous shaded-pole motor. IET Electric Power Applications. Jan. 2013;**7**(1):16, 26

[9] Zhou J, Tseng K-J. Performance analysis of single-phase line-start permanent-magnet synchronous motor. IEEE Transactions on Energy Conversion. Dec. 2002;**17**(4):453-462

[10] Honsinger VB. Permanent magnet machines: Asychronous operation. IEEE Transactions on Power Apparatus and Systems. Jul. 1980;**PAS-99**(0018-9510):1503-1509

[11] Miller TJE. Synchronization of line-start permanent-magnet ac motors. IEEE Power Engineering Review. Jul. 1984;**PER-4**(0272-1724):57-58

[12] Nam H, Jung SK, Kang GH, Hong JP, Jung TU, Baek SM. Design of pole-change single-phase induction motor for household appli- ances. IEEE Transactions on Industry Applications. May 2004;**40**(7989338):780-788

[13] Lee B-H, Hong J-P, Lee J-H. Optimum design criteria for maximum torque and efficiency of a line-start permanent-magnet motor using response surface methodology and finite element method. IEEE Transactions on Magnetics. Jan. 2012;**48**(12487177):863-866

[14] Popescu M, Miller TJE, McGilp M, Strappazzon G, Trivillin N, Santarossa R. Line-start permanent-magnet motor: Single-phase starting performance analysis. IEEE Transactions on Industry Applications. July 2003;**39**(7709854):1021-1030

[15] Behbahanifard H, Sadoughi A. Line start permanent magnet synchronous motor performance and design; a review. Journal of World's Electrical Engineering and Technology. 2015; **2322**:5114

[16] Sarac V, Atanasova-Pacemska T. Simulation model for prediction of transient performance characteristics of single-phase shaded pole motor. Journal of Electrical Engineering. 2016; **67**(4):253-260

[17] Fang L, Lee BH, Hong J-P, Nam H, Ha S-H. Study on the permanent magnet reduction design in single-phase line-start permanent magnet motor for household appliance. In: 2008 International Conference on Electrical Machines and Systems; Oct 2008. pp. 3289-3292

[18] Miller TJE, Popescu M, Cossar C, McGilp M, Strappazzon G, Trivillin N, Santarossa R. Line-start permanent-magnet motor single-phase steady-state performance analysis. IEEE Transactions on Industry Applications. March 2004;**40**(7943119):516-525

[19] Popescu M, Miller TJE, McGilp M, Strappazzon G, Trivillin N, Santarossa R. Asynchronous performance analysis of a single-phase capacitor-start, capacitor-run permanent magnet motor. IEEE Transactions on Energy Conversion. Feb. 2005;**20**(1):142-150

[20] Rahman MA, Zubayer HM, Wang K, Kurihara K, Osheiba AM, Jabbar MA, Hoque MA. Single phase line-start high efficiency interior permanent magnet motors. In: 2011 IEEE International Electric Machines Drives Conference (IEMDC); May 2011. pp. 19-28

[21] Iepure LI, Tutelea L, Boldea I. Fem analysis and control of a tapered airgap single phase PMSM. In: 2008 11th International Conference on Optimization of Electrical and Electronic Equipment; May 2008. pp. 241-248

[22] Marcic T, Stumberger B, Stumberger G, Hadziselimovic M, Virtic P, Dolinar D. Line-starting three-and single-phase interior permanent magnet synchronous motors;direct comparison to induction motors. IEEE Transactions on Magnetics. Nov 2008;**44**(11):4413-4416

[23] Attaianese C, Pizzo AD, Pagano E. Optimization and modelling of single-phase capacitor motors (in Italian) [PhD thesis]

[24] Huang H, Fuchs EF, White JC. Optimization of single-phase induction motor design: The maximum efficiency and minimum cost of an optimal design. IEEE Transactions on Energy Conversion. June 1988;**3**(2):357-366

[25] Miller TJE. Single-phase permanent magnet motor analysis. In IEEE Transactions on Industry Applications. May 1985;**IA-21**(3):651-658

[26] Popescu M, Miller TJE, McGlip MI, Strappazzon G, Trivillin N, Santarossa R. Line start permanent magnet motor: Single-phase starting performance analysis. IEEE Transactions on Industry Applications. Jul./Aug. 2003;**39**(4)

[27] Popescu M, Miller TJE, McGlip MI, Cossar C, Strappazzon G, Trivillin N, Santarossa R. Line start permanent magnet motor: Single-phase steady-state performance analysis. IEEE Transactions on Industry Applications. Mar./Apr. 2004;**40**(2)

[28] Popescu M, Miller TJE, McGlip M, Strappazzon G, Trivillin N, Santarossa R. Asynchronous performance analysis of a single phase capacitor-start, capacitor-run permanent magnet motor. In IEEE Transactions on Energy Conversion. March 2005;**20**(1):142-150

5

Start-Up of a PID Fuzzy Logic-Embedded Control System for the Speed of a DC Motor using LabVIEW

Ruthber Rodriguez Serrezuela,
Jorge Luis Aroca Trujillo,
Roberto Sagaro Zamora and Vadim Azhmyakov

Abstract

This work explains the speed control design for a DC motor using fuzzy logic with LabVIEW software. It is also a literature review about the design and the implementation environment and is presented using fuzzy logic to describe the materials and methods used. Various processes on the subject highlight the idea, creation, development, and implementation of intelligent control, and the results considering the application and development for this purpose are presented.

Keywords: fuzzy logic, LabVIEW, MATLAB, rules, industrial applications

1. Introduction

In the last decade, the use of various control techniques such as hybrid [1, 2] or artificial intelligence has grown exponentially in the fields of power electronics, control systems, and positioning systems. To carry out these development methods, a large database and understanding of control methods are necessary to describe the analysis system to facilitate decision-making. One of these methods is fuzzy logic [3, 4], which is necessary to realize improvements over classical logic and is basically a statement that can be understood as a logical value of 1 or 0.

In the control field, specifically in the DC motor, control can be applied by intelligent control techniques such as fuzzy logic, which is further applied in parallel with conventional control techniques. In fuzzy logic there are two well-known algorithms: logarithm Mamdani and Takagi Sugeno; these are useful in applications such as microelectronics and power electronics [5, 6].

Since it is a more compact and computationally efficient representation than a Mamdani system, the Sugeno system lends itself to the use of adaptive techniques for constructing fuzzy models [7, 8]. These adaptive techniques can be used to customize the membership functions so that the fuzzy system models the data with a significant improvement [9, 10].

2. Materials and methods

2.1. Process description

The device on which the fuzzy logic control will be applied is a 12 V DC motor as shown in **Figure 1**.

The block diagram of **Figure 2** shows the control process. In the first stage, the input is the set-point or reference rate to be applied to the motor. Next is the analog-digital converter, which is distinguished as a data acquisition card. The second stage is the fuzzy logic control developed in LabVIEW. The data acquisition card is used once again since this should act as both analog-digital converter at the input and digital-analog converter at the output. Hereafter, the analog signal passes through the actuator block, which is H-bridge pulse-width modulation (PWM) modified to the motor speed [11]. The next block in the plant is the DC motor. The tachometer, which is built into the DC motor, is a feedback system that checks whether the engine speed has altered and eliminates unwanted phenomena.

LabVIEW [12] (**Figure 3**) is the software used to develop the controller through its fuzzy logic blocks. To facilitate programming the LabVIEW toolbox contains toolkits. This software is mainly used for data acquisition, mathematical analysis, instrument control, and controller design. In 2008 the software was used to control the Large Hadron Collider [13], a particle accelerator designed by about 2000 scientists from 34 countries around the world. This appliance is the largest of its kind and is located near Geneva.

Figure 1. 12 V DC motor.

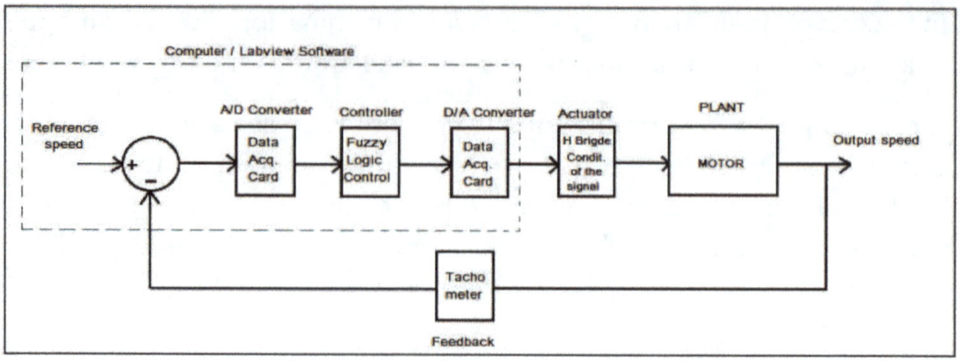

Figure 2. Process control (closed loop).

To interact with LabVIEW, the user needs a data acquisition card, e.g., a National Instruments NI myDAQ (**Figure 4**). The function of the data acquisition card is to collect physical signals during the process and transform them into electrical voltages that are subsequently scanned on a PC and can be processed. To make the transformation to a digital signal the card is needed at the adaptation stage [14–16]

Within the data acquisition process is the transducer, which is an element that converts the physical quantity to be measured into an output signal, usually a voltage or current that is then processed on a PC or other system.

Signal conditioning is the stage that is responsible for filtering and adapting the outgoing signal transducer, which is then sent to the analog-digital converter, which as we know is the stage responsible for converting an analog signal to the digital output stage to connect the data acquisition system to a PC (**Figure 5**).

In the initial window for the design of a new fuzzy controller, data such as name, designer name, date, and time are observed for the project. In the window in **Figure 6**, the fuzzy set editor and its consequences are introduced.

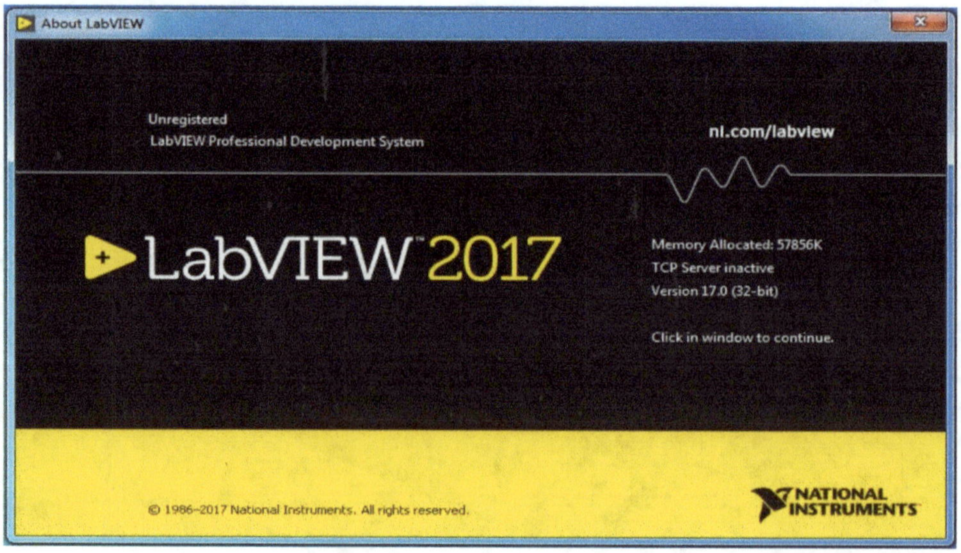

Figure 3. LabVIEW software programming.

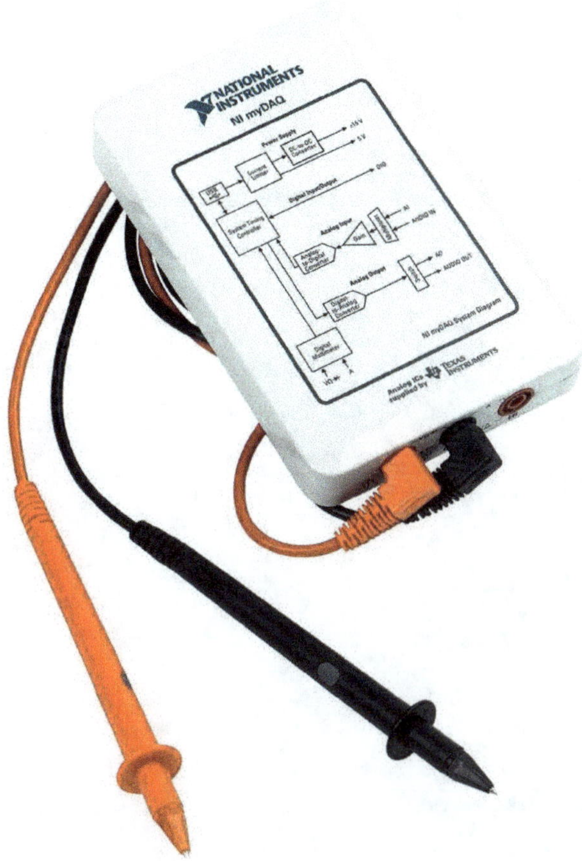

Figure 4. Data acquisition card.

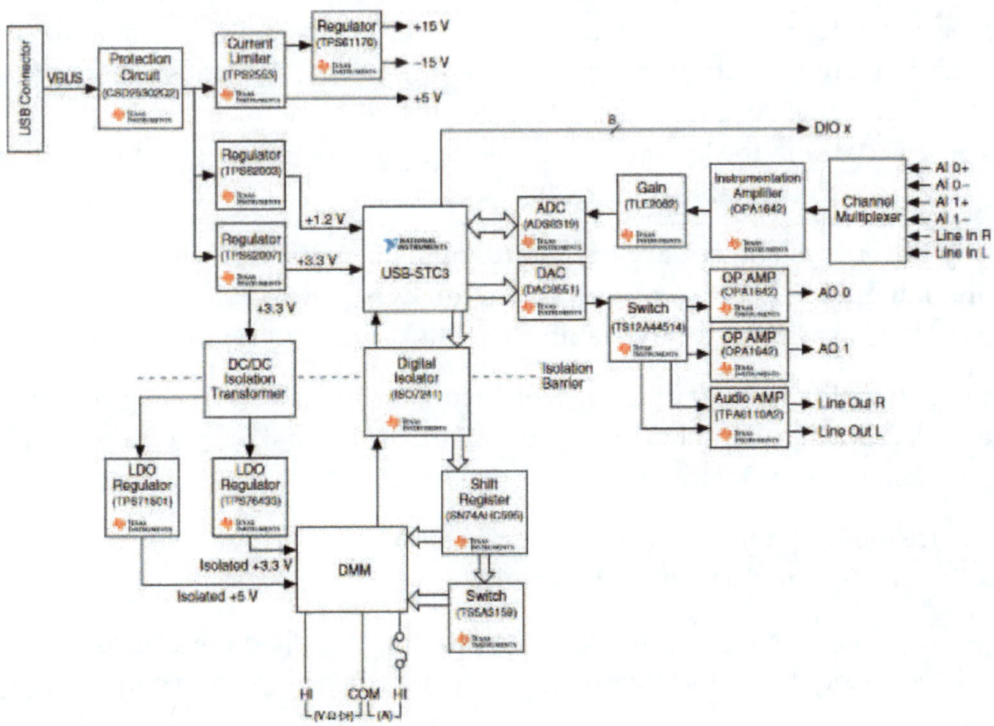

Figure 5. NI MyDAQ components may be changed or substituted without notice.

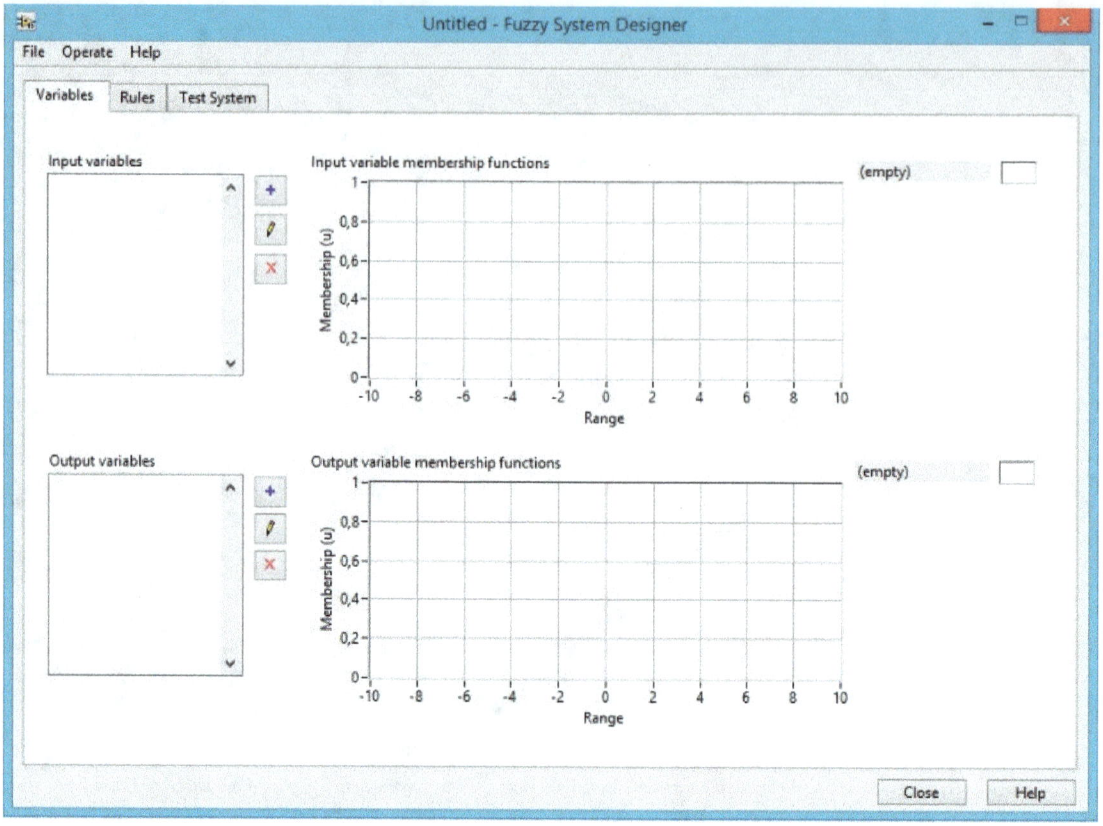

Figure 6. Main window to start a new fuzzy controller in LabVIEW.

The input and output linguistic variables and their ranges are entered functionally. This takes into account the fuzzy set editor, which starts with a default configuration. The advantage over a conventional proportional-integral-derivative (PID) controller is that a fuzzy PD (proportional and derivative) controller can implement nonlinear control strategies, which uses linguistic rules. The rules evaluate the difference between the measured value and the set value, the error signal, and the tendency of the error signal to determine whether to increment or decrement the control variable (**Figure 7**).

By deploying the box, it is necessary the input linguistic variables, where each one contains the following linguistic terms: NEx (term negative), ZOx (zero), POx (term positive), default data range is −1 to 1, variable name default: INX with x = 1, …, 4 (**Figure 8**).

The result is output linguistic variables, which contain the following linguistic terms: NEo (term negative), ZOo (zero) and POo (term positive), default data range is from −1 to 1, variable name default: Output (**Figure 9**).

The interference method, the method of defuzzification and output default configuration, is determined because of the rule editor (**Figure 10**).

The block features input and output characteristics that simulate the controller and graphically display the output behavior regarding parameter rules in the rule editor (**Figure 11**).

At the end of this process, the fuzzy controller with all the information on the rules, which is observed in the graphical environment of LabVIEW as a block, is obtained (**Figure 12**).

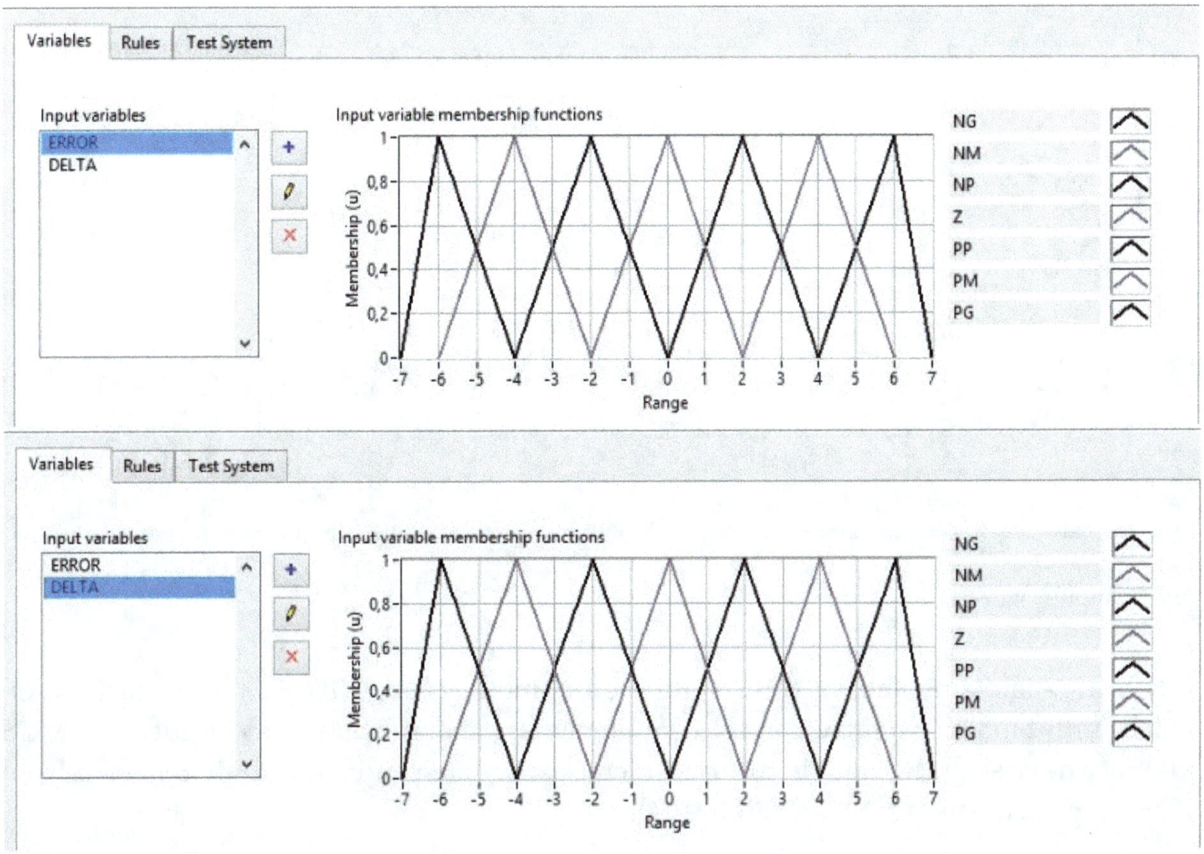

Figure 7. Fuzzy default settings set editor.

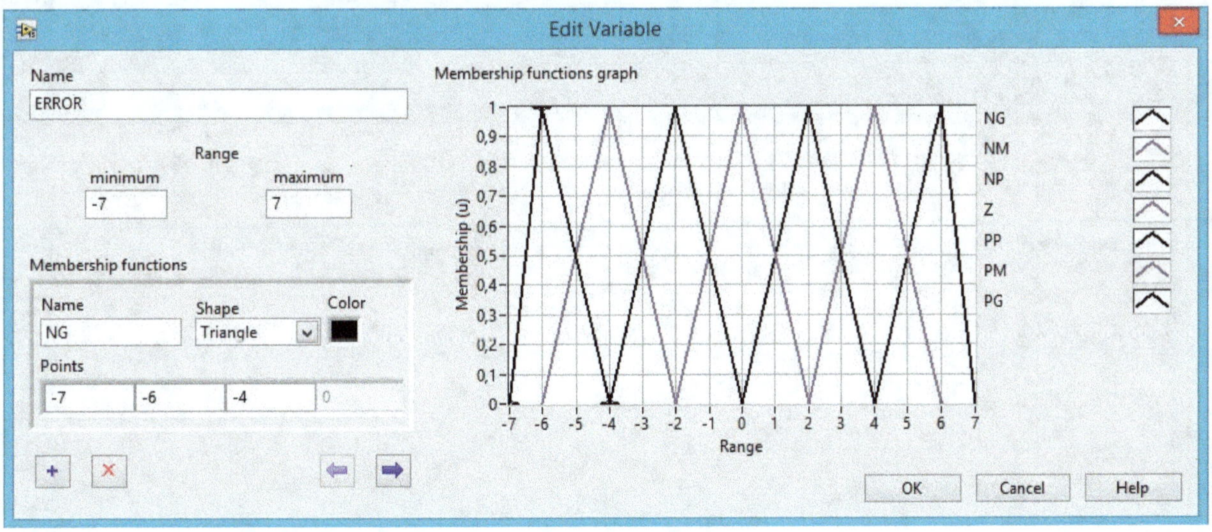

Figure 8. Input linguistic variables.

Pulse width modulation (PWM) is a process in which the duty cycle of a periodic signal, possibly to transmit information, is modified through a communications channel or to control the amount of power sent to a load (**Figure 13**).

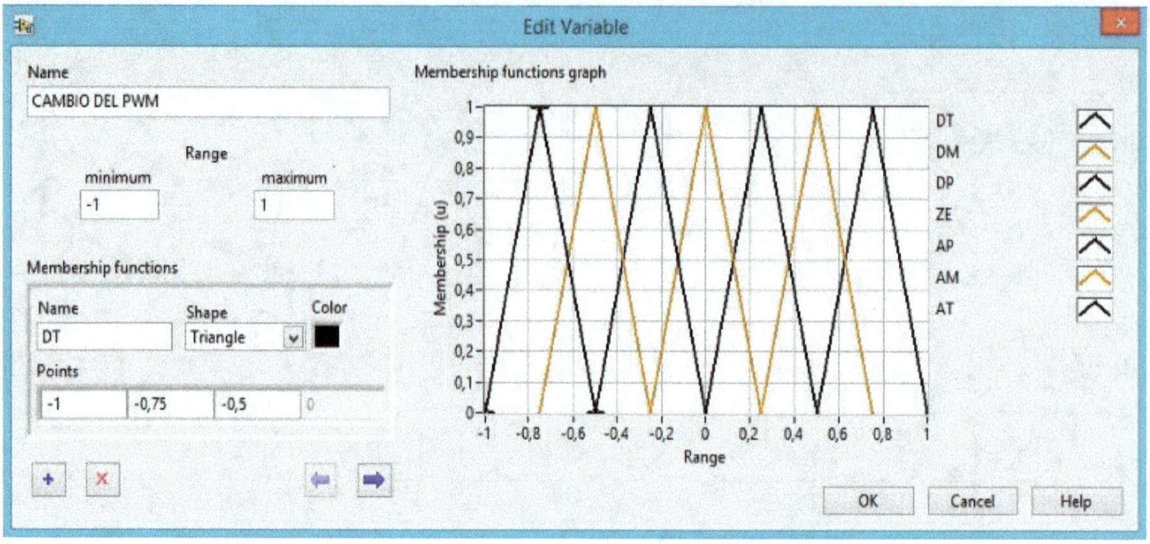

Figure 9. Output linguistic variables.

The H-bridge implemented for controlling the DC motor is an LMD18200 [5], which is constructed with bipolar circuitry control complementary metal-oxide-semiconductor (CMOS) field-effect devices and a double-diffused metal-oxide-semiconductor (DMOS), and accommodates a peak output up to 6 A (**Figure 14**).

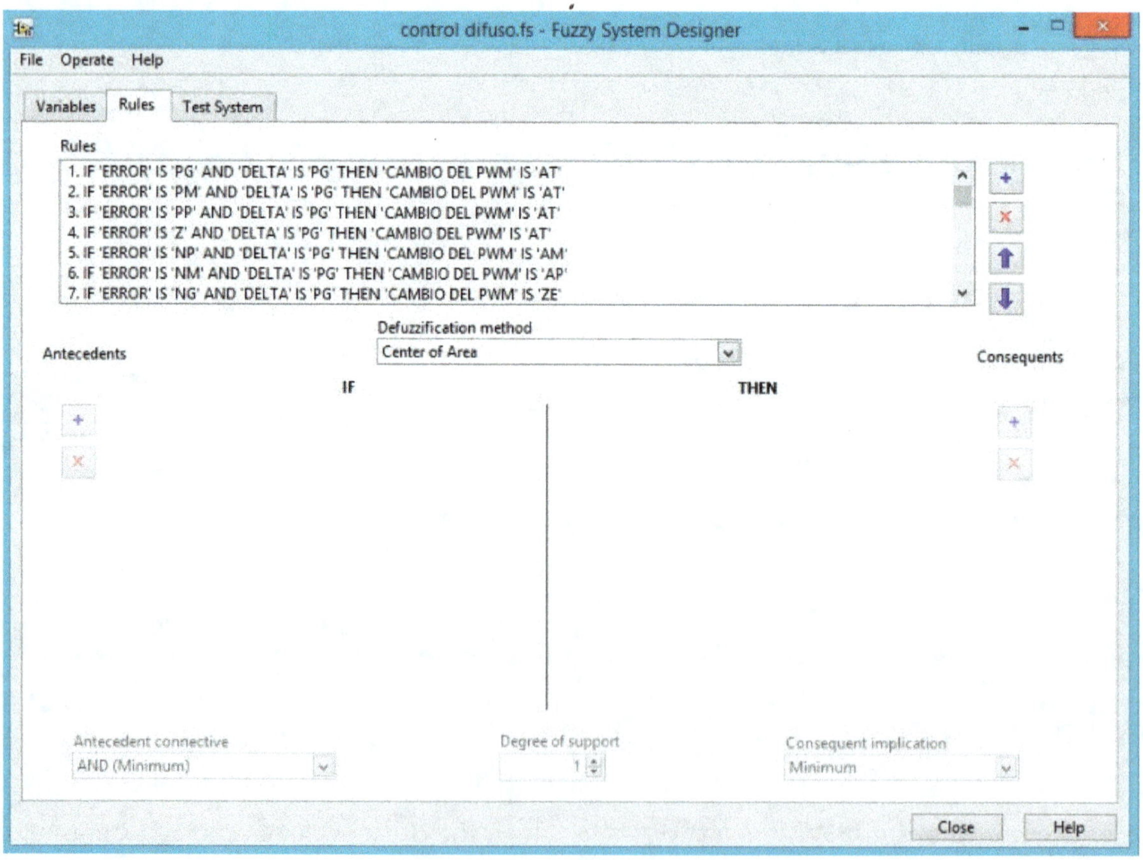

Figure 10. Rule editor.

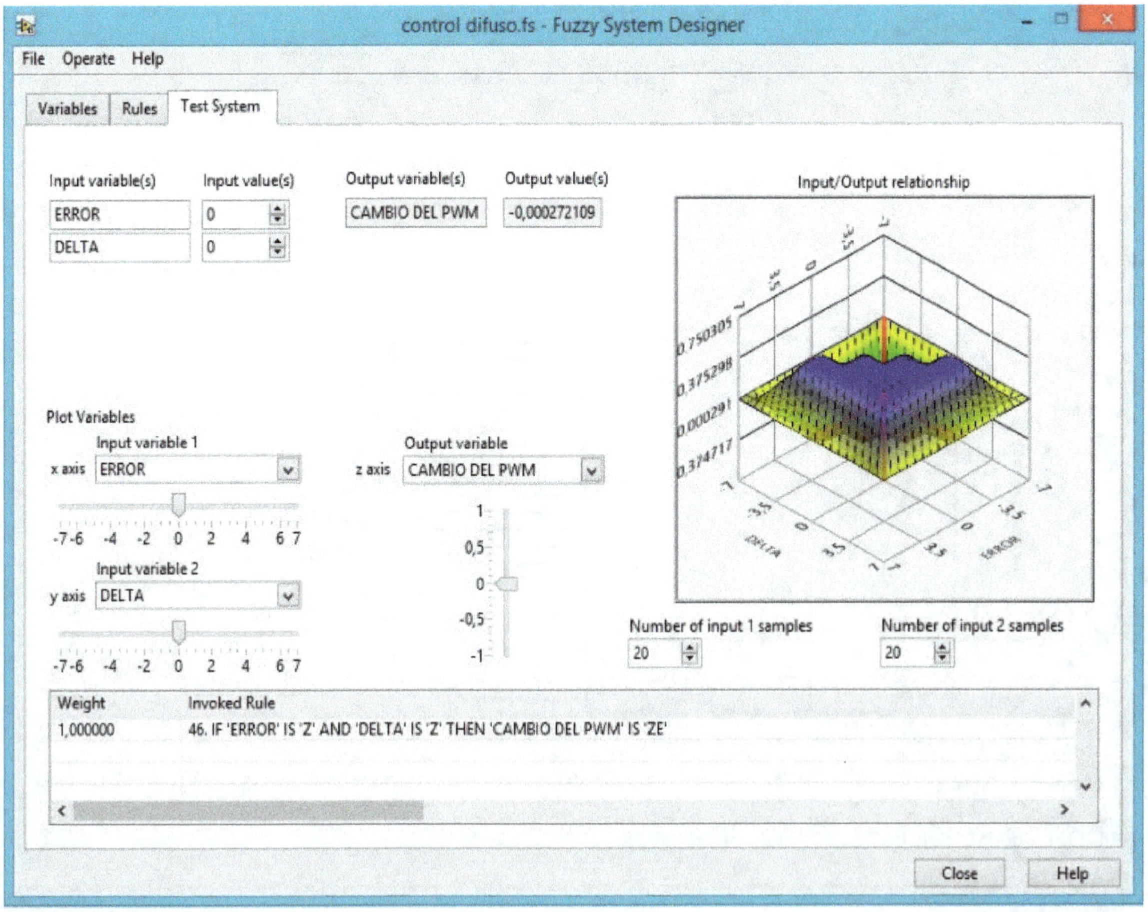

Figure 11. Input and output characteristics.

Feedback should be between 0 and 5 V as it is the voltage that supports the data acquisition card; a tachometer generates voltages from 0 to 12 V. This does not serve as an input to the program, thus it designs and implements an actuator to condition the signal, taking into account that the behavior guarantees a range between 0 and 5 V, which is the optimal voltage for scanning the signal (**Figures 15** and **16**).

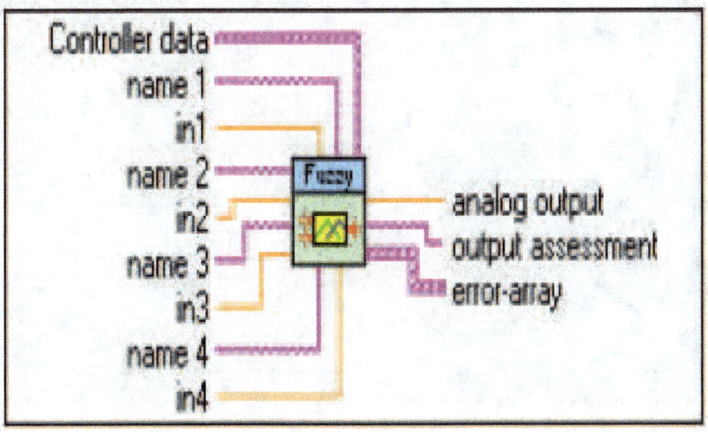

Figure 12. Fuzzy controller in LabVIEW graphical interface.

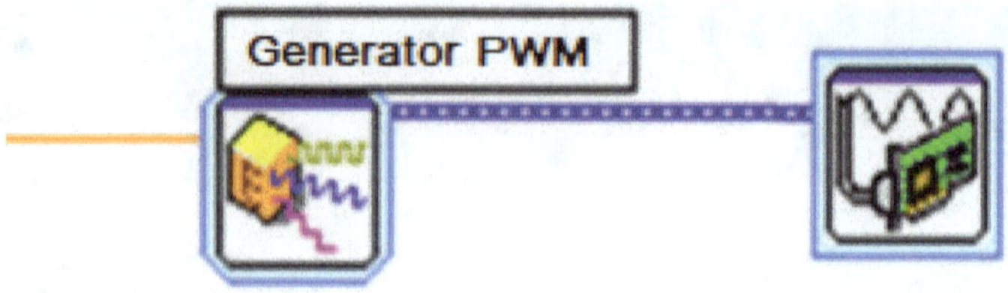

Figure 13. PWM output in LabVIEW.

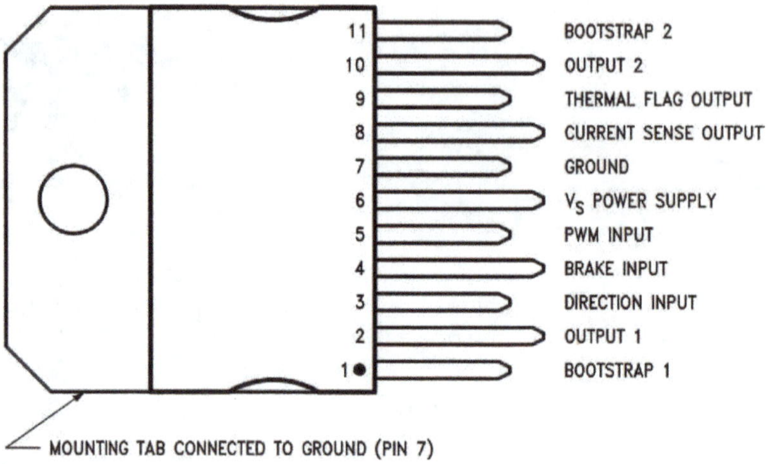

Figure 14. Datasheet for an LM18200T H-bridge.

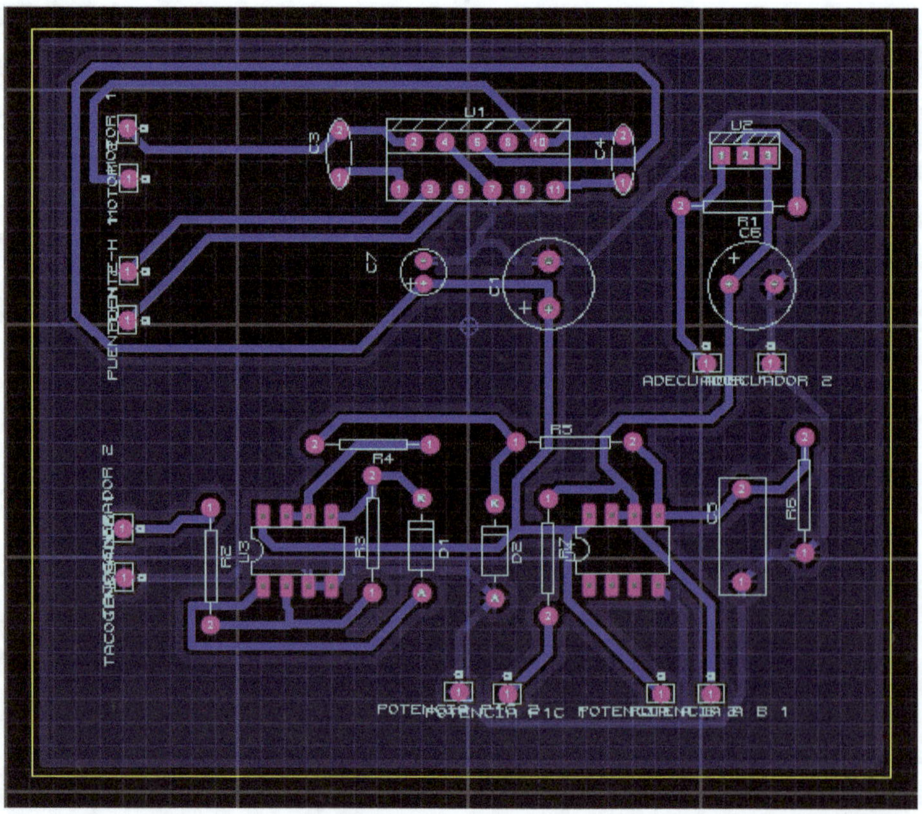

Figure 15. Design of the transducer in PROTEUS software.

Figure 16. Card and bridge transducer circuit H.

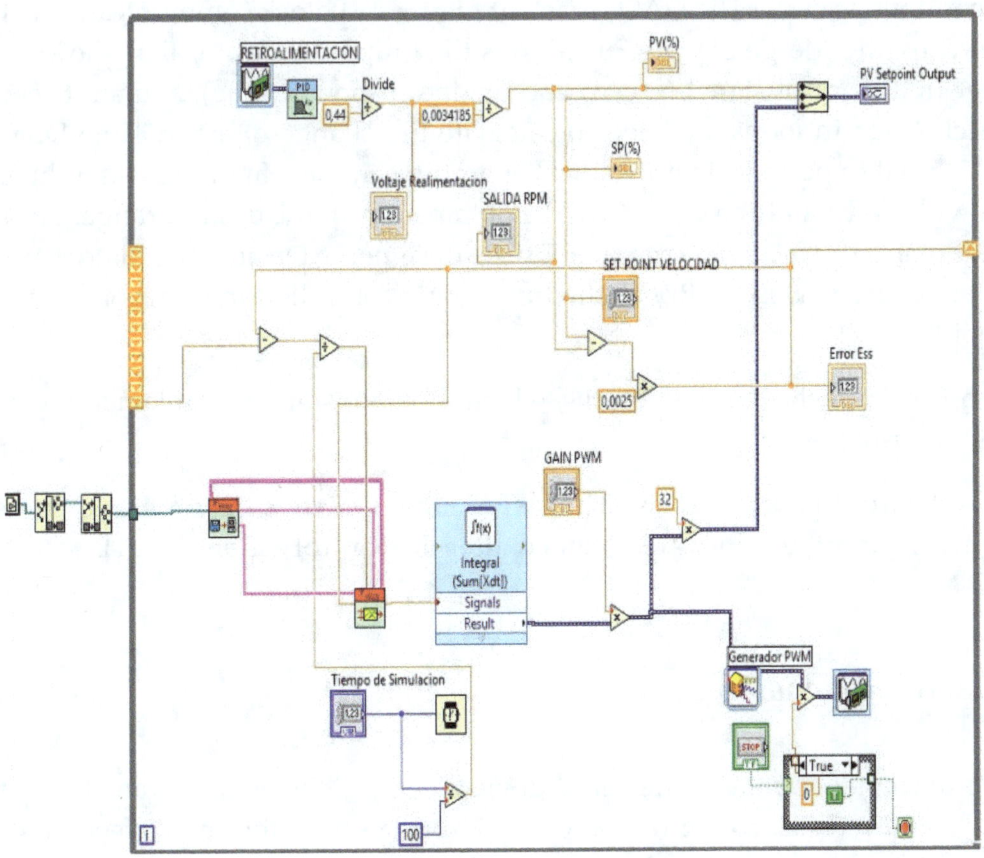

Figure 17. Block diagram for a DC motor controller in LabVIEW.

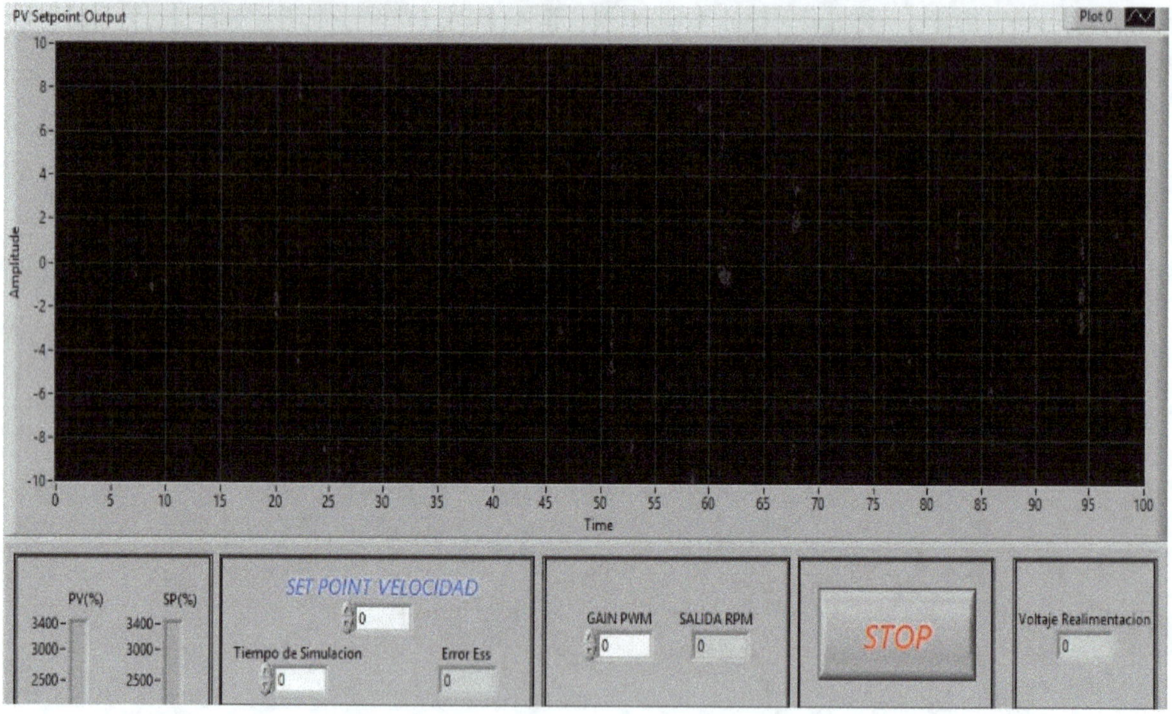

Figure 18. Front panel for a DC motor controller in LabVIEW.

Figure 17 shows the blocks of indicators and controls, such as the while loop; all the blocks work in a continuous cycle. The DAQ Assistant Express VI block communicates with the data acquisition card, the Merge Signals Function is the graphic indicator that shows the sign of the setpoint (red) control signal (green) and engine response (blue), and shift registers and the Feedback Node in loops are used to calculate the change of error. Time Domain Math Express VI is used to perform mathematical operations in time, in this case the change will be integrated with PWM and the output will be obtained from the desired range. The strip path function maintains the file path generated by the designer extension. Simulate Signal Express VI simulates square waves for PWM, the fuzzy logic controller, the controller, and the load fuzzy setpoint [7, 17].

A while loop encloses the set of all blocks, which is responsible for implementing everything inside continuously [18, 19].

Figure 18 is the front panel view, which is where the user runs the program, the results are seen graphically, and the values are obtained after the control action [20, 21].

3. Results and discussion

At the time of connecting the entire set of devices such as the engine, transducer, font, procuring card data, and PC, configuration of the H-bridge pins and motor feedback are taken into account. After verifying that the assembly has the correct speed, control is executed in LabVIEW and the system is energized.

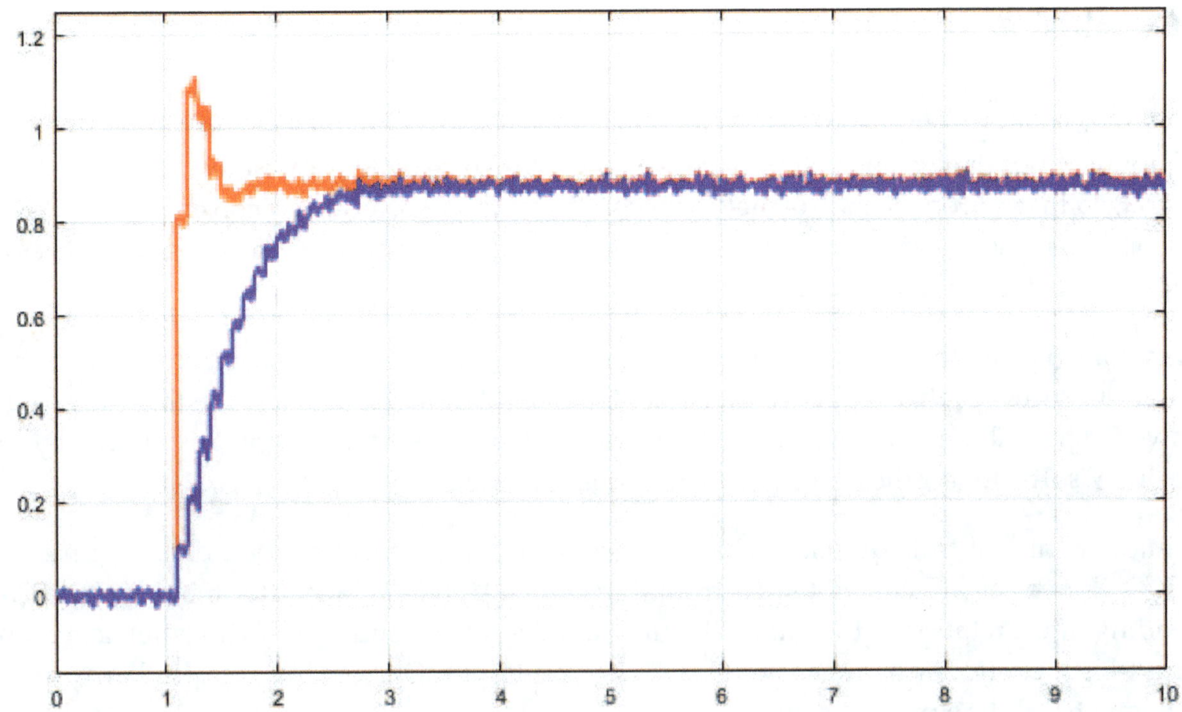

Figure 19. Front panel showing the resulting graph of the fuzzy controller.

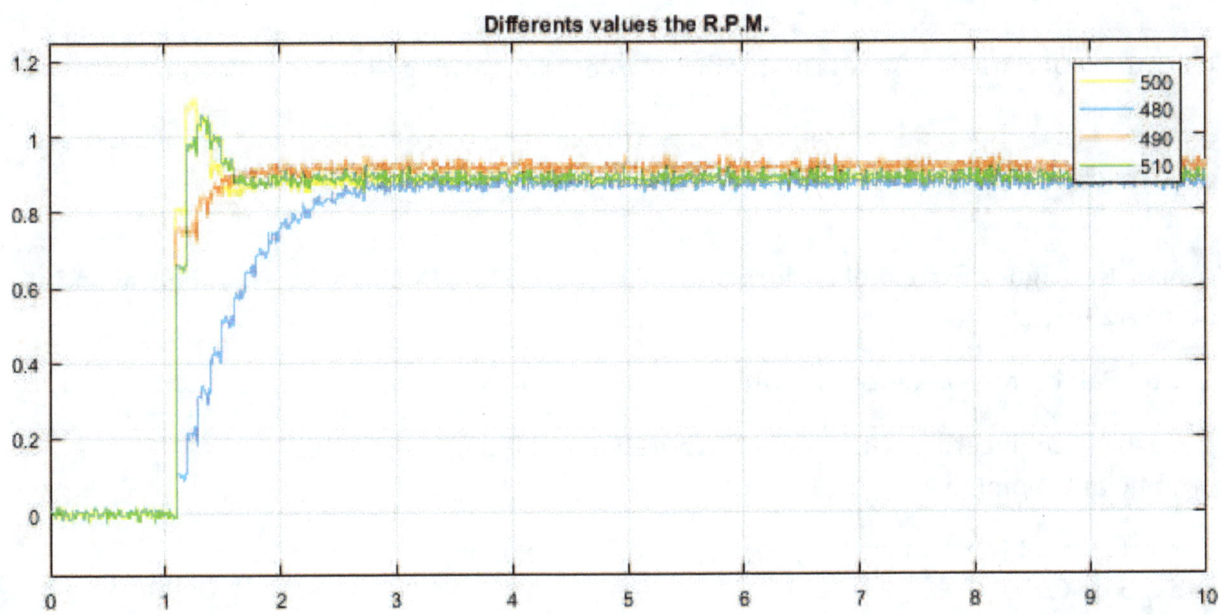

Figure 20. Diffuse logic PID control response at different engine speeds in RPM.

Within the controller, a value of 1600 RPM is entered to start the control engine speed, and on the graphic belonging to the GUI LabVIEW variation, PWM is observed in terms of engine speed in RPM. These results can be compared with a classic PID control showing that our fuzzy logic controller meets the design requirements. Achieving error values of 3% is acceptable in a practical manner (**Figures 19** and **20**).

4. Conclusion

The foregoing is accomplished by designing a control system to understand alternative control applications that can used, for example, in an industrial sector. It is noted that fuzzy logic is a fast, solid, and robust control method for a system that is affected by constant instabilities, which in this case could be disturbances affecting a DC motor by applying opposing forces or weights within its axis rotation.

Fuzzy logic is a method of computing based on decision-making and is used for expert systems and process control. Fuzzy logic differs from traditional Boolean logic in that fuzzy logic allows for partial membership in a set. The great advantages and possibilities offered by the LabVIEW software for these applications are also stressed.

To implement real-time decision making or control of a physical system, data acquired by an A/D Converter must be obtained from a fuzzy controller. Which after making the corresponding interpretation of the linguistic rules and the corresponding defusification leaves the diffuse logic controller by means of a D/A Converter with DAQ analog output hardware to implement real-time process control.

Conflict of interest

The authors declare that there is no conflict of interest regarding the publication of this chapter.

Author details

Ruthber Rodriguez Serrezuela[1*], Jorge Luis Aroca Trujillo[1], Roberto Sagaro Zamora[2] and Vadim Azhmyakov[3]

*Address all correspondence to: ruthber.rodriguez@corhuila.edu.co

1 Industrial Engineering, University Corporation of Huila, CORHUILA, Neiva, Republic of Colombia

2 Department of Mechanical Engineering and Design, Universidad de Oriente, Santiago de Cuba, Cuba

3 Faculty of Basic Sciences, University of Medellin, UdeM, Medellin, Republic of Colombia

References

[1] Azhmyakov V, Rodriguez Serrezuela R, Ríos Gallardo AM, Gerardo Vargas W. An approximation based approach to optimal control of switched dynamic systems. Mathematical Problems in Engineering. 2014;**2014**

[2] Rojas JHC, Serrezuela RR, López JAQ, Perdomo KLR. LQR hybrid approach control of a robotic arm two degrees of freedom. International Journal of Applied Engineering Research. 2016;**11**(17):9221-9228

[3] Serrezuela RR, Chavarro AFC. Multivariable control alternatives for the prototype tower distillation and evaporation plant. International Journal of Applied Engineering Research. 2016;**11**(8):6039-6043

[4] Thepsatorn P, Numsomran A, Tipsuwanporn V, Teanthong T. DC motor speed control using fuzzy logic based on LabVIEW. In: SICE-ICASE 2006 International Joint Conference. IEEE; 2006. pp. 3617-3620

[5] Horng JH. SCADA system of DC motor with implementation of fuzzy logic controller on neural network. Advances in Engineering Software. 2002;**33**(6):361-364

[6] Deka N, Hazarika PJ. Liquid level control using fuzzy logic in LabVIEW with Arduino. International Journal of Advance Research, Ideas and Innovations in Technology. 2017;**3**(1)

[7] Ammar HH, Azar AT, Tembi TD, Tony K, Sosa A. Design and implementation of fuzzy PID controller into multi agent smart library system prototype. In: International Conference on Advanced Machine Learning Technologies and Applications. Cham: Springer; 2018. pp. 127-137

[8] Martinez DA, Poveda JD, Montenegro D. Li-ion battery management system based in fuzzy logic for improving electric vehicle autonomy. In: 2017 IEEE Workshop on Power Electronics and Power Quality Applications (PEPQA). IEEE; 2017. pp. 1-6

[9] Dhimish M, Holmes V, Mehrdadi B, Dales M. Comparing Mamdani Sugeno fuzzy logic and RBF ANN network for PV fault detection. Renewable Energy. 2018;**117**:257-274

[10] Yilmaz U, Kircay A, Borekci S. PV system fuzzy logic MPPT method and PI control as a charge controller. Renewable and Sustainable Energy Reviews. 2018;**81**:994-1001

[11] Safwat IM, Wu X, Zhao X, Li W. Adaptive fuzzy logic control of boost converter fed by stand-alone PEM fuel cell stack. In: 2017 IEEE Transportation Electrification Conference and Expo, Asia-Pacific (ITEC Asia-Pacific). IEEE; 2017. pp. 1-6

[12] Buele J, Espinoza J, Pilatásig M, Silva F, Chuquitarco A, Tigse J, Guerrero L. Interactive system for monitoring and control of a flow station using LabVIEW. In: International Conference on Information Theoretic Security. Cham: Springer; 2018. pp. 583-592

[13] Noor IMM, Kamal M. Control of wastewater treatment by using the integration MATLAB and LabVIEW. Pertanika Journal of Science & Technology. 2017;**25**

[14] Nolasco JJM, Medina JAP. LabVIEW-based classic, fuzzy and neural controllers algorithm design applied to a level control prototype. IEEE Latin America Transactions. 2017;**15**(6):1154-1162

[15] Dhimish M, Holmes V, Mehrdadi B, Dales M, Mather P. Photovoltaic fault detection algorithm based on theoretical curves modelling and fuzzy classification system. Energy. 2017;**140**:276-290

[16] Oswald R. Fuzzy control of inverted pendulum swing-up and balancing [Doctoral dissertation]. Northridge: California State University; 2017

[17] Aroca Trujillo JL, Pérez-Ruiz A, Serrezuela RR, Garavito ECDIJ. Generation and control of basic geometric trajectories for a manipulator robot using CompactRIO. Journal of Robotics. 2017;**3**:1-11

[18] Aroca Trujillo JL, Rodríguez Serrezuela R, Ramírez Zarta JB, Navarrete Ramos AM. Direct and inverse kinematics of a manipulator robot of five degrees of freedom implemented in embedded system—CompactRIO. In: Ajaykumar BS, Sarkar D, editors. Advanced Engineering Research and Applications. Nueva Deli, India: Research India Publication; 2018. pp. 405-419

[19] Rodríguez Serrezuela R, Aroca Trujillo JL, Navarrete Ramos AM, Ramírez Zarta JB. Applications alternatives of multivariable control in the tower distillation and evaporation plant. In: Ajaykumar BS, Sarkar D, editors. Advanced Engineering Research and Applications. Nueva Deli, India: Research India Publication; 2018. pp. 452-465

[20] Serrezuela RR, Cardozo MAT, Chavarro AFC. Design and implementation of a PID fuzzy control for the speed of a DC motor. Journal of Engineering and Applied Sciences. 2017;**12**(8):2655-2660

[21] Serrezuela RR, Chavarro AFC, Cardozo MAT, Toquica AL, Martinez LFO. Kinematic modelling of a robotic arm manipulator using Matlab. Journal of Engineering and Applied Sciences. 2017;**12**(7):2037-2045

6

Advanced Control of the Permanent Magnet Synchronous Motor

Marian Găiceanu

Abstract

The electrical machines are the core of the electrical drives. By introducing the vector control techniques for the alternative current machines, the high performances in drive systems are attained. One on the alternative current machines is the permanent magnet synchronous motor (PMSM). Due to their advantages, it becomes a very popular solution in the electrical drive field. In this chapter, an optimal control solution applied on the PMSM based on the Riccati solution is developed by the author. The objectives of the optimal control drive system are regulation, stability, robustness to the load disturbance variation and the energy reduction. Comparative with the conventional cascaded control, the proposed solution conducts up to 10% to energy efficiency improvement in transient regimes. The efficiency improvement depends on the chosen weighted matrices. Both the conventional and optimal controllers are implemented in Matlab-Simulink. The real-time solution based on the dSpace platform is provided.

Keywords: PMSM, PI control, quadratic control, Riccati, Matlab-Simulink, dSpace

1. The origins of the optimal control

Variational calculus uses multivariable functions and has the objective of obtaining a function (maximum or minimum) of a functional. The optimal command consists of finding time-varying functions or gain factors placed on the state feedback in order to find the extreme point of a functional cost. Therefore, optimal control derives from variational calculus. Optimization is a collection of methods and techniques for obtaining optimal solutions to automation problems.

The issue of optimization has encountered several stages of development, marked by the emergence of research areas. The optimization problem has been initiated since antiquity, continuing in the Renaissance period (with remarkable progress in the development of the optimization problem through the *variational calculus*), respectively, and its expansion to the *optimal command* (born in 1697) by solving the brachistochrone problem [1]. The foundation of the optimal control theory has been done by Semionovici Pontryagin's maximum principle (1956).

Thus, Johann Bernoulli (1667–1748) launched a challenge to the problem of the minimum time ("brachistochrone problem", the problem of determining a curve to ensure the fastest descent of a mass on its surface). By using variational calculation, Johann Bernoulli solved this problem in 1697, originally formulated in 1638 by Galileo Galilei (1564–1642).

In June 1696, Bernoulli formulated the issue of the minimum time in *Acta Eruditorum* journal: two points A and B are to determine the AMB curve of a moving point M, with the initial velocity $v_A = 0$, which under its own weight will move from A to B in the shortest possible time. The shortest distance is the straight segment AB, but it is not the one that can be traversed in the shortest possible time (**Figure 1**).

The Pontryagin maximum principle led to the birth of optimal control, a wider theory than variational calculus.

The origins of the variational calculation, after Herman H. Goldstine (1980), originated in 1662, when Pierre de Fermat (1601–1665), by posturing the Fermat's principle by specifying that the fastest path of a light beam passing through a single optical medium is the one in which light passes "in minimum time" (Heron's interpretation).

In consultation with Leibniz (Gottfried Wilhelm Freiherr von Leibniz, 1646–1716), the founder of variational calculus alongside Newton, Bernoulli sets the deadline to refer to the solution in the summer of 1697. Thus, besides Johann's solution, the following mathematicians answer this challenge: (1) Leibniz presents his solution in a letter addressed to Bernoulli in June of the same year, (2) Jakob Bernoulli (1654–1705, with probabilistic theory distributions that bear his name) offers another solution to the larger brother's solution, (3) Tschirnhaus, (4) L'Hopital and (5) Newton presents his solution in February 1697 to the Royal Society.

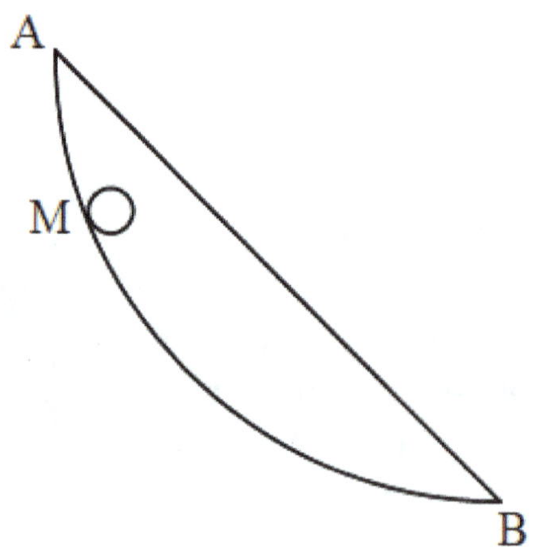

Figure 1. Brachistochrone problem.

2. Modern optimal control

The optimal command appeared with the publication of Bernoulli's solution in *Acta Eruditorum*. The optimal control theory started with the Pontryagin maximum principle (USSR, 1956). Jacopo Francesco Riccati (1676–1754) presented a scalar form of the equation that bears his name, an equation used to solve the linear quadratic problems of today. The varied engineering calculations were developed by Placido Cicara (Torino, 1957), G. Newton and Gould Kaiser (1957), Derek F. Lawden (London, 1963), Athans F. (1966) and D. McRuer (1973). The design of the control by using a quadratic performance index was studied by G. Newton and Gould Kaiser (the square error integrator); Kalman (1960) introduces quadratic penalty on control and output error, obtaining optimal control by solving the inverse Riccati equation; and Chang (1961) introduces a restriction, by a K2 weighting factor, into the composition of the quadratic performance index. Athans had named this control type, both for monovariable linear systems and for multivariable systems, as control based on a *linear quadratic regulator.*

Bellman (1950) developed dynamic programming by introducing optimal nonlinear control. Bellman used the Hamilton-Jacobi theory and discrete systems. The disadvantage of dynamic programming is the limitation of the variables used due to the need for an important memory space to determine the solution.

Numerical solutions of optimal control were determined by Kalman and Athans (1960) for linear dynamic systems. Moreover, Kalman (1963) determined solutions of multivariable, linear quadratic optimal problems using Riccati matrix equations.

An optimization of dynamic systems has been strongly developed in the field of frequencies by the emergence of robust H-infinite (H∞) and H2 control. Also their name, these optimal commands are more robust to disturbance (due to high-frequency unmodified dynamics) and variations in process parameters. Optimization research is growing fastly seeking to develop modern theories that began in the 1980s, establishing robust optimal control laws for process parameter variations and modeling uncertainties.

3. The problem of optimal control

The problem formulation of the optimization of a dynamic system should include [2–4]:

1. The dynamics of the system.
2. The objectives of the problem have to be formulated accordingly. Mandatory, one of the objectives should be related to the specific energy of the discussed system (mechanical or electrical in this case) to be related.
3. Determining the set of allowable control.
4. The performance criterion or index (mathematically is the functional cost), which integrates the proposed objectives by combining the states and controls of the dynamic system.

The problem of the optimal control is to find a solution, being an admissible control that minimizes or maximizes the chosen performance criterion. That solution is called *optimal control*.

Depending on the objectives imposed, the following types of optimal problems can be highlighted:

A. *Minimum time problem*: the goal is to achieve the desired final state of a dynamic system in minimum time.

B. *The minimum fuel consumption* aims to minimize the fuel consumed by a system to move to the desired final state with minimum fuel consumption.

C. *Minimum energy problem*: if there is an electrical signal as control signal, part of the system energy, by applying this control, the system energy will be minimized.

D. *The optimal control problem of the final state* will contain the final state error in the chosen performance index. The resulting control will minimize the error.

Due to its applicability in the technique, the performance index based on quadratic energy criteria is highly used.

Linear quadratic problems lead to minimal energy consumption and can be of three types: state feedback control, output control problem and optimal tracking problem.

Linear quadratic problem leads to solving Riccati equations: algebraic and differential (continuous time and discrete time).

The optimal control of the dynamic system involves the synthesis of the optimal control directly from the extremes of the functional cost.

Depending on the final time, dynamic optimization problems can be:

- Infinite final time (infinite horizon), for *linear time-invariant* systems with state constraints. In order to solve this problem type, the algebraic Riccati equation (ARE) is used.
- Finite final time (finite horizon) for *linear time-variant* systems. This type of optimization problem requires the solution of the matrix Riccati differential equation (MRDE).

By taking into account the process constraints, optimization can be without constraints or free state, with constraints or with free time.

4. Control of the synchronous machine with permanent magnets

The mathematical models of the PMSM in stator/rotor reference frame as well as a vector control solution are presented succinctly.

4.1. The mathematical model of the synchronous machine with permanent magnets in fixed reference frame

In order to determine the mathematical model of the PMSM in a fixed reference frame, the following hypotheses are considered: the iron core is non-saturated (applying the superposition effect is valid); the symmetrical three-phase stator windings, the infinite permeability of the air gap, constant air-gap width, the sinusoidal distribution of the magnetic field in the air gap and iron losses are neglected; the permanent magnets are placed on the rotor; and the isotropic machine is considered.

For the stator windings (**Figure 2a**), the second Kirchhoff Law conducts to

$$u_A = R_1 i_A + \frac{d\psi_A}{dt}, \quad u_B = R_1 i_B + \frac{d\psi_B}{dt}, \quad u_C = R_1 i_C + \frac{d\psi_C}{dt} \tag{1}$$

where u_A, u_B and u_C represent the phase voltages; i_A, i_B and i_C the phase currents; ψ_A, ψ_B and ψ_C the magnetic fluxes; and R_1 the resistance of the stator phase.

By taking into consideration the magnetic flux ψ_0 of the rotor permanent magnet, the magnetic flux of each stator winding is as follows (L_{AA} self-inductance and mutual inductances $L_{AB} = L_{BC} = L_{CA}$):

$$\begin{aligned}
\psi_A &= L_{AA} i_A + L_{AB} i_B + L_{AC} i_C + \psi_0 \cos\theta \\
\psi_B &= L_{BA} i_A + L_{BB} i_B + L_{BC} i_C + \psi_0 \cos\left(\theta - \frac{2\pi}{3}\right) \\
\psi_C &= L_{CA} i_A + L_{CB} i_B + L_{CC} i_C + \psi_0 \cos\left(\theta - \frac{4\pi}{3}\right)
\end{aligned} \tag{2}$$

or in the form:

$$\begin{aligned}
\psi_A &= L_1 i_A + \psi_0 \cos\theta \\
\psi_B &= L_1 i_B + \psi_0 \cos\left(\theta - \frac{2\pi}{3}\right) \\
\psi_C &= L_1 i_C + \psi_0 \cos\left(\theta - \frac{4\pi}{3}\right)
\end{aligned} \tag{3}$$

where $L_1 = \frac{3}{2} L_0$ is the synchronous inductance.

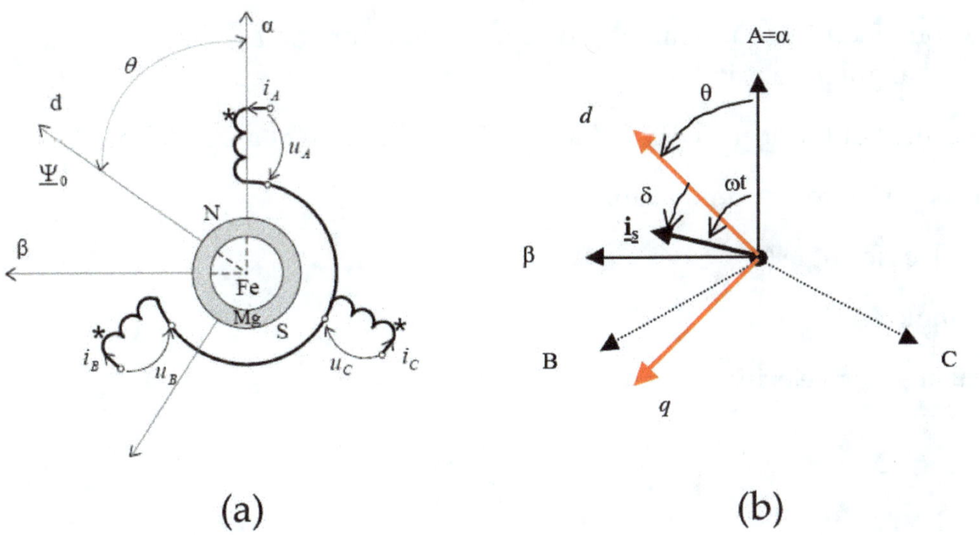

Figure 2. (a) The three-phase PMSM in stator reference frame and (b) the synchronous reference frame.

The stator and rotor phasor equations:

$$\underline{u}_1^s = R_1\underline{i}_1^s + \frac{d}{dt}\left(L_1\underline{i}_1^s + \psi_0 e^{j\theta}\right) \tag{4}$$

$$\underline{u}_1^r = R_1\underline{i}_1^r + L_1\frac{d\underline{i}_1^r}{dt} + j\theta\left(L_1\underline{i}_1^r + \psi_0\right) \tag{5}$$

in which the connection between the fixed reference frame and synchronously one is described by the following transformation:

$$\underline{i}_1^r = \underline{i}_1^s e^{-j\theta}, \underline{u}_1^r = \underline{u}_1^s e^{-j\theta} \tag{6}$$

By knowing the electromagnetic energy between the stator and rotor, the electromagnetic torque can be found:

$$T_e = p\frac{\partial\left(\psi_0 i_A \cos\theta + \psi_0 i_B \cos\left(\theta - \frac{2\pi}{3}\right) + \psi_0 i_C \cos\left(\theta - \frac{4\pi}{3}\right)\right)}{\partial\theta}\Bigg|_{\text{iA, iB, iC=const}} \tag{7}$$

Taking into consideration that the stator windings are star connected:

$$i_A + i_B + i_C = 0 \tag{8}$$

after some manipulations, the final form of the electromagnetic torque is obtained:

$$T_e = \frac{3}{2}p\psi_0\left[-i_\alpha \sin\theta + i_\beta \cos\theta\right] = \frac{3}{2}p\psi_0 i_q \tag{9}$$

The circular motion equation:

$$J\frac{d\Omega}{dt} = T_e - T_l - T_v \tag{10}$$

where the combined inertia moment reduced to the rotor shaft J, the viscous torque T_v, load torque T_l and angular velocity Ω.

The significances of the specific PMSM angles from **Figure 1(b)** are as follows:

- Angle of the stator current vector ωt.
- d-axis electric angle θ.
- Load angle δ.

Mechanical angular velocity:

$$\Omega = \frac{\dot{\theta}}{p}[rad/s] \tag{11}$$

p—pole pair number.

4.2. Vector control of the synchronous machine with permanent magnets

Vector control in a synchronous reference frame is the best way to control the permanent magnet synchronous machine (PMSM). The field-oriented control is used to control the spatial vectors of magnetic flux, stator current and voltage. As with field-oriented vector control of the three-phase asynchronous machine, the synchronous reference system chosen to drive PMSM allows the stator current vector to decompose into the field-generating component and torque component. The structure of the vector control leads to almost identical operation with the self-excited DC machine, which simplifies the structure of the PMSM control. In order to achieve the high dynamic performances in the PMSM drive, the vector control has been chosen. In this way, the parameters of the PMSM become constant. The *d*-axis stator component determines the magnetic flux changes, and the q-axis stator component is responsible to control the mechanical units: torque and speed.

The block diagram for vector control of the permanent magnet synchronous machine is shown in **Figure 3** [5].

In order to make the vector control, the following methodology should be applied:

1. Measurement of the three-phase stator currents.

2. By using the direct (forward) Clarke transformation (in order to obtain the equivalent two-phase PMSM model in fixed coordinates):

$$\begin{aligned} i_\alpha(t) &= \frac{i_A(t)}{\sqrt{2}} \\ i_\beta(t) &= \frac{i_A(t) + 2i_B(t)}{\sqrt{6}}. \end{aligned} \tag{12}$$

The above-deducted equation is based on the existence of the isolated stator windings neutral (8).

3. Direct Park transformation (based on the measured or estimated rotor angle)

$$\begin{aligned} i_d(t) &= i_\alpha(t)\cos\theta + i_\beta(t)\sin\theta \\ i_q(t) &= -i_\alpha(t)\sin\theta + i_\beta(t)\cos\theta \end{aligned} \tag{13}$$

4. The cascaded loop control of the two main axes: the mechanical and magnetic field.

5. The inverse (reverse) Park (IPARK) and Clarke (ICLARKE) transformations in order to deliver the reference switching states to the three-phase power inverter:

$$\begin{aligned} i_\alpha(t) &= i_d(t)\cos\theta - i_q(t)\sin\theta \\ i_\beta(t) &= i_d(t)\sin\theta + i_q(t)\cos\theta \end{aligned} \tag{14}$$

$$\begin{aligned} i_A(t) &= i_\alpha(t)\sqrt{2} \\ i_B(t) &= \frac{\sqrt{6}i_\beta(t) - i_A(t)}{2} \\ i_C(t) &= -[i_A(t) + i_B(t)] \end{aligned} \tag{15}$$

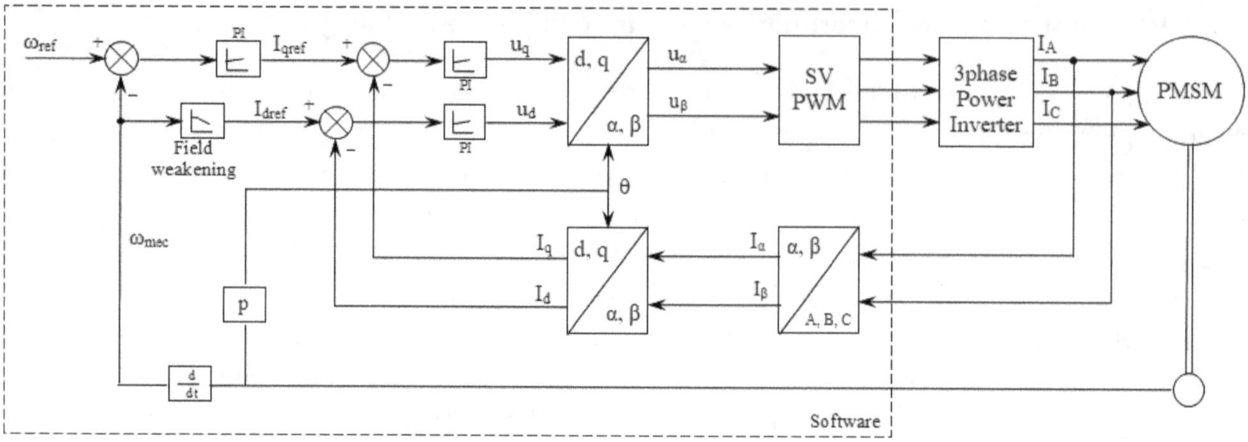

Figure 3. Block diagram for vector control of PMSM.

The voltage source power inverter supplies the stator windings of the PMSM with the adequate reference voltages such that the objectives of the control are attained.

The general mathematical model of PMSM in a synchronous reference frame (d, q) is as follows:

$$\begin{cases} \dfrac{di_d}{dt} = -\dfrac{R}{L_d} i_d + \dfrac{L_q}{L_d}\omega_r i_q + \dfrac{1}{L_d} u_d \\ \dfrac{di_q}{dt} = -\dfrac{\dot{\theta}_r L_d}{L_q} i_d - \dfrac{R}{L_q} i_q - \dfrac{\psi_0}{L_q}\omega_r + \dfrac{1}{L_q} u_q \\ \dfrac{d\omega_r}{dt} = \dfrac{1}{J} T_e - \dfrac{F}{J}\omega_r - \dfrac{p}{J} T_l \\ \dfrac{d\theta_r}{dt} = \omega_r \end{cases} \tag{16}$$

with the electromagnetic torque $T_e = \frac{3p\psi_0}{2} i_q + \frac{3p(L_q - L_d)}{2} i_d i_q$.

Synchronous motors with permanent magnets (with permanent magnet rotor inside, anisotropic PMSM) have different inductances q and d ($L_{sq} > L_{sd}$). For an isotropic PMSM, the inductances are equal: $L_{sq} = L_{sd}$.

The used symbols in the mathematical PMSM model are: J [kg m^2], the reduced inertia to the rotor shaft; T_e [Nm], T_l [Nm], the electromagnetic torque and load torque; u_d, the direct axis component of the stator voltage [Vrms]; u_q, the quadrature axis component of the stator voltage [Vrms]; i_d, the direct axis component of the stator current [Arms]; i_q, the quadrature axis component of the stator current [Arms]; θ, electric rotor angle [rad elect]; Ω, angular mechanical velocity [rad/s]; p, pole pairs.

4.2.1. *Design of the current controller*

In order to design the current control, the second Kirchhoff Law is applied to the stator equivalent circuit:

$$L_{ph}\frac{di}{dt} = -R_{ph}\cdot i + u \tag{17}$$

The corresponding transfer function is as follows:

$$G(s) = \frac{I}{U^*} = \frac{1}{L_{ph}s + R_{ph}} = \frac{g}{1 + s\tau} \tag{18}$$

where $\tau = \frac{L_{ph}}{R_{ph}}$ and $g = \frac{1}{R_{ph}}$.

By taking into consideration the step reference signal of the current loop, the proportional integral control is involved:

$$C(s) = \frac{U^*}{E_I} = K_p + \frac{K_I}{s} \tag{19}$$

The performance specifications of the PI control are:

- The bandwidth of the current loop (Hz) f_c.
- The damping coefficient ζ.

The open loop transfer function is

$$L(s) = C(s)G(s), \tag{20}$$

and the closed loop transfer function is

$$F(s) = \frac{L(s)}{1 + L(s)} = \frac{g(K_I + K_P s)}{\tau s^2 + s(gK_p + 1) + gK_I} = \frac{gK_I}{\tau}\frac{1 + s\frac{K_P}{K_I}}{s^2 + s\frac{gK_P + 1}{\tau} + \frac{gK_I}{\tau}}. \tag{21}$$

The denominator of the closed loop transfer function could be written in the standard form of the second-order system:

$$s^2 + 2\zeta\omega_0 s + \omega_0^2 \tag{22}$$

where the natural pulsation is denoted by ω_0 (**Table 1**).

Switching frequency (kHz)	The damping coefficient	The bandwidth (Hz)
8	1.5	1200
4	1.5	850
2	1.5	600

Table 1. The bandwidth of the current loop at different switching frequency.

By comparing Eq. (21) with Eq. (22), the parameters of the PI current control are obtained:

$$K_I = \frac{\tau}{g}\omega_0^2 = L_{ph}\omega_0^2$$
$$K_P = \frac{2\zeta\omega_0\tau - 1}{g} = 2\zeta\omega_0 L_{ph} - R_{ph} \quad (23)$$

The calculated poles of the closed loop current system are as follows:

$$s_{1,2} = \omega_0\left(-\zeta \pm \sqrt{\zeta^2 - 1}\right) \quad (24)$$

The connection between the bandwidth and the natural pulsation is based on the below-mentioned equation:

$$\omega_0 = 2\pi\left(\frac{f_c}{2\zeta}\right) \quad (25)$$

4.2.2. Design of the speed controller

Taking into account that the reference speed signal is the ramp type, the speed control design is based on the symmetrical optimum, the PI controller can be deducted:

$$C_v(s) = \frac{T_e^*}{E_v} = K_{pv} + \frac{K_{Iv}}{s} \quad (26)$$

$$\tau_3 = 4T_{\Sigma n}, \quad (27)$$

$$\tau_4 = \frac{8T_{\Sigma n}{}^2 k_{TT}}{k_{TI}J}, \quad (28)$$

with the specific constants [6]:

$$T_{\Sigma n} = 2T_{\Sigma i} + T_{TN}, k_m = \frac{3}{2}p\psi_0, k_{TT} = \frac{10}{1.2\cdot\Omega_r}\left[\frac{rot/\min}{V}\right], k_{TI} = \frac{10}{I_{q\max}}\left[\frac{V}{A}\right]. \quad (29)$$

The equivalent controller parameters are as follows:

$$K_{pv} = \frac{\tau_3}{\tau_4}$$
$$K_{Iv} = \frac{1}{\tau_4} \quad (30)$$

5. Optimal control of the electrical drives with PMSM

The high-efficiency PMSM during steady state is obtained at the rated load. During the transient regime, the main loss component into the PMSM drive system is the stator copper losses. For the PMSM operating in often dynamic regimes, the optimal control based on the energetic criteria should be taken into account.

5.1. Optimal control problem statement

The optimal control problem statement consists of defining the dynamic system, the objectives, the initial and final conditions and the functional. The functional or the performance index includes the combination of the independent states weighted by the matrices defined by the designer of the drive system. In order to include the energy terms, the quadratic performance index is chosen. The optimal control problem consists in finding the control which minimizes the performance index. The optimal control problem is unconstrained. By adequately choosing the weighted matrices, the boundary restriction of the states is provided.

The *objectives* of the optimal electric drive based on the PMSM are regulation (by obtaining zero steady-state error), stability (admissible electrical and mechanical limits), rejection of the perturbation and energy reduction.

The dynamic regimes of PMSM drive (starting, braking and reversing) are characterized by the absorption of large currents, high slipping and a significant reduction in conversion efficiency (below 50%).

Other side effects are increased power loss in the PMSM, therefore worsening of its thermal regime, reducing life span and increasing the harmonic pollution of the power supply network.

For these reasons it is proposed an optimal control based on energy minimization. The optimal controller has in view the minimization of the absorbed energy necessary to fulfil the required dynamic regimes.

5.1.1. *The dynamic system*

The constant flux is considered. Therefore, the imposed reference of the d-axis stator current is fixed at zero: $i^*_{sd} = 0$. Taking into account that the speed of the magnetic field wave is the same as of the mechanical rotor in synchronous machine case, to implement the FOC, it is necessary only to measure the mechanical speed, delivering the necessary angle for inverse Park transformation (**Figure 3**).

The state-space mathematical model of the PMSM is as follows [7]:

$$\begin{bmatrix} \overset{o}{\omega}_r(t) \\ \overset{o}{i}_{sq}(t) \end{bmatrix} = \begin{bmatrix} -\dfrac{F_v}{J} & \dfrac{3p\Psi_0}{2J} \\ -\dfrac{\Psi_0}{L_s} & -\dfrac{R_s}{L_s} \end{bmatrix} \cdot \begin{bmatrix} \omega_r(t) \\ i_{sq}(t) \end{bmatrix} + \begin{bmatrix} 0 \\ \dfrac{1}{L_s} \end{bmatrix} \cdot u_{sq}(t) + \begin{bmatrix} -\dfrac{1}{J} \\ 0 \end{bmatrix} \cdot T_l(t) \tag{31}$$

By noting the state vector $\mathbf{x}(t) = \begin{bmatrix} \omega_m(t) \\ i_{sq}(t) \end{bmatrix}$, **u** is the control vector; $\mathbf{u}(t) = [u_{sq}(t)]$, and the perturbation vector is $\mathbf{w}(t) = [T_l(t)]$; the mathematical model of the PMSM in the standard state space can be configured:

$$\dot{\mathbf{x}}(t) = \mathbf{A}\mathbf{x}(t) + \mathbf{B}\mathbf{u}(t) + \mathbf{G}\mathbf{w}(t) \tag{32}$$

The initial and final conditions are stated for a starting:

$$\mathbf{x}_0 = \begin{vmatrix} \omega_r(t_0) \\ i_{sq}(t_0) \end{vmatrix} = \begin{vmatrix} 0 \\ 0 \end{vmatrix}, t_0 = 0 \text{ [s]} \tag{33}$$

$$\mathbf{x}_1 = \begin{vmatrix} \omega_{rN} \\ 0 \end{vmatrix}, t_1 = 0.4 \text{ [s]} \tag{34}$$

The *performance criterion* is based on the quadratic functional having in view both the steady-state and dynamic-state performances.

The chosen performance criterion [8] is as follows:

$$J = \underbrace{\frac{1}{2}[\mathbf{x}(t_1) - \mathbf{x}_1]^T \mathbf{S}\ [\mathbf{x}(t_1) - \mathbf{x}_1]}_{\Lambda|_{t=t_1}} + \frac{1}{2}\int_{t_0}^{t_1} [\mathbf{x}^T(t)\ \mathbf{Q}\mathbf{x}(t) + \mathbf{u}^T(t)\ \mathbf{R}\mathbf{u}(t)]dt \tag{35}$$

in which $\Lambda|_{t=t_1}$ is the *terminal cost.*

The weighting matrices are designed such that S ≥ 0 and Q ≥ 0 are positive semidefinite; **S** has in view the minimization of the square error between the obtained final state and the imposed one at the final time t_1; Q has in view the smooth angular speed without oscillations and minimization of the stator copper losses.

R > 0 is positive definite, in order to assure the minimum condition existence and to limit the magnitude of the control.

The *optimal problem* consists of determining the control, $\mathbf{u}^*(t)$, which minimizes the quadratic functional (35):

$$J_{\min} = J(\mathbf{u}^*) \tag{36}$$

The electrical drive system based on the PMSM is completely controllable and observable. Therefore, the existence of the minimum is assured.

By taking into consideration the above-mentioned conditions, the optimal problem is unconstrained, fixed time and free-end point.

5.2. The solution of the optimal control problem

The solution of the optimal control problem exists, and it is unique thanks to the complete controllability and observability of the dynamic system by adequate choosing of the weighting matrices: Q, S $\geq$ 0 and R > 0 [8].

5.2.1. Hamiltonian of the optimal problem

By knowing the dynamic system and the integral part of the performance index, the Hamiltonian of the proposed problem can be determined:

$$H(p,x,u,t)=\frac{1}{2}\left[\mathbf{x}^{\mathbf{T}}(\mathbf{t})\ \ \mathbf{Qx}(\mathbf{t})+\mathbf{u}^{\mathbf{T}}(\mathbf{t})\mathbf{Ru}(\mathbf{t})+\mathbf{p}^{\mathbf{T}}(\mathbf{t})\ \ \cdot\overset{\circ}{\mathbf{x}}(\mathbf{t})\right] \tag{37}$$

Through the Hamiltonian (37) the unknown co-state vector is introduced $\mathbf{p}(t)\in\mathfrak{R}^2$.

By using the necessary conditions of the optimality:

$$\left.\frac{\partial\Lambda}{\partial\mathbf{x}}\right|_{\mathbf{x}^*(t_1)}-\mathbf{p}(t_1)=0 \tag{38}$$

$$\left.\frac{\partial H}{\partial\mathbf{x}}\right|_*+\dot{\mathbf{p}}(t)=0 \tag{39}$$

$$\left.\frac{\partial H}{\partial\mathbf{u}}\right|_*=0, \tag{40}$$

both the *canonical system* and the optimal control are obtained:

$$\begin{aligned}\dot{\mathbf{x}}^*(t)&=\frac{\partial H}{\partial\mathbf{p}}[\mathbf{x}^*(t),\mathbf{p}(t),\mathbf{u}^*(t),t]\\ \dot{\mathbf{p}}(t)&=-\frac{\partial H}{\partial\mathbf{x}}[\mathbf{x}^*(t),\mathbf{p}(t),\mathbf{u}^*(t),t]\end{aligned} \tag{41}$$

By using Eq. (40), the *optimal control* is obtained:

$$u^*(t)=-\mathbf{R}^{-1}\mathbf{B}^{\mathbf{T}}\mathbf{p}(\mathbf{t}) \tag{42}$$

The canonical system can be solved by knowing the initial and final conditions: $\mathbf{x}(t_0)=\mathbf{x}_0$ and $\mathbf{p}(t_1)$.

The final condition for the co-state vector is obtained by the *transversality condition*:

$$\mathbf{p}^*(t_1)=\frac{\partial\Lambda}{\partial\mathbf{x}}[\mathbf{x}^*(t_1)] \tag{43}$$

$$\mathbf{p}^*(t_1)=\mathbf{S}[\mathbf{x}(\mathbf{t}_1)-\mathbf{x}_1] \tag{44}$$

The co-state and state vectors are the solutions of the canonical system (41).

The adopted nonrecursive solution of the MRDE in numerical form [8] at any instant t has been found:

$$u_{sq}{}^*(t)=\mathbf{R}^{-1}\mathbf{B}^T[-\mathbf{P}(t_1-t)\mathbf{x}(t)+\mathbf{K}_1(t_1-t)\mathbf{x}_1+\mathbf{K}_2(t_1-t)\mathbf{w}(t)], \tag{45}$$

where the feedback gain matrix $\mathbf{P}(t_1-t)$ is computed as the solution of the nonrecursive MRDE, the matrix $\mathbf{K}_1$ takes into consideration the zero steady-state error at the final time t_1 (i.e. to achieve the imposed reference $\mathbf{x}_1$ at the final time t_1) and the matrix K_2 has in view the feedforward compensation of the load torque. The matrices $\mathbf{K}_1$ and $\mathbf{K}_2$ are calculated via $\mathbf{P}(t_1-t)$, where t_1-t means the time remaining until the final time [8].

6. Numerical results

Both the vector control and optimal control (45) of the PMSM drive system (31) have been implemented in Matlab-Simulink, and numerical results based on the discretization using Z transform and zero order hold for a starting of a 2.81 [kW] and 1420 [rpm] PMSM under a rating load of 19 [Nm].

6.1. Vector control of the PMSM

For operation at rated regime, the magnetizing flux is maintained constant. Therefore, the *d*-axis stator component is set to zero value, $i_{sd}^{*}(t) = 0$ [A].

By using an elevator prototype, the experimental results are provided. The optimal control has been implemented on ControlDesk, the dSpace real-time interface.

The Matlab-Simulink vector control implementation of the PMSM is shown in **Figure 4**. The performance of the speed controller is depicted in **Figure 5** under a rated load torque (**Figure 6**). The PMSM drive is tested in both dynamic and steady-state regimes. The starting, breaking and speed reversal are shown in **Figure 5**. In **Figures 5–10**, the appropriate signals of the PMSM drive system based on the vector control are shown.

In **Figures 7** and **8**, the stator current controller performances are shown. In **Figure 9**, the two-phase stator currents are obtained. At the same time, in **Figures 9** and **10**, the appropriate *d* and *q* stator voltage components are depicted.

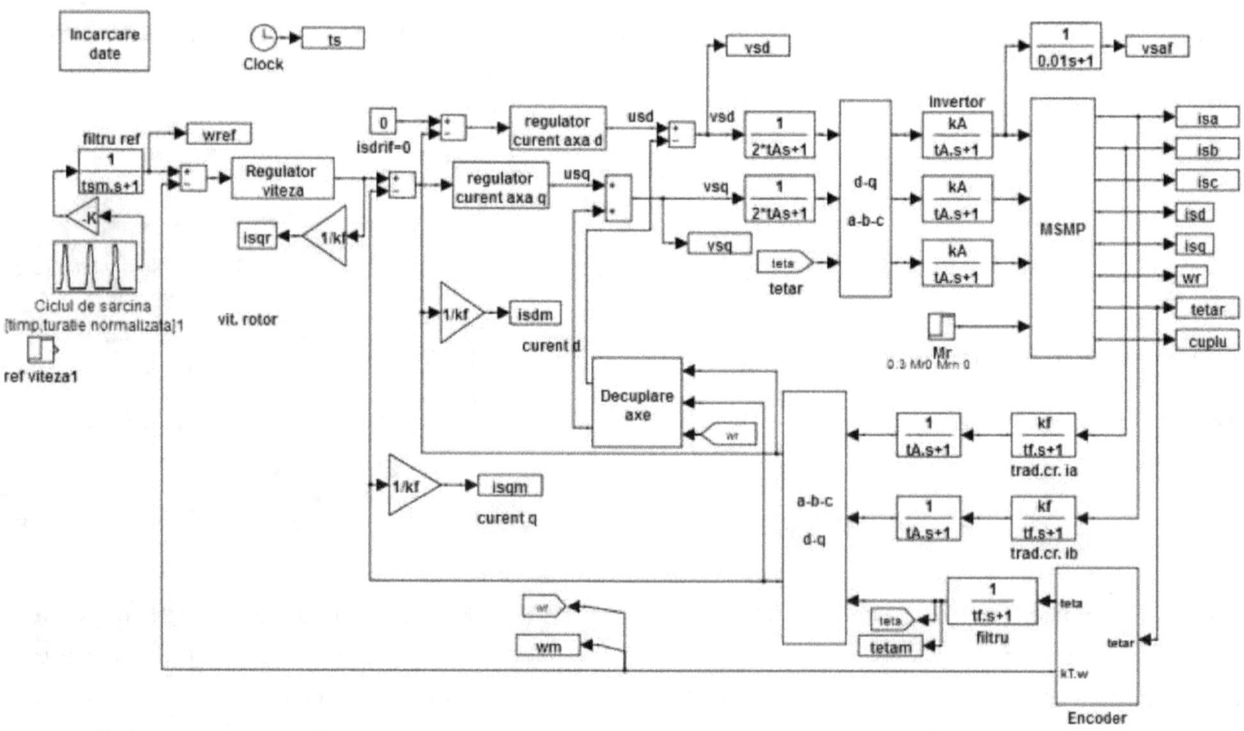

Figure 4. The Simulink implementation of the PMSM vector control.

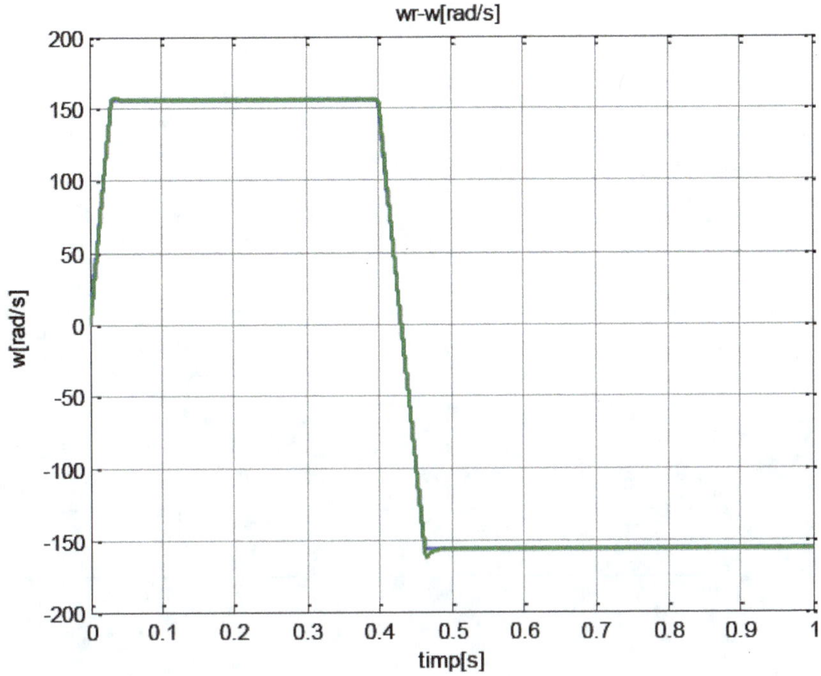

Figure 5. Speed reference and the actual speed.

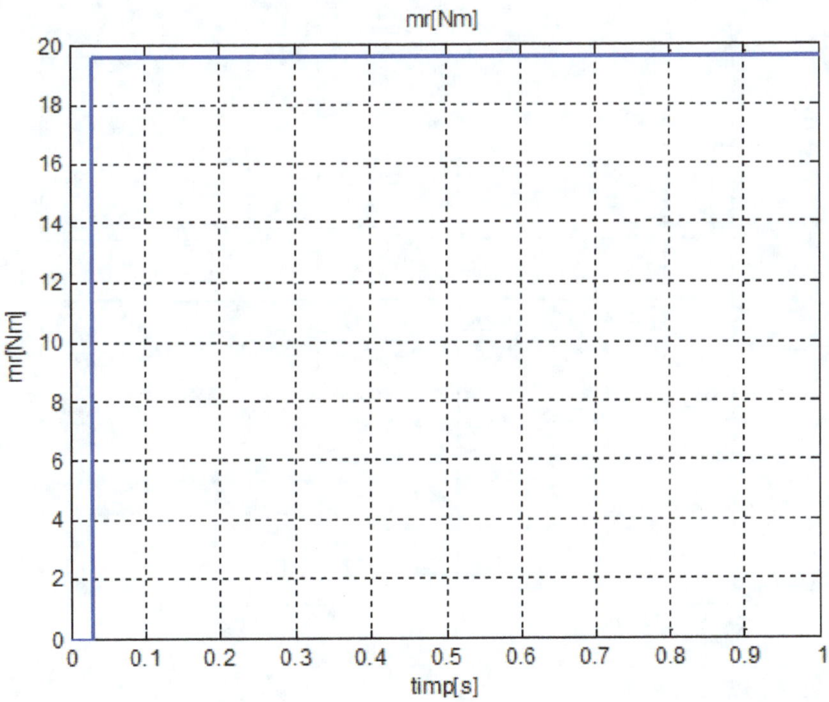

Figure 6. The rated load torque of the elevator.

6.2. Optimal control of the PMSM

The problem formulation supposes the initial and final condition specification. The initial conditions are formulated for a starting: $t_o = 0$ [s], $n(t_o) = 0$ and $i_{sq}(t_o) = 0$. The final conditions are $t_1 = 0.4$ [s], $n_1 = 1420$ [rpm] and $i_{sq}(t_1) = 0$. The implemented optimal control of the PMSM

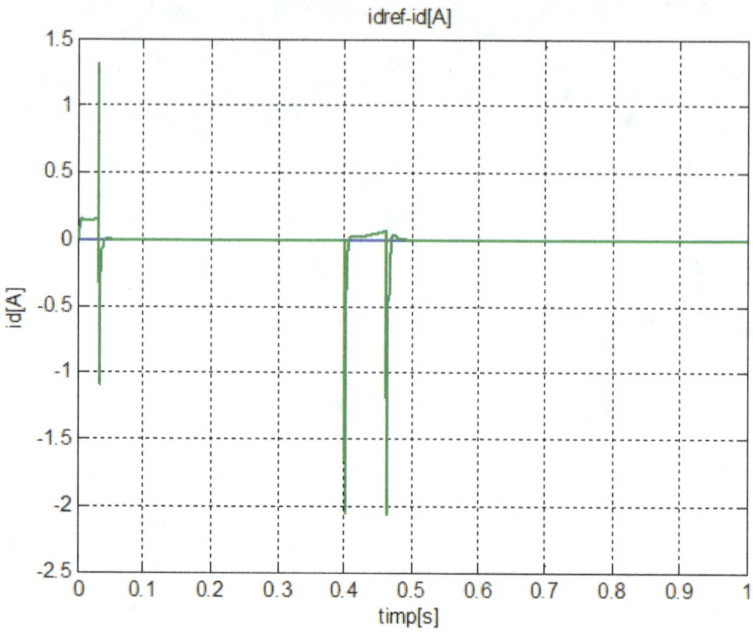

Figure 7. The d-axis stator current controller.

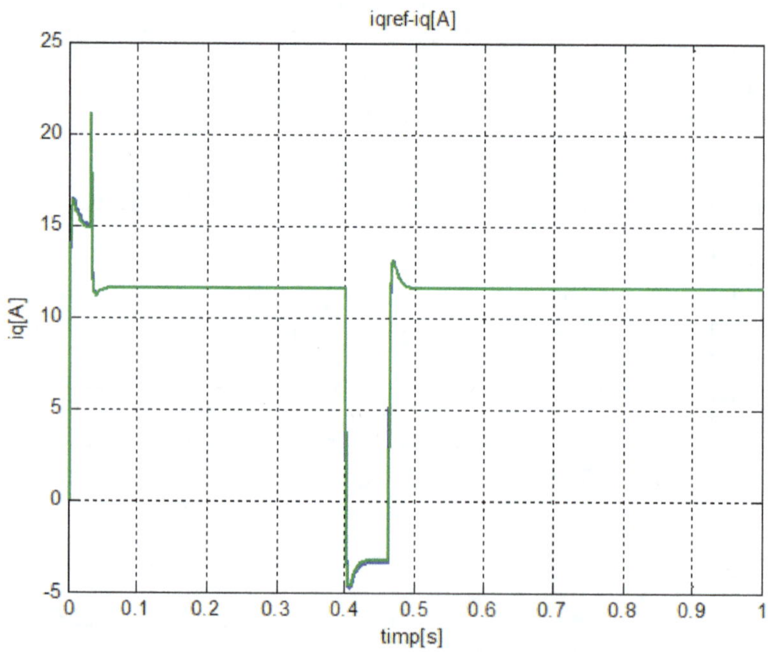

Figure 8. The q-axis stator current controller.

conducts to the specific speed reference as in **Figure 11**. At the same time, the performances of the speed controller are shown in **Figure 11**.

The performances of the speed and current (d-axis, q-axis) controllers are depicted in **Figures 11–13**. The shape of the stator phase current (**Figure 14**) and the developed electromagnetic torque (**Figure 15**) depend on the reference speed (**Figure 11**).

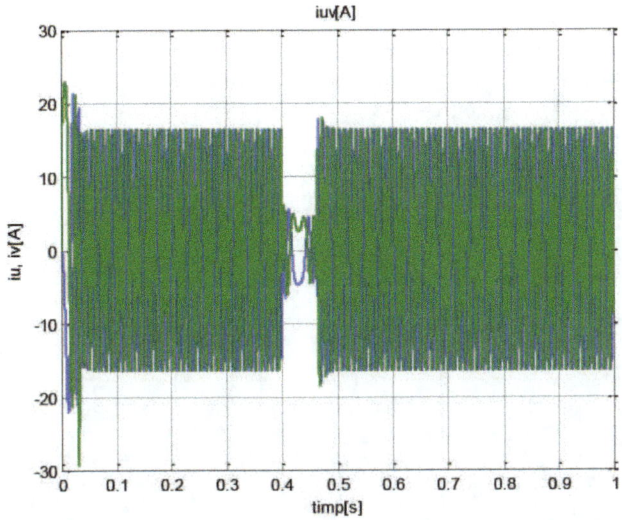

Figure 9. The phase stator currents.

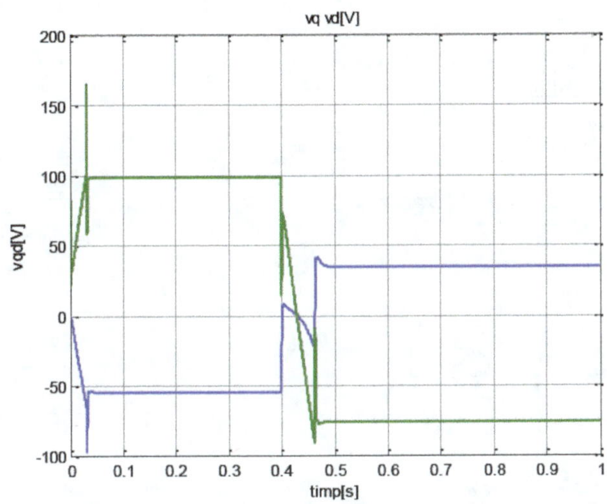

Figure 10. The stator voltage components.

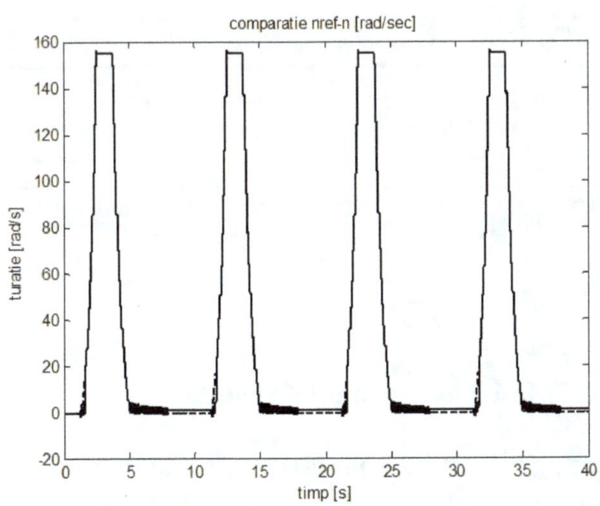

Figure 11. The performances of the optimal speed controller.

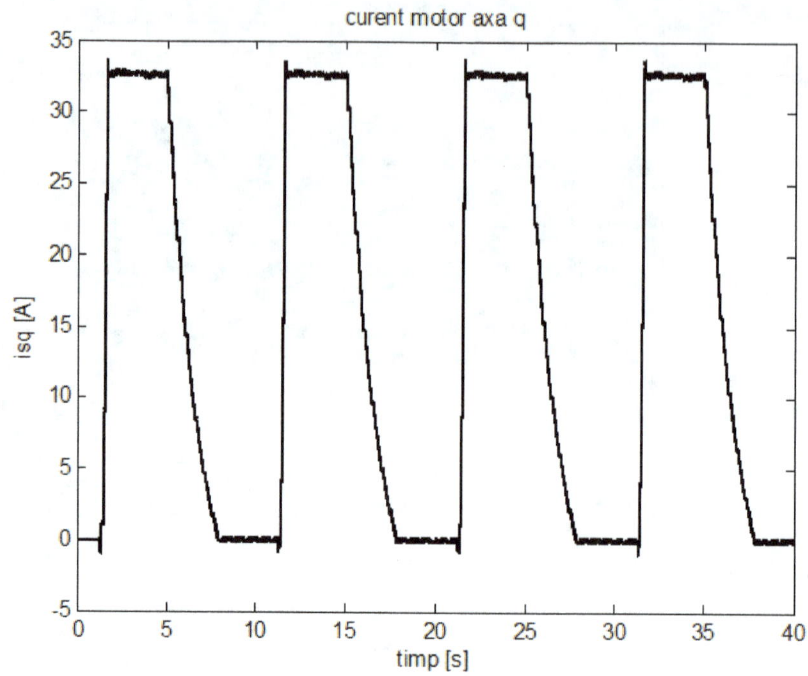

Figure 12. The *q*-axis current controller.

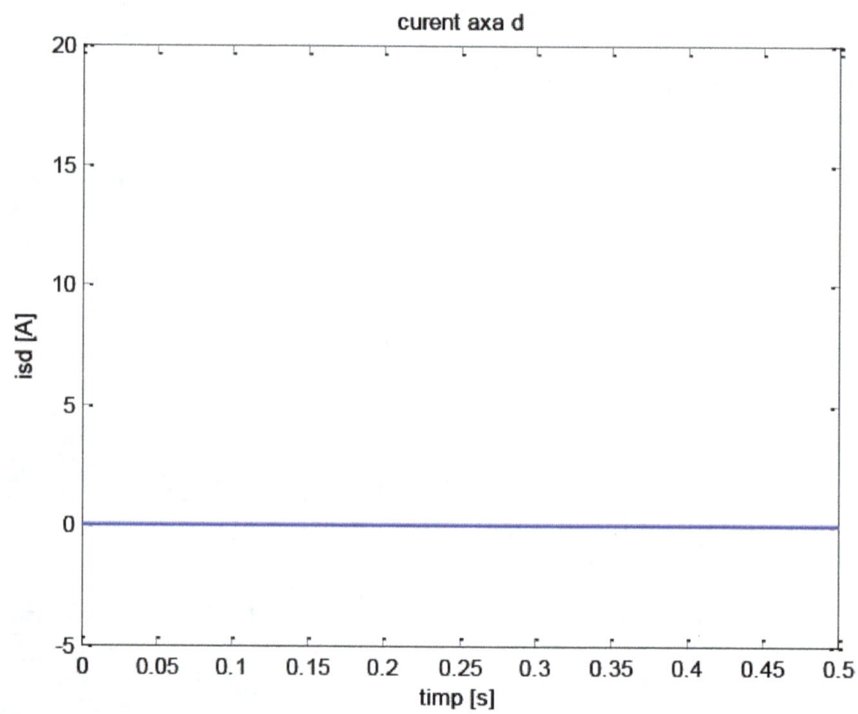

Figure 13. The *d*-axis current controller.

6.3. dSpace control of the PMSM: experimental results

For the PMSM the same speed profiles have been implemented as in simulation tool. The optimal control has been implemented on the dSpace platform in order to control an elevator (**Figures 16–18**).

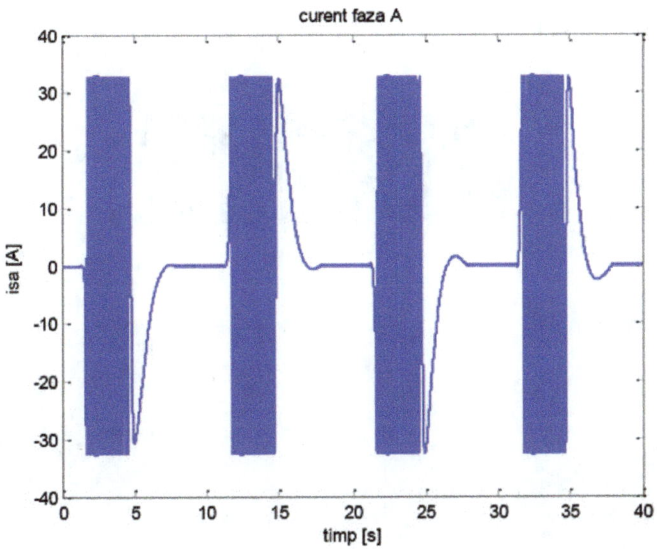

Figure 14. The stator phase current.

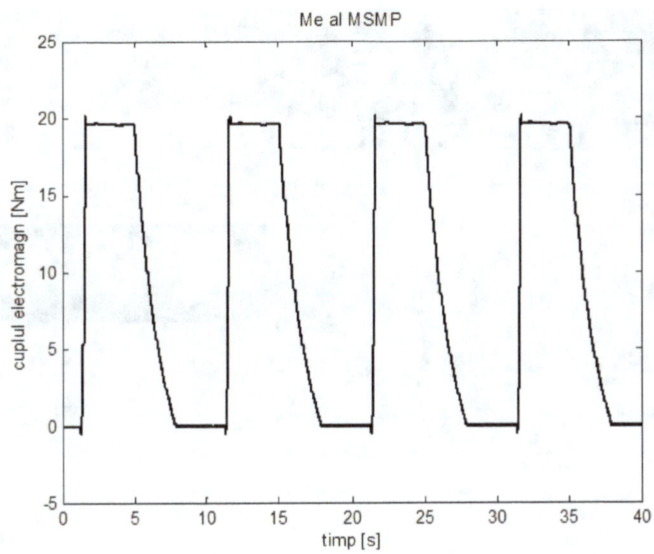

Figure 15. The electromagnetic torque.

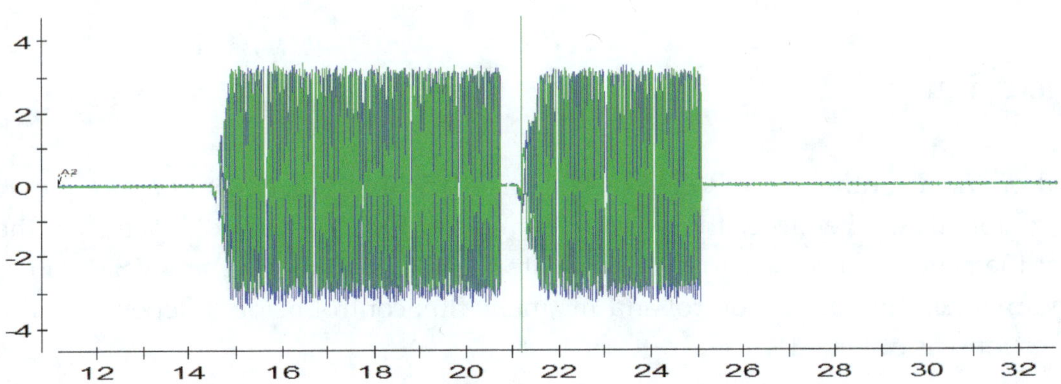

Figure 16. The stator phase currents (U,V) for the PMSM [9].

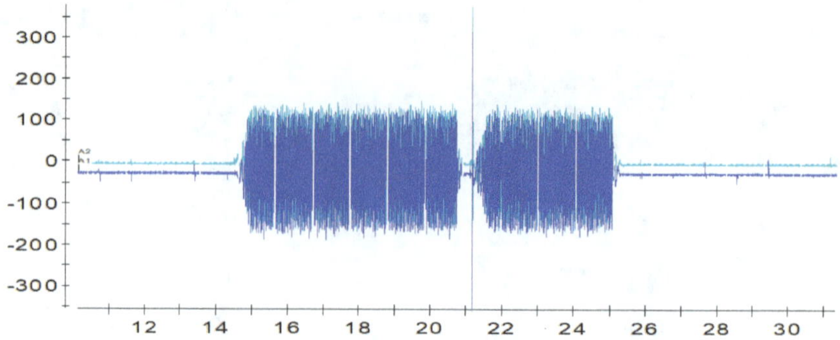

Figure 17. The stator voltage of the PMSM [9].

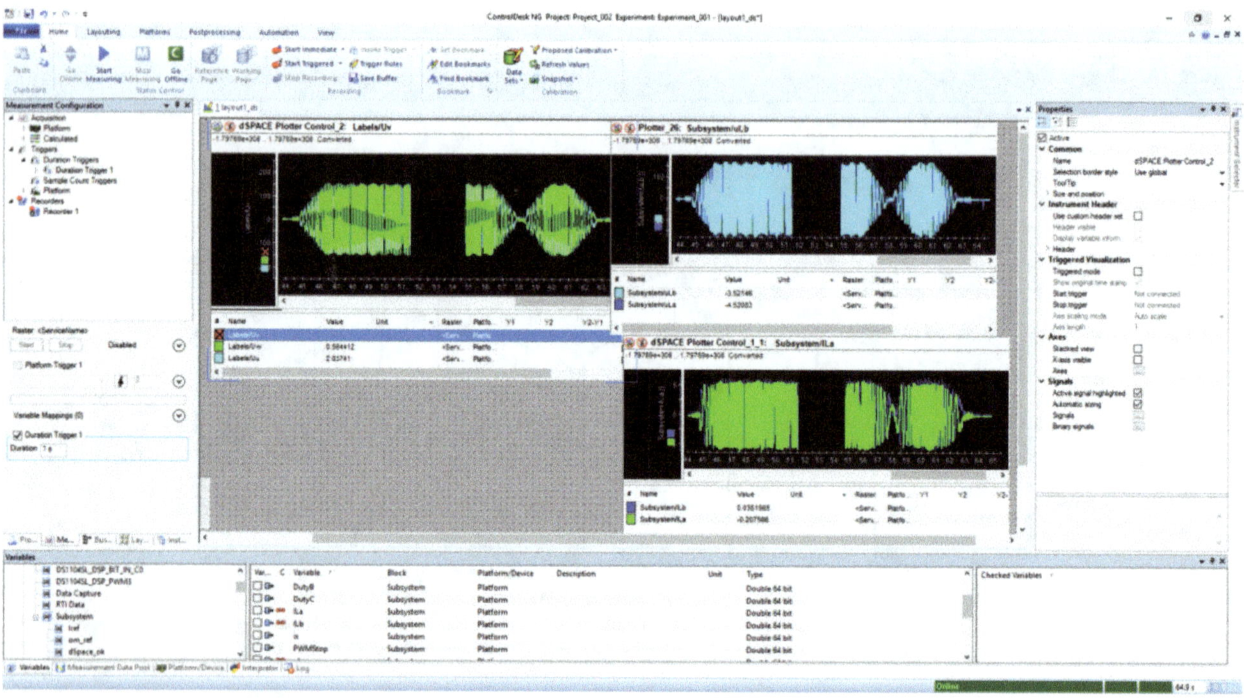

Figure 18. The ControlDesk real-time interface for controlling PMSM [9].

7. Conclusions

The field-oriented control of the PMSM has the main advantage of decoupling torque from the magnetization flux. Therefore, the parameters of the PMSM do not depend on the rotor position; the stator and rotor currents are treated as DC quantities. The PMSM control is in cascaded manner, the torque control and magnetic flux control being independent; the quick torque response is obtained.

The synchronous machine with permanent magnets is a serious contender for their use in specific applications: irrigation, wind turbines, conveyor belts and elevators.

It is noted that by introducing an optimal regulator, the state vector is limited due to the proper choice of the weighting matrices. By minimizing the quadratic performance index, energy saving is initiated.

In **Figure 6**, the removal of the system disturbance at t = 0.02 s in a relatively small period compared to the evolution of the process is shown.

The reduction of the energy assures either an increase of the operational period of the electrical drive components or the permanent magnet synchronous motor overload permission.

Acknowledgements

This work was supported by a grant of the Romanian National Authority for Scientific Research, CNDI-UEFISCDI, project number PN-II-PT-PCCA-2011-3.2-1680.

Author details

Marian Găiceanu

Address all correspondence to: marian.gaiceanu@ieee.org

Integrated Energy Conversion Systems and Advanced Control of Complex Processes Research Center, Dunarea de Jos University of Galati, Galati, Romania

References

[1] Freguglia P, Giaquinta M. The Brachistochrone Problem: Johann and Jakob Bernoulli, Book: The Early Period of the Calculus of Variations. Switzerland: Birkhäuser Basel, Springer International Publishing; 2016. DOI: 10.1007/978-3-319-38945-5_2

[2] Matychyn I, Onyshchenko V. On time-optimal control of fractional-order systems. Journal of Computational and Applied Mathematics. 2018;**339**:245-257

[3] Frank B, Kleinert J, Filla R. Optimal control of wheel loader actuators in gravel applications. Automation in Construction. 2018;**91**:1-14

[4] Rojas RA, Carcaterra A. An approach to optimal semi-active control of vibration energy harvesting based on MEMS. Mechanical Systems and Signal Processing. 2018;**107**:291-316

[5] Leonhard W. Control of Electrical Drives. Berlin: Springer-Verlag; 1996

[6] Găiceanu M. Matlab-Simulink-Based Compound Model Reference Adaptive Control for DC Motor, Adaptive Robust Control Systems Anh Tuan Le. United Kingdom

[7] Gaiceanu M, Rosu E. Analysis of the nonrecursive advanced optimal control of the permanent magnet synchronous motor drive. Applied Mechanics and Materials. 2013; **367**:194-198

[8] Athans M, Falb PL. Optimal Control. New York: McGraw-Hill; 1966

[9] Gaiceanu M, Epure S. Improvements on the electric drive elevator prototype. Part II Operating Tests. April 2018. DOI: 10.1515/sbeef-2017-0022

7

Robust Diagnosis by Observer Using the Bond Graph Approach

Abderrahmène Sallami

Abstract

In this chapter, we have proposed a Luenberger observer linear systems diagnostic technique using the bond graph. Indeed, an observer can reconstruct or estimate the current state of a real system using the available measurements, without prior knowledge of the initial conditions. In addition, it allows to estimate the nonmeasurable states of a system. The design of the observer is carried out using graphical methods from the structural properties of the model bond graph becomes simple and practical to build. We presented the bond graph approach for the construction of a full-order observer and proposed a new BG-based observer diagnostic method. Subsequently, we presented the uncertain parameter systems modeled by the bond graph approach, and we also proposed a new method for diagnosing systems with uncertain parameters by Luenberger observer. In the last part of this chapter, we developed and proposed an observer bench diagnosis technique (BG-DOS/BG-GOS) to detect and locate defects.

Keywords: robust diagnosis, bond graph, Luenberger observer, DC motor, generation of robust residues

1. Introduction

Nowadays, the engineering sciences rely heavily on the estimation of the state of the systems. Indeed, the complete knowledge of the state of a system is often necessary for the elaboration of a control law or the setting up of a strategy of monitoring or diagnosis. In practice, the state of a system is not always available and the input and output signals are the only quantities accessible by measurement. The most widespread solution to this problem consists in coupling to the system another auxiliary system, called estimator or observer of state. The observer provides an estimate of the state of the system based on its model and measurements of its inputs and outputs. The observer conventionally used, within the framework of linear systems, is said to have proportional gain or Luenberger. In recent years, observer diagnosis is used to estimate shareholder defects and sensor defects [1].

In this chapter, we show how the leap graph model can be used for modeling, simulation and residual determination by Luenberger observers for diagnosis system.

2. Observer diagnosis

2.1. Observational diagnosis using the analytical model

The principle of diagnosis consists in estimating, by appropriate techniques, all the components of the state vector or, more generally, the output of the process, using the error of estimation as a residue [2].

This operation is carried out by means of a proportional observer. The functional diagram of such a method is given in **Figure 1**.

The residue equations are defined as follows:

State estimation residue:

$$r_x = \tilde{X} = x - \hat{x} \tag{1}$$

Exit estimate residue:

$$r_y = \tilde{Y} = y - \hat{y} = C(x - \hat{x}) \tag{2}$$

2.1.1. *Residual in the case of normal operation*

We consider a linear system whose equations of states are defined as follows:

$$\begin{cases} \dot{x}(t) = Ax(t) + Bu(t) \\ y(t) = Cx(t) \end{cases} \tag{3}$$

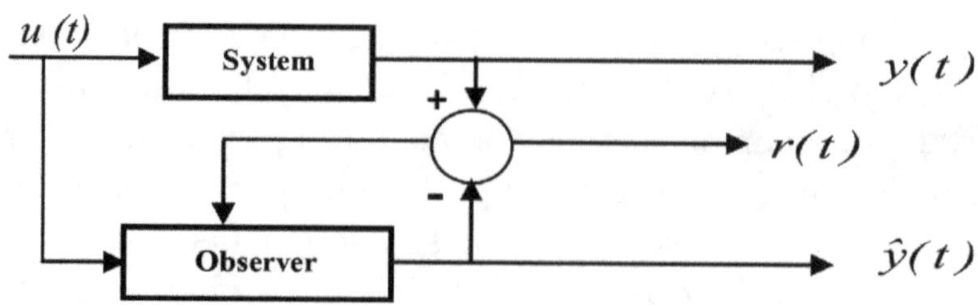

Figure 1. Observational diagnosis.

The proportional state observer is defined by equations:

$$\begin{cases} \dot{\hat{x}}(t) = A\hat{x}(t) + Bu(t) + K(y - \hat{y}) \\ \hat{y}(t) = C\hat{x}(t) \end{cases} \tag{4}$$

The state estimation or state reconstruction residual is by definition:

$$r_x = \tilde{X} = x - \hat{x} \tag{5}$$

For a proportional observer, the evolution of the residue r(t) is:

$$\dot{r}_x = (A - KC)r_x \tag{6}$$

The Laplace transform of Eq. (6) is written:

$$r_x(p) = (pI - A + KC)^{-1}r_0 \tag{7}$$

The output reconstruction residue is defined by:

$$r_y = \tilde{Y} = y - \hat{y} = C(x - \hat{x}) = C\tilde{X} = Cr_x \tag{8}$$

Hence using the previous result, Eq. (8) is written:

$$r_y(p) = C(pI - A + KC)^{-1}r_0 \tag{9}$$

2.1.2. Residual with a sensor failure

We consider again the same linear system in which the measurements made are subjected to a defect which we denote by fc(t) of unknown amplitude and appearing at an unknown time (**Figure 2**):

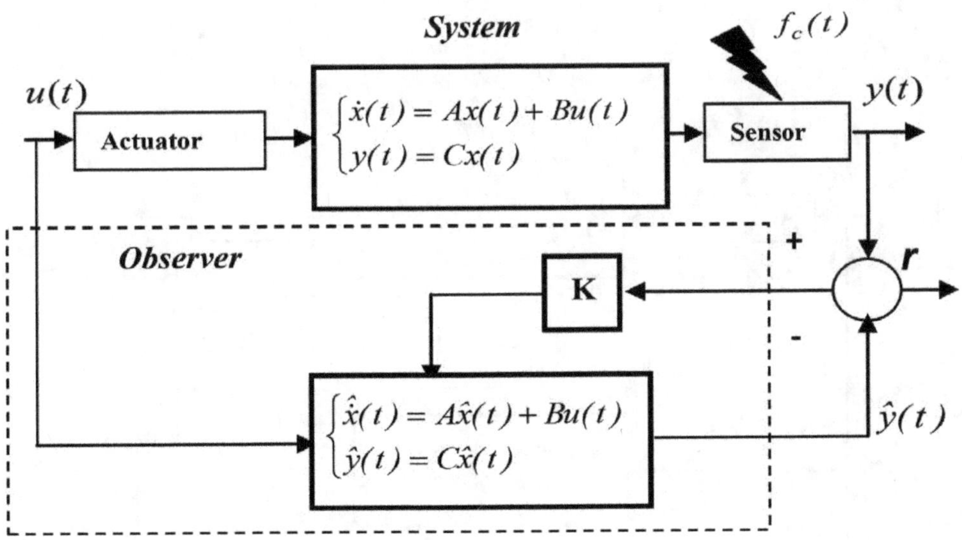

Figure 2. Proportional observer with sensor fault.

The equations of states become:

$$\begin{cases} \dot{x}(t) = Ax(t) + Bu(t) \\ y(t) = Cx(t) + Ff_c(t) \end{cases} \tag{10}$$

In order to diagnose the state of operation of the system, the states reconstruct or must be sensitive to this bias and must highlight this sensor defect, isolate it and possibly quantify it. The residue r(t) is:

$$\dot{r}_x(t) = (A - KC)r_x(t) + Ff_c(t) \tag{11}$$

The equation is written using the Laplace transformation:

$$r_x(p) = (pI - A + KC)^{-1}(r_0 + Ff_c(p)) \tag{12}$$

To see the error of reconstruction of output according to the biases of the sensors, we do not take into account the initial conditions of or:

$$r_x(p) = \left[I - C(pI - A + KC)^{-1}K)\right]Ff_c(p) \tag{13}$$

2.1.3. Residual with an actuator failure

In the case where the control applied to the system is misinterpreted by the actuator, the latter introduces a fault which we denote by $f_a(t)$ (**Figure 3**).

The equation describing the faulty system is:

$$\begin{cases} \dot{x}(t) = Ax(t) + Bu(t) + Df_a(t) \\ y(t) = Cx(t) + Ff_c(t) \end{cases} \tag{14}$$

For zero initial conditions, the expression of the state reconstruction residue is of the form:

$$r_x(p) = \left[I - C(pI - A + KC)^{-1}K)\right]Ff_c(p) + C(pI - A + KC)^{-1}Df_a(p) \tag{15}$$

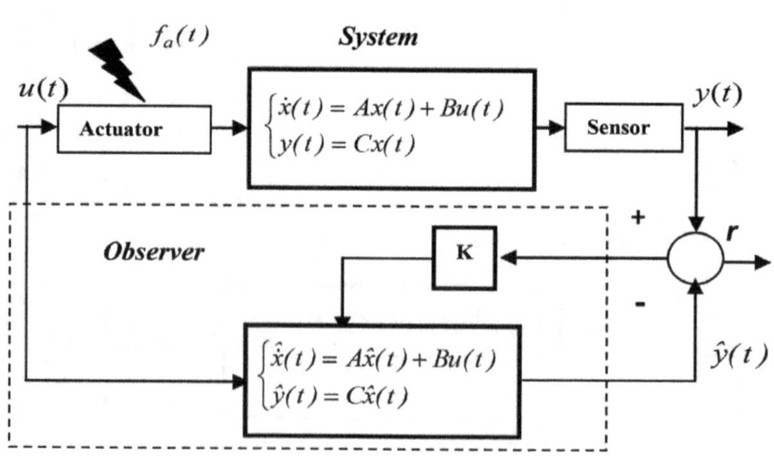

Figure 3. Proportional observer with actuator fault.

The output reconstruction residue is defined by equation:

$$r_x(p) = C(pI - A + KC)^{-1}Df_a(p) \tag{16}$$

2.1.4. Residual with sensor and actuator failures

In this case, we have the addition of two defects at the system level:

- Fault at sensor $fc(t)$;
- Fault at actuator $fa(t)$.

The equations of the system are then written as follows:

$$\begin{cases} \dot{x}(t) = Ax(t) + Bu(t) + Df_a(t) \\ y(t) = Cx(t) + Ff_c(t) \end{cases} \tag{17}$$

Applying the principle of superposition, since the system is linear, we can determine the output reconstruction residue which is the sum of the residuals caused by the sensor alone and the actuator alone. The expression of this residue is as follows:

$$r_x(p) = \left[I - C(pI - A + KC)^{-1}K)\right]Ff_c(p) + C(pI - A + KC)^{-1}Df_a(p) \tag{18}$$

This residue is sensitive to sensor and actuator defects.

2.2. Observational diagnosis using the bond graph model

2.2.1. Construction of a Luenberger observer based on model bond graph

To construct the observers, one must check the observability of the system. From a Bond Graph perspective, proposed by [3], a system modeled by leap graph is structurally observable, if the following two conditions are met:

- *First condition*: There is at least one causal path linking a sensor to each dynamic element I or C in the integral causality when the hop graph is the preferred integral causal model.
- *Second condition*: All elements I or C admitting a derivative causality when we put the leap graph model in derivative causality, and dualize the sensors.

Figure 4 presents, respectively, the proportional observer using the bond graph for elements I and C [4].

2.2.2. Proportional observer diagnosis using the bond graph model

The observer provides an estimate of the state of the system based on its model and measurements of its inputs and outputs. The observer conventionally used in linear systems is said to have Proportional (P) or Luenberger gain [5].

Consider a continuous system described by the equation of state using the hop graph represented in **Figure 5**:

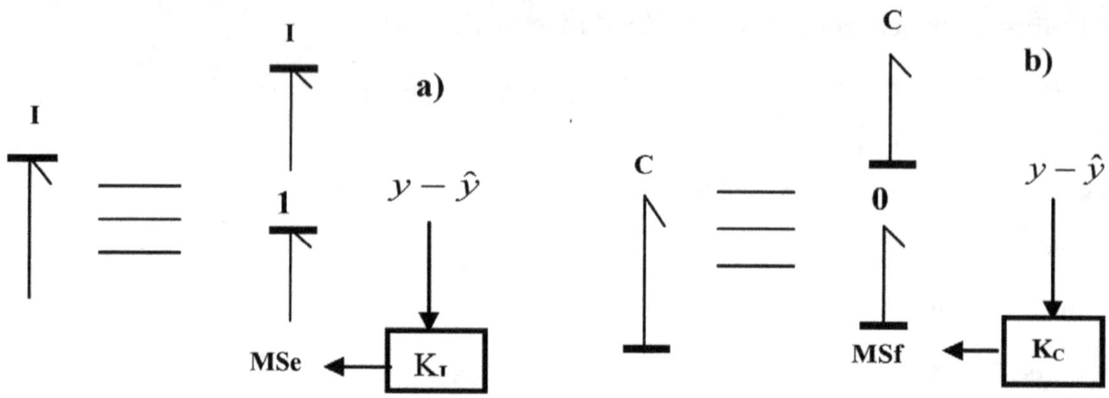

Figure 4. Construction of a proportional observer case of: (a) element I and (b) element C.

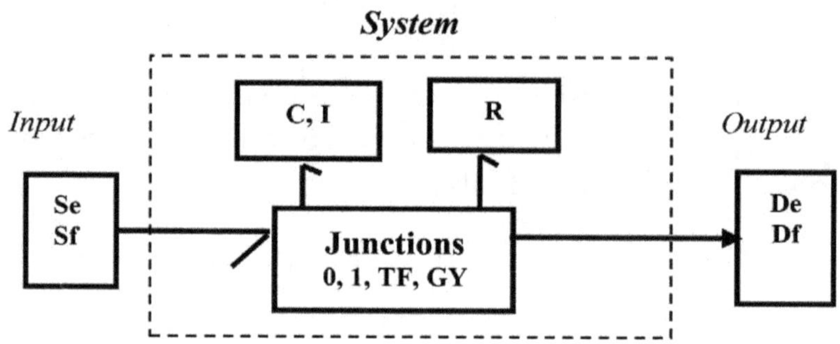

Figure 5. Schema of the continuous system described by bond graph.

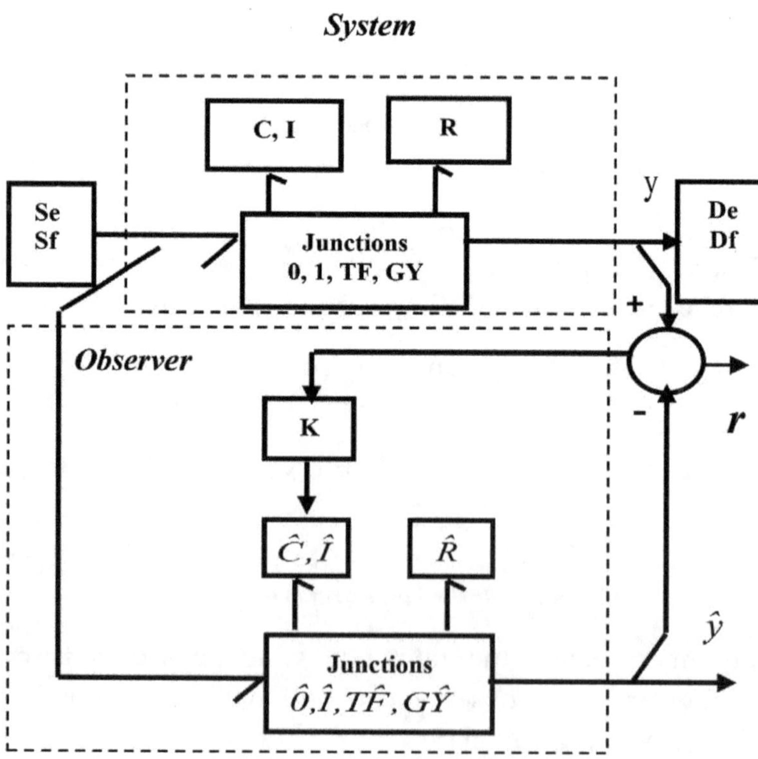

Figure 6. Structure of a Luenberger observer using graph graph.

The state observer of proportional type is represented by **Figure 6**. The equation of state is of the following form:

$$\begin{cases} \dot{\hat{x}}(t) = \begin{pmatrix} \dot{\hat{p}}_I \\ \dot{\hat{q}}_c \end{pmatrix} = A\begin{pmatrix} p_L \\ q_C \end{pmatrix} + Bu(t) + K(y(t) - \hat{y}(t)) \\ \hat{y} = C\begin{pmatrix} \hat{p}_L \\ \hat{q}_C \end{pmatrix} \end{cases}$$

2.2.3. *Generation of residues*

The generation of residues from a proportional observer using the bond graph model is summarized by the following steps:

Verify that the bond graph model of the system is structurally observable, if so, then continue the next steps;

Construction of the observer using the graph;

The symbolic expression of residue is deduced from equation:

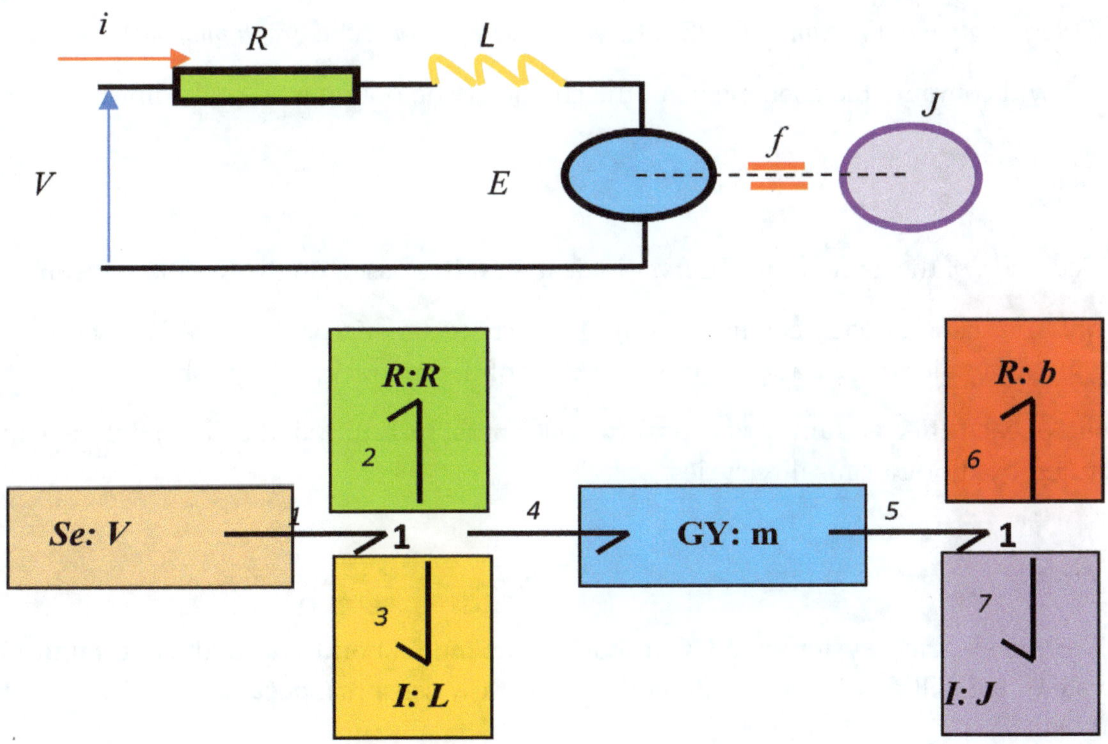

a) b)

Figure 7. (a): Direct current motor, (b): Bond model of the DC motor.

$$r = \dot{y} - \hat{y} \tag{19}$$

After calculation, the residue is in the form:

$$r : \Phi(R, I, C, TF, GY) \tag{20}$$

2.3. Determination of the gain of the Luenberger observer

2.3.1. Determination of the gain of the Luenberger observer by the analytical method

To guarantee the asymptotic convergence toward zero of the estimation error, choose the gain K to stabilize the matrix (A-KC).

The gain K can be determined by two methods:

- Placement of the poles: This method makes it possible to have a stable observer and allows playing on the criterion of rapidity of the convergence of the observer by choosing eigenvalues greater or less in absolute values.
- LYAPUNOV criterion: This method guarantees the stability of the observer but gives no idea about the speed of convergence of the observer since it does not allow to choose the eigenvalues.

2.3.2. Determination of the gain of the Luenberger observer by the bond graph method

First, we will compute the coefficients of the characteristic polynomial of the linear system:

$$p_A = p^n + a_1 \cdot p^{n-1} + \ldots\ldots + a_{n-1} \cdot p + a_n \tag{21}$$

The calculation of the gain of the observer by leap graph is based on the Rahmani theorem, [6].

Theorem: The value of each coefficient (ai) of the characteristic polynomial PA isequal to the total gain of families of causation cycles of order i of the model graph graph:

The gain of each affected family of causality cycles must be multiplied by (-1) d if the family is constituted by disjoint causality cycles.

2.4. Example

Consider the electrical system of a DC motor and its model bond graph given in **Figure 7**. On this system, we will detect and locate faults at the flow sensors (speed sensor Df_1 and current sensor Df_2).

2.4.1. State equation

The equation of state of the DC motor is as follows:

$$\begin{cases} \begin{pmatrix} \dot{p}_3 \\ \dot{p}_7 \end{pmatrix} = \begin{bmatrix} -\frac{R}{L} & -\frac{m}{J} \\ \frac{m}{L} & -\frac{b}{J} \end{bmatrix} \begin{pmatrix} p_3 \\ p_7 \end{pmatrix} + \begin{bmatrix} 1 \\ 0 \end{bmatrix} U \\ y = \begin{bmatrix} \frac{1}{L} & 0 \\ 0 & \frac{1}{J} \end{bmatrix} \begin{pmatrix} p_3 \\ p_7 \end{pmatrix} \end{cases} \tag{22}$$

With R: armature resistance, L: armature inductance, m: torque coefficient, J: moment of inertia, b: coefficient of friction.

R = 1 Ω; L = 5 mH; b = 10^{-4} Nm/rd. S^{-1}; J = 10^{-3} Kg.m^2, m = 0.2 Nm/A.

$$\begin{cases} \begin{pmatrix} \dot{p}_3 \\ \dot{p}_7 \end{pmatrix} = \begin{bmatrix} -200 & -200 \\ 40 & -0.1 \end{bmatrix} \begin{pmatrix} p_3 \\ p_7 \end{pmatrix} + \begin{bmatrix} 1 \\ 0 \end{bmatrix} U \\ y = \begin{bmatrix} 200 & 0 \\ 0 & 1000 \end{bmatrix} \begin{pmatrix} p_3 \\ p_7 \end{pmatrix} \end{cases}$$

2.4.2. Determination of the gain of an observer Luenberger by the analytical method

It is desired that the poles have the following values s1 = –160 and s1 = –75, then the characteristic polynomial:

Let us calculate the polynomials characteristic of the observer $P_1(A\text{-}K_1C_1)$ and $P_2(A\text{-}K_2C_2)$

- For the characteristic polynomial $P_1(A\text{-}K_1C_1)$

$$\begin{cases} C_1 = \begin{bmatrix} \frac{1}{L} & 0 \end{bmatrix} = [200 \quad 0] \\ K_1 = \begin{bmatrix} k_{11} \\ k_{12} \end{bmatrix} \end{cases} \rightarrow A - K_1C_1 = \begin{bmatrix} -200 - 200k_{11} & -200 \\ 40 - 200k_{12} & -0.1 \end{bmatrix}$$

$$\det(sI - (A - K_1C_1)) = \begin{bmatrix} s + 200 + 200k_{11} & 200 \\ -40 + 200k_{12} & s + 0.1 \end{bmatrix} = s^2 + (200.1 + k_{11})s + 8020 + 40000k_{22}$$

$$\rightarrow \begin{cases} 200k_{21} + 200.1 = 235 \\ 40000k_{22} + 8020 = 12000 \end{cases} \rightarrow \begin{cases} k_{11} = 0.1745 \\ k_{12} = -0.099 \end{cases}$$

- For the characteristic polynomial $P_2(A\text{-}K_2C_2)$

$$\begin{cases} C_2 = \begin{bmatrix} 0 & \frac{1}{J} \end{bmatrix} = \begin{bmatrix} 0 & 1000 \end{bmatrix} \\ K_1 = \begin{bmatrix} k_{21} \\ k_{22} \end{bmatrix} \end{cases} \rightarrow \det(sI - (A - K_2C_2)) = \begin{bmatrix} s+200 & 200+1000k_{21} \\ -40 & s+0.1+1000k_{22} \end{bmatrix}$$

$$\rightarrow \begin{cases} k_{21} = -0.0750 \\ k_{22} = 0.0349 \end{cases}$$

2.4.3. Determination the gain of a Luenberger observer by bond graph model

Using the theorem of Rahmani [6]:

- The coefficient a_1 is equal to the total gain of families of first order causality cycles in the jump graph model of **Figure 8**.
- The coefficient a_2 is equal to the total gain of second-order families of the causation cycles in the jump graph model of **Figure 8**.

To calculate the gains k_{11} and k_{12}, it is necessary to use **Table 1** that indicates the causality cycles in the observer's bond graph model.

- For the BGO1 model

The coefficient a1 is equal to the sum of the gains of the first order:

$a_1 = G_1 + G_2 + G_3$

$$200\, k_{11} = a_1 - (R/L) + (b/J) = 235 - (200 + 0.1)$$

So $k_{11} = 0.1745$.

To calculate k_{12}, it is necessary to determine the coefficient of a_2 (a_2 equal to the sum of the gains of the second order):

- For the BGO2 model

The coefficient a1 is equal to the sum of the gains of the first order:

$a_1 = G_1 + G_2 + G_4$

$$1000\, k_{22} = a_1 - (R/L) + (b/J) = 235 - (200 + 0.1)$$

So $k_{22} = 0.0349$

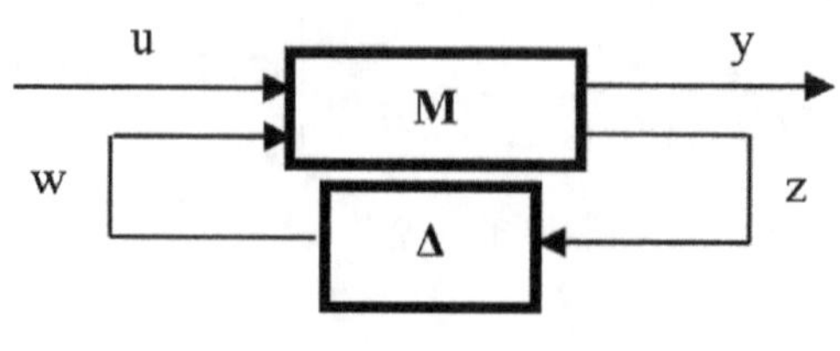

Figure 8. Schema of the LFT principle.

First order of the causation cycle	Gain
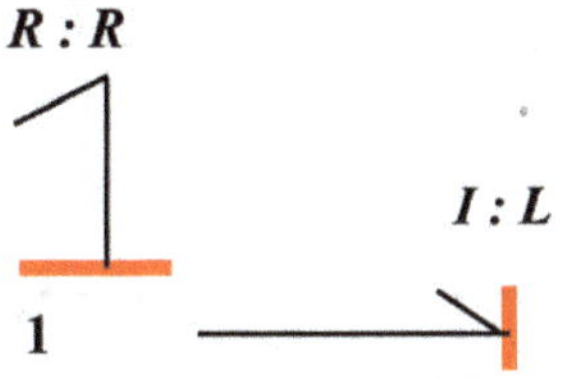	$G_1 = (-1)(-R/Ls)$
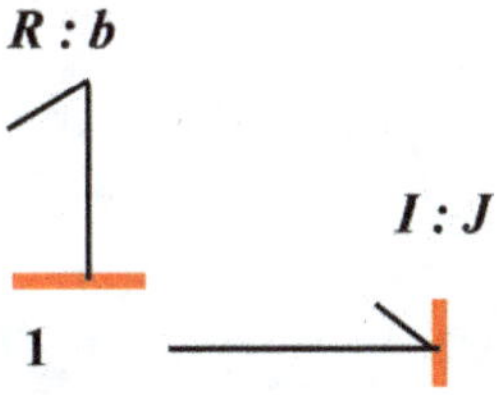	$G_2 = (-1)\ (-b/Js)$
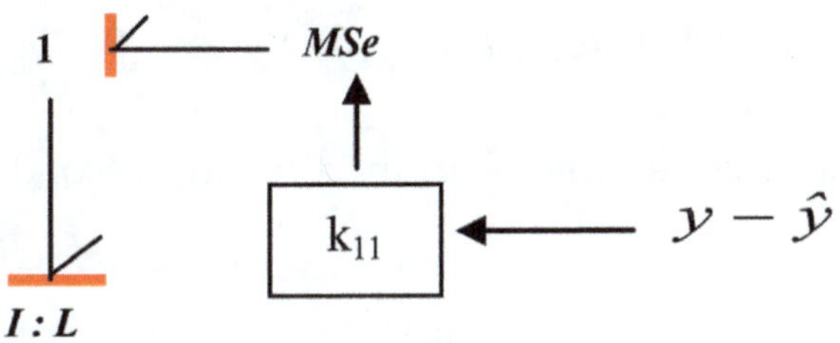	$G_3 = (k_{11})\ (1/Ls)$
Second order of causality cycle	**Gain**
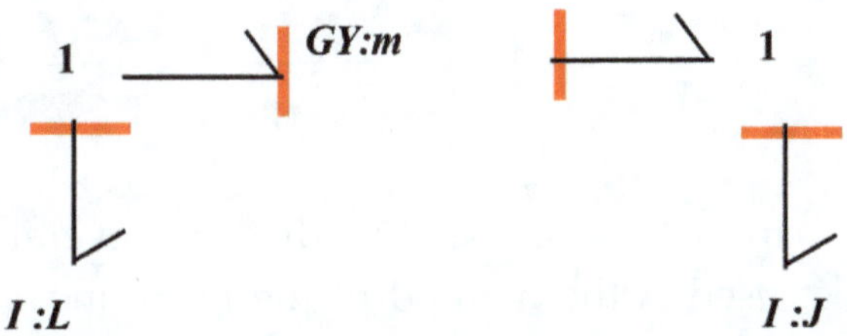	$G_4 = (m^2/JLs^2)$
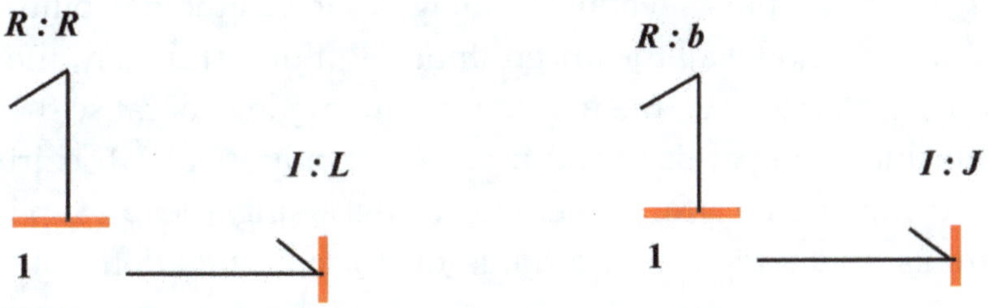	$G_5 = (Rb/JLs^2)$
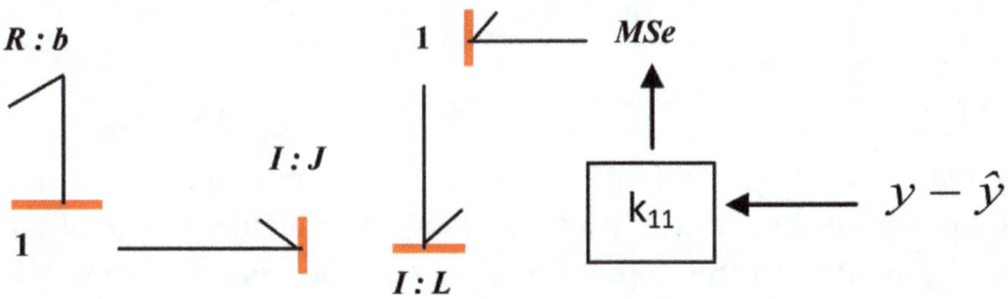	$G_6 = (bk_{11}/JLs^2)$
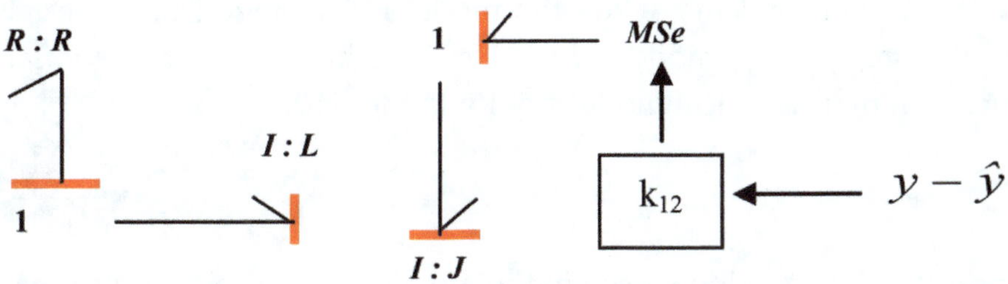	$G_7 = (Rk_{12}/JLs^2)$

Table 1. Causation cycles in the bond graph model of observers BGO1 and BGO2.

Causal of order i	Family of cycles	
	Total gain in BGO1	**Total gain in BGO2**
1	200.1 + 200 k_{11}	200.1 + 1000 k_{22}
2	8020 + 20 k_{11} + 4000 k_{12}	8020 + 40,000 k_{21} + 200,000 k_{22}

Table 2. Total gains of families of causal cycles in the bond graph model of observers BGO_1 and BGO_2.

To calculate k_{21}, it is necessary to determine the coefficient of a_2 (a_2 equal to the sum of the gains of the second order):

$$a_2 = G_5 + G_6 + G_8 + G_{10} = \left(m^2/JL\right) + (Rb/JL) + (\mathrm{m}k_{21}/JL) + (Rk_{22}/JL)$$

So $$k_{21} = JL\left(a_2 - \left[\left(m^2/JL\right) + (Rb/JL) + k_{22}(R/JL)\right]\right)/m = -0.0750$$

Table 2 represents the total earnings of i-order causal cycle families in the first and second observer graph (BGO1 and BGO2).

3. Robust diagnosis using bond graph model

3.1. Introduction

The diagnosis of industrial systems with uncertain parameters has been studied by several researchers in recent years. Indeed, the method most used is the method of the form linear fractional transformations (LFT), this method offers several advantages such as the flexibility of the diagnosis point of view that allows to model all the uncertainties. But unfortunately, the transition to the LFT form is not always possible (e.g., the nonlinear state models) because the separation of the nominal part from its uncertain part is very difficult (even impossible). Djeziri [7] opted for bond graph modeling using the LFT form method, using a single leap graph model in LFT form generates residuals and adaptive thresholds of normal operation with perfect separation.

3.2. Building a bond graph

Two methods are proposed by Dauphin-Tanguy [8] and Sié Kam [9] to construct parametric uncertainties by BG. The first is to represent uncertainty on a leap graph element as another element of the same type, causally related to the nominal element or the rest of the model. These uncertainties are kept in derived causality when the model is in integral preferential causality so as not to modify the order of the model. The second method is linear fractional transformation (LFT) introduced on mathematical models by Redheffer [10].

3.3. Representation LFT

Linear fractional transformations (LFTs) are widely used in the modeling of uncertain systems. The universality of LFT is due to the fact that any rational expression can be written in this form according to Alazard [11]. This form of representation is widely used for the synthesis of the control laws of uncertain systems using the principle of μ-analysis. It consists of separating the nominal part of a model from its uncertain part as illustrated in **Figure 8**.

The nominal values are grouped in an augmented matrix denoted M, supposed to be proper, and the uncertainties whatever their type (structured and unstructured parametric uncertainties, modeling uncertainties, measurement noises, etc.) are combined in a matrix Δ of structure diagonal. In the linear case, this standard form leads to a state representation of the form (**Figure 8**):

$$\begin{aligned} \dot{x} &= A \cdot x + B_1 \cdot w + B_2 \cdot u \\ z &= C_1 \cdot x + D_{11} \cdot w + D_{12} \cdot u \\ y &= C_2 \cdot x + D_{21} \cdot w + D_{22} \cdot u \end{aligned} \quad (23)$$

- $x \in R^n$: the state vector of the system;
- $u \in R^m$: the vector combining the control inputs of the system;
- $y \in R^p$: the vector grouping the measured outputs of the system;
- $w \in R^l$ and $z \in R^l$: group respectively the inputs and the auxiliary outputs;
- n, m, l and p are positive integers.

The matrices (A, B_1, B_2, C_1, C_2, D_{11}, D_{12}, D_{21}, D_{22}) are matrices of suitable dimensions.

3.4. Formatting assumptions LFT

Forming LFT requires that the model be clean and observable [12]. The bond graph methodology allows by causal manipulations to check these properties directly on the bond graph model.

Property 1: A leap graph is unique if and only if it contains no derivative causal dynamic component when it is a preferential integral causality, and reciprocally.

Property 2: A leap graph is structurally observable if and only if the following conditions are met:

- On the bond graph model in integral causality, there exists a causal path between all dynamic elements I and C in integral causality and a detector De or Df;
- All dynamic elements I and C admit causality derived on the leap graph model in preferential derived causality. If dynamic elements I or C remain in integral causality, the dualization of detectors De and Df must make it possible to put them into derivative causality.

3.5. Modeling of BG elements by LFT

The modeling of linear systems with uncertain parameters has been developed in Djeziri's thesis [13], the modeling of uncertain graph leap elements (R, I, C, TF and GY) has been determined. We will therefore limit ourselves in this section to show the two methods of modeling uncertain BG elements, as well as the advantages of BG-LFT for robust diagnosis.

3.5.1. BG element with a multiplicative uncertainty

The introduction of a multiplicative uncertainty on, for example, the element R in causality resistance gives Eq. (25):

$$e_R = R_n \cdot (1 + a_R) \cdot f_R = R_n \cdot f_R + a_R \cdot R_n \cdot f_R = e_n + a_R \cdot e_n = e_n + e_{inc} \tag{24}$$

- R_n: The nominal value of the R element;
- a_R: The multiplicative uncertainty on the parameter;
- e_R and f_R: represent the force and the flux in the element R, respectively;
- e_n and e_{inc}: represent respectively the effort provided by the nominal parameter and the effort introduced by the multiplicative uncertainty.

The bond graph model equivalent to the mathematical model of Eq. (24) is given in **Figure 9**.

3.5.2. Construction of a BG-LFT model

The complete BG-LFT model is represented by the diagram in **Figure 10**.

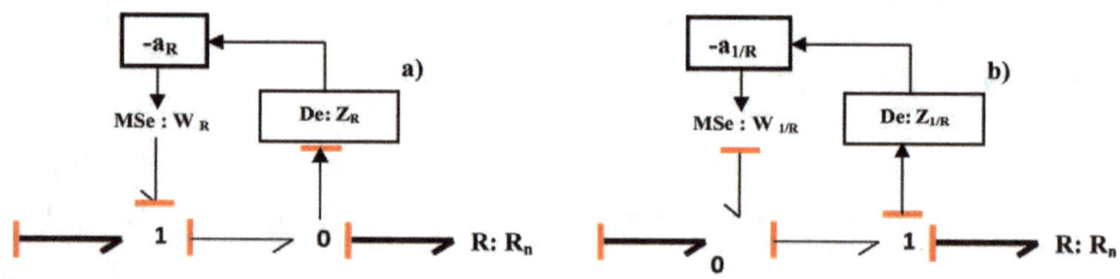

Figure 9. (a): Model BG-LFT of an element R causality resistance with multiplicative uncertainty, (b): Model BG-LFT of an element R with conductance causation with multiplicative uncertainty.

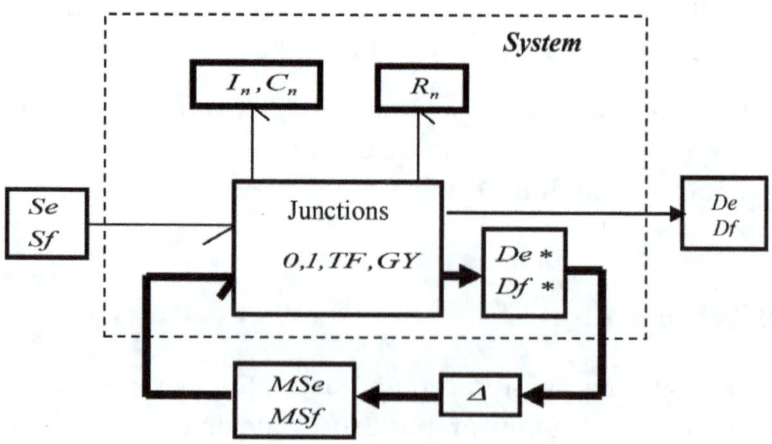

Figure 10. Representation in form BG-LFT.

3.5.3. Rugged diagnosis by ARRs

The generation of robust analytical redundancy relationships (ARRs) from a clean, observable and overdetermined leap graph is summarized by the following steps:

Step 1: Checking the state of the coupling on the model graph deterministic preferential derived causality; if the system is overdetermined, then proceed with the following steps;

Step 2: The leap graph template is set to LFT;

Step 3: The symbolic expression of the RRA is deduced from the equations at the junctions. This first form will be expressed by:

$$\text{For junction } 0: \sum b_i \cdot f_{inc} + \sum \dot{Sf} + \sum w_i = 0 \tag{25}$$

$$\text{For junction } 1: \sum b_i \cdot e_{inc} + \sum \dot{Se} + \sum w_i = 0 \tag{26}$$

With $\sum \dot{Sf}$ the sum of the flux sources linked to the junction 0, the $\sum \dot{Se}$ sum of the force sources linked to the junction 1, b = ± 1 according to whether the half-arrow enters or leaves the junction, e_{in} and f_{in} are the unknown variables .

Is the $\sum \dot{w}_i$ sum of the modulated inputs corresponding to the uncertainties on the elements related to the junction.

Step 4: Unknown variables are eliminated by traversing causal paths between detectors or sources and unknown variables;

Step 5: After removing the unknown variables, the uncertain RRAs are in the form:

$$RRA : \Phi\left(\sum \text{Se}, \sum \text{Sf}, De, Df, \sum w_i, R_n, I_n, C_n, TF_n, GY_n\right) \tag{27}$$

TF_n and GY_n are, respectively, the nominal values of the modules of the elements TF and GY, also R_n, C_n and I_n are the nominal values of the elements R, C and I.

3.5.4. Robust diagnosis by Luenberger observer using bond graph

The model BG-LFT of the system with the observer of Luenberger is represented by the diagram in **Figure 11**.

3.5.5. Location of defects per observer bench

A single residue allows the detection of a fault. However, the location of a defect requires a set of structured residues. These residues must be designed to be sensitive to certain defects and insensitive to others. The use of observer banks constructed from only part of the inputs and/or outputs of the system makes it possible to respond to this problem. The symptoms generated are compared with fault signatures.

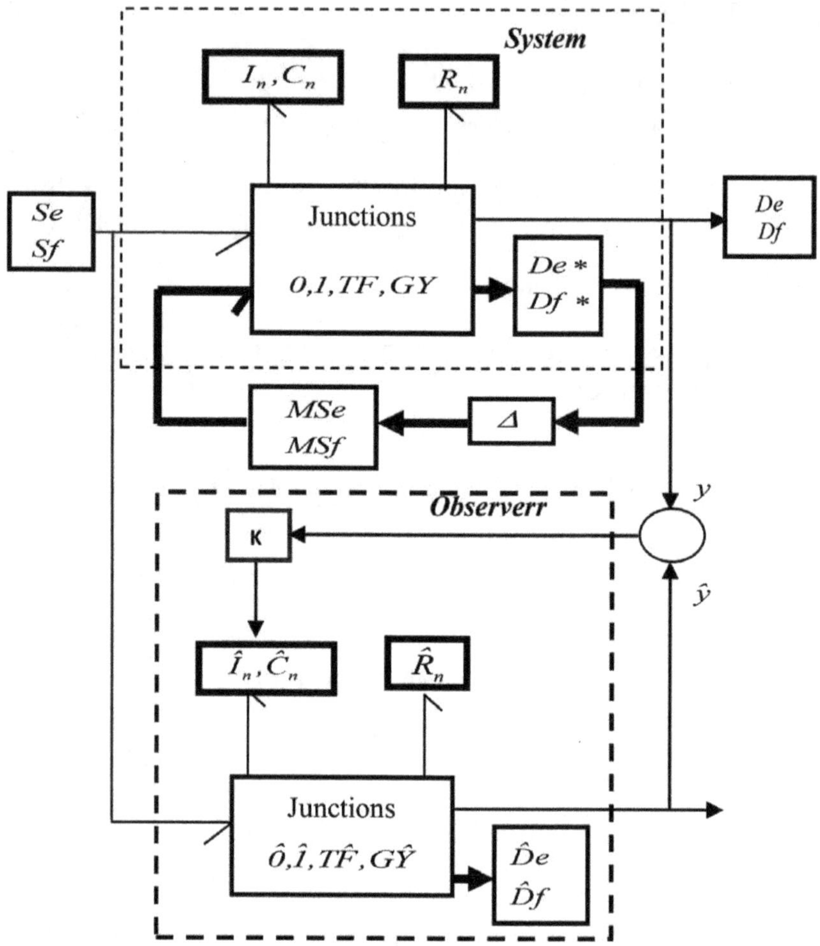

Figure 11. Representation in BG-LFT form of the system with observer from Luenberger.

There are two observer banks:

- The Dedicated Observer Scheme (DOS): the ith observer is controlled by the ith output (input) and all inputs (outputs). The other outputs (inputs) are considered unknown.
- The Generalized Observer Scheme (GOS) structure: the ith observer is controlled by all outputs (inputs) except the ith and all inputs (outputs).

For a given observer bank, the signature of the various defects is defined in the signature table. The rows and columns of this table correspond, respectively, to defects and symptoms. The cells in the table are filled with binary values. A zero (0) means that the symptom is not sensitive to the defect. One (1) means that the symptom is sensitive to this defect.

Figures 12 and **13**, respectively, present the principle of detection and localization of faults by observers (DOS) and (GOS).

Tables 3 and **4** illustrate the signatures of the various defects as a function of the residuals.

3.6. Example

Consider the same example (DC motor) given in **Figure 7**.

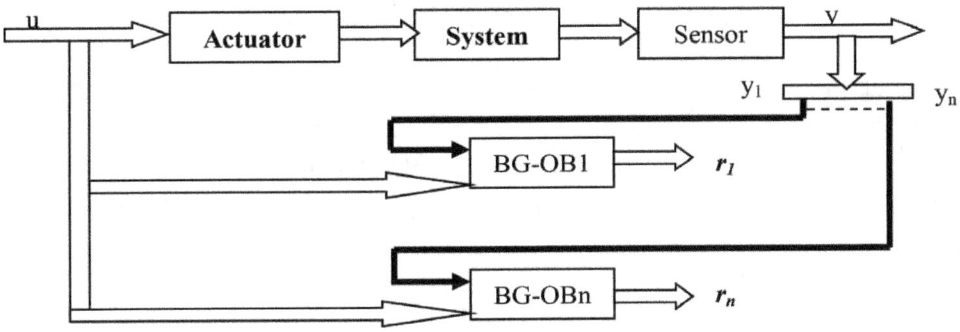

Figure 12. Sensor faults with DOS structure.

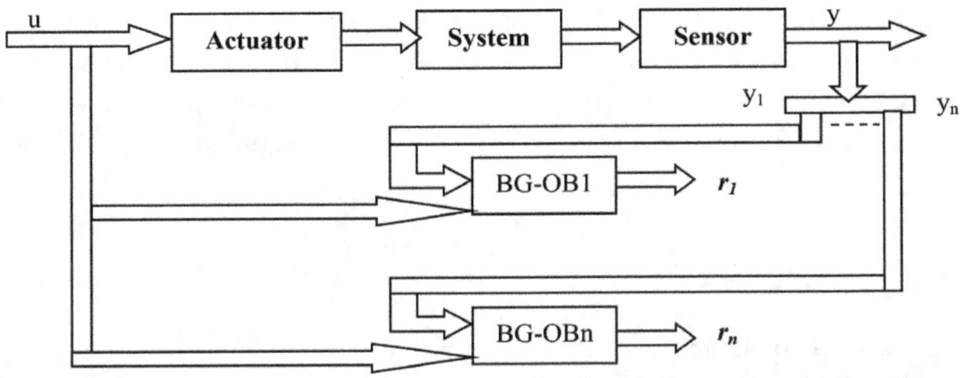

Figure 13. Sensor faults with GOS structure.

	F_{s1}	F_{s2}	....	F_{sn}
r_1	1	0	0	0
r_2	0	1	0	0
...	0	0	1	0
r_n	0	0	0	1

Table 3. Signature of sensor faults with DOS structure.

	F_{s1}	F_{s2}	....	F_{sn}
r_1	0	1	1	1
r_2	1	0	1	1
...	1	1	0	1
r_n	1	1	1	0

Table 4. Signature of sensor faults with GOS structure.

3.6.1. Proportional observer diagnosis using the bond graph model

The application of leap graph for the diagnosis of industrial systems is mainly justified by the fact that the model can be fine-tuned by adding or removing jump graph elements (graphically) according to simplifying assumptions. It is therefore also desirable that for the state estimation, the observer design is carried out using graphic methods and taking advantage of the properties.

3.6.2. Requirements

For observer construction, we need to check the following conditions:

The first condition is satisfied: When the bond graph model of the DC motor is put into a preferred integral causality, it is clear that there exists a causal path linking the sensors Df_1 and Df_2 to each dynamic element I in the integral causality (**Figure 14(a)**). The second condition is satisfied: when we put the jump engine model of the DC motor in derived causality, all the elements I admit a derivative causality, and the sensors Df_1 and Df_2 are dualized (**Figure 14(b)**).

3.6.3. Calculation of residues

Figure 15 represents a construction of the proportional observer of the system using the bond graph approach.

We simulated the system with the 20sim software. **Figure 16** shows the evolution of the real and estimated state variables.

From model BG of **Figure 15**, the residues r_1(t) and r_2(t) can be deduced.

Residue r_1: $f_2 - f_{\hat{2}} = Df_2 - f_{\hat{2}} = 0$.

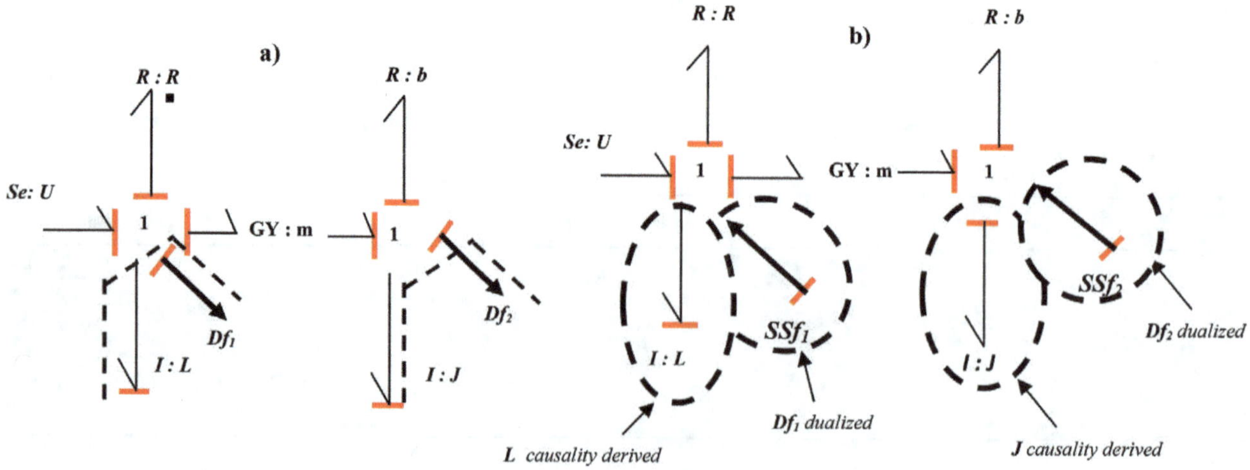

Figure 14. (a): Bond model of the DC motor in integral causality, (b): bond model of the DC motor in derived causality.

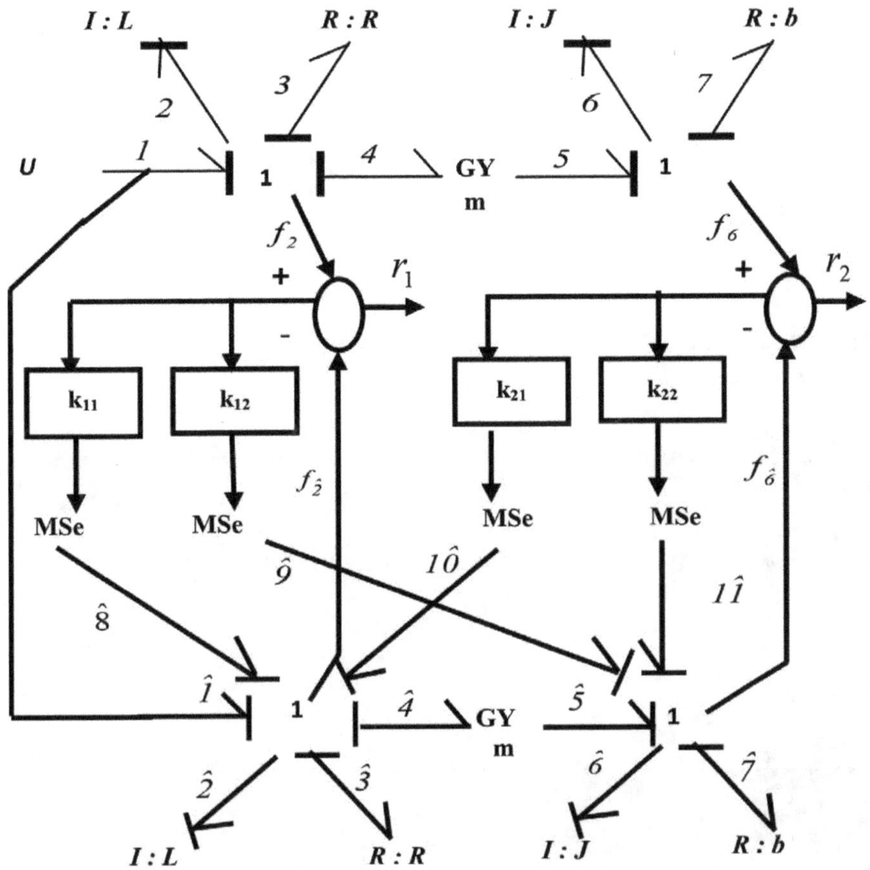

Figure 15. Proportional observer using bond graph of DC motor monitor.

Therefore,

$$r_1(t)\left(m^2Rb + m^2RK_{11} + mK_{12}\right) + \frac{dr_1(t)}{dt}\left(m^2RJ + m^2Lb + m^2LK_{11}\right) + \frac{d^2r_1(t)}{dt}\left(m^2LJ\right) \\ +r_2(t)\left(m^2RK_{21} + mK_{22}\right) + \frac{dr_2(t)}{dt}\left(m^2LK_{21}\right) = 0 \tag{28}$$

Residue r_2: $f_6 - f_{\hat{6}} = Df_1 - f_{\hat{6}} = 0$

We proceed in the same way, we end:

$$r_2(t)\left(m^2Rb + m^2bK_{22} + mK_{21}\right) + \frac{dr_2(t)}{dt}\left(m^2RJ + m^2Lb + m^2JK_{21}\right) + \frac{d^2r_2(t)}{dt}\left(m^2LJ\right) \\ r_1(t)\left(m^2bK_{12} + mK_{11}\right) + \frac{dr_2(t)}{dt}\left(m^2JK_{12}\right) = 0 \tag{29}$$

Figure 17 shows the convergence of residuals toward zero.

3.6.4. Génération des résidus avec défauts capteurs

The sensors Df_1 and Df_2 are affected by defects (F_{C1} and F_{C2}), then:

Residue r_1: $\left(Df_1 + F_{C1}\right) - f\hat{6} = 0$

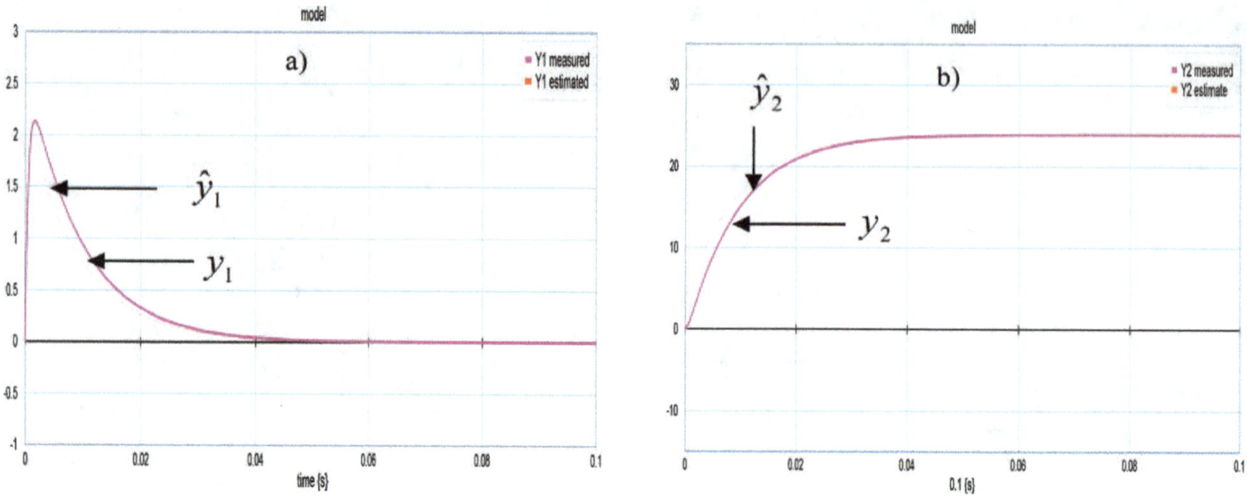

Figure 16. Output variables, (a): output variables y_1 and $\hat{y}_1$, (b): output variables y_2 and $\hat{y}_2$.

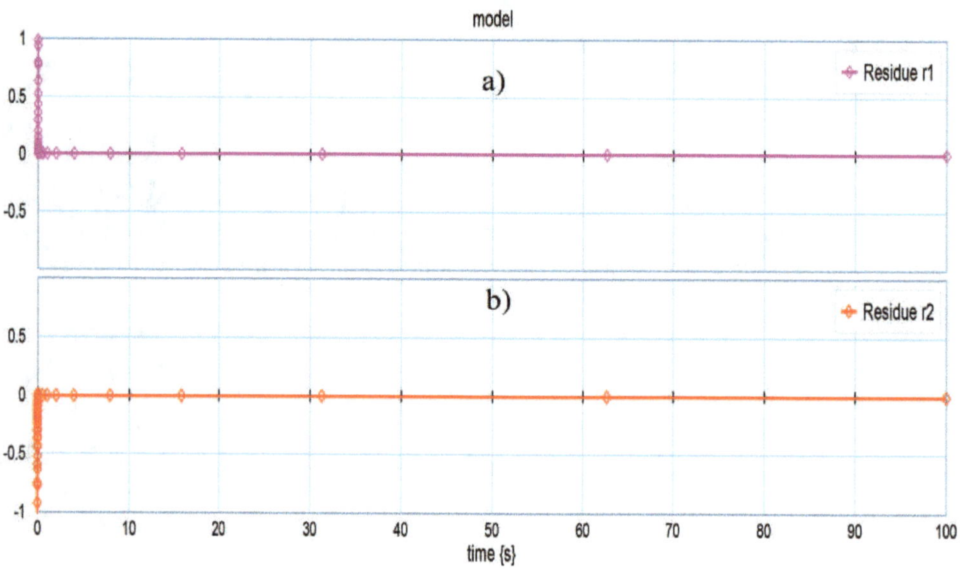

Figure 17. (a): Residue $r_1(t)$, (b): Residue $r_2(t)$ in the case of normal operation.

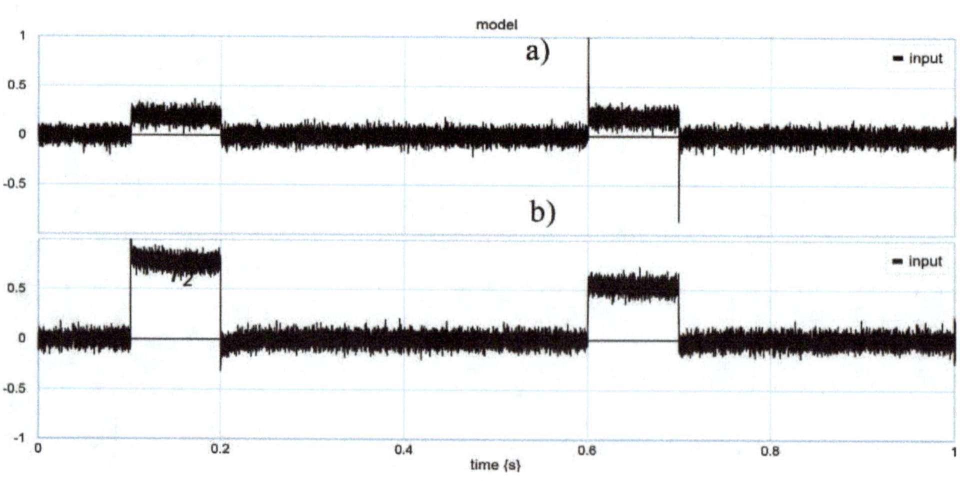

Figure 18. (a): Residue $r_1(t)$ with defects of the sensors Df_1 and Df_2, (b): Residue $r_2(t)$ with defects of the sensors Df_1 and Df_2.

$$\begin{aligned}
&r_1(t)\left(m^2Rb+m^2RK_{11}+mK_{12}\right)+\frac{dr_1(t)}{dt}\left(m^2RJ+m^2Lb+m^2LK_{11}\right)\\
&+\frac{d^2r_1(t)}{dt}\left(m^2LJ\right)+r_2(t)\left(m^2RK_{21}+mK_{22}\right)+\frac{dr_2(t)}{dt}\left(m^2LK_{21}\right)=F_{C1}(t)\left(bm^2\right)\\
&+\frac{dF_{C1}(t)}{dt}\left(Lm^2\right)-F_{C2}(t)(m)
\end{aligned} \tag{30}$$

Residue r_2: $(Df_2+F_{S2})-f\hat{2}=0$

$$\begin{aligned}
&r_2(t)\left(m^2Rb+m^2bK_{22}+mK_{21}\right)+\frac{dr_2(t)}{dt}\left(m^2RJ+m^2Lb+m^2JK_{21}\right)\\
&+\frac{d^2r_2(t)}{dt}\left(m^2LJ\right)+r_1(t)\left(m^2bK_{12}+mK_{11}\right)+\frac{dr_2(t)}{dt}\left(m^2JK_{12}\right)=\\
&F_{C2}(t)\left(bm^2\right)+\frac{dF_{C2}(t)}{dt}\left(Jm^2\right)-F_{C1}(t)(m)
\end{aligned} \tag{31}$$

Eqs. (30) and (31) show that residues $r_1(t)$ and $r_2(t)$ are sensitive to sensor defects. **Figure 18** confirms the sensitivity of the residuals to the defects of the sensors Df_1 and Df_2.

3.6.5. Robust diagnosis by observer from Luenberger

Figure 19 shows the Luenberger observer of the DC motor using the BG-LFT model.

From the BG-LFT model (**Figure 19**), residues $R_1(t)$ and $R_2(t)$ can be deduced.

Residue R_1: $f_2-f_{\hat{2}}=0$

$$\begin{aligned}
&\left[r_1(t)\left(m^2Rb+m^2RK_{11}+mK_{12}\right)+\frac{dr_1(t)}{dt}\left(m^2RJ+m^2Lb+m^2LK_{11}\right)\right.\\
&\left.+\frac{d^2r_1(t)}{dt}\left(m^2LJ\right)+r_2(t)\left(m^2RK_{21}+mK_{22}\right)+\frac{dr_2(t)}{dt}\left(m^2LK_{21}\right)\right]\\
&-\left[w_J(t)\left(m^2R_n\right)+\frac{dw_J(t)}{dt}\left(m^2L_n\right)+w_L(t)(m)\right]=0
\end{aligned} \tag{32}$$

Eq. (32) is composed of two parts: the first part corresponds to the evolution of the normal residue (r_{1n}) and the second part represents the evolution related to the uncertainty of the parameters (d_1).

$$\begin{cases}
R_1=r_{1n}+d_1\\
r_{1n}=\left[r_1(t)\left(m^2Rb+m^2RK_{11}+mK_{12}\right)+\frac{dr_1(t)}{dt}\left(m^2RJ+m^2Lb+m^2LK_{11}\right)\right.\\
\left.+\frac{d^2r_1(t)}{dt}\left(m^2LJ\right)+r_2(t)\left(m^2RK_{21}+mK_{22}\right)+\frac{dr_2(t)}{dt}\left(m^2LK_{21}\right)\right]\\
|d_1|=w_J(t)\left(m^2R_n\right)+\frac{dw_J(t)}{dt}\left(m^2L_n\right)+w_L(t)(m)
\end{cases}$$

Residue R_2: $f_6-f\hat{6}=0$

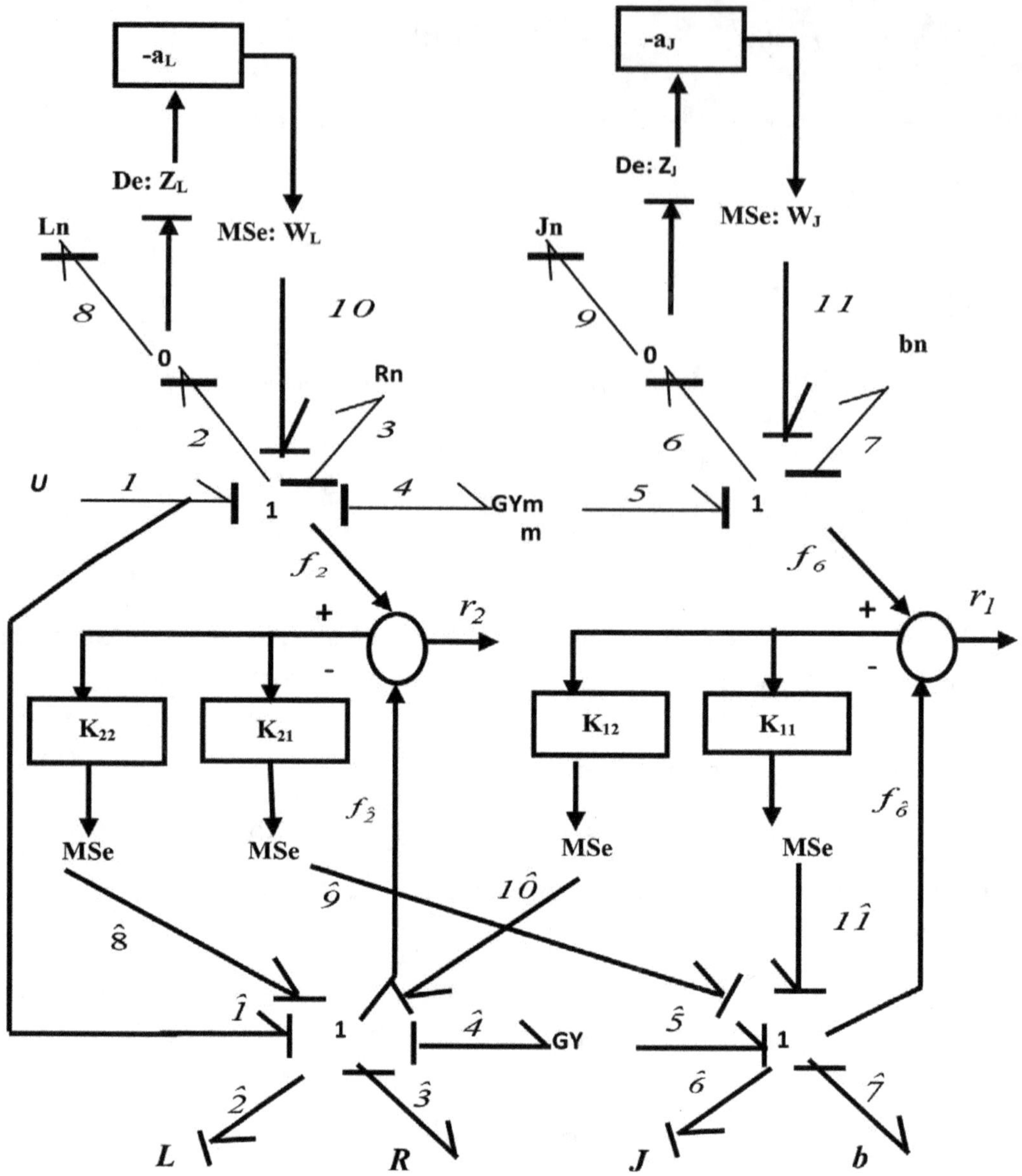

Figure 19. Luenberger observer of the DC motor using the BG-LFT model.

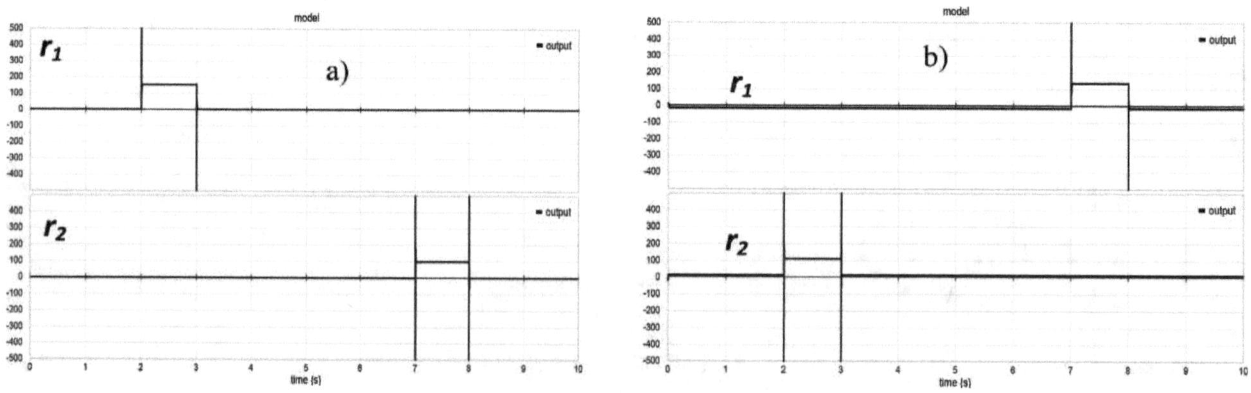

Figure 20. (a): Residue $r_1(t)$ and $r_2(t)$ for the DOS structure, (b): residue $r_1(t)$ and $r_2(t)$ for the DOS structure.

Structure DOS			Structure GOS		
	Df_1	Df_2		$Df1$	$Df2$
r_1	1	0	r_1	1	0
r_2	0	1	r_2	0	1

Table 5. Signatures of faults. (a) Structure DOS and (b) structure GOS.

$$\begin{aligned}&[r_2(t)(m^2Rb+m^2bK_{22}+mK_{21})+\frac{dr_2(t)}{dt}(m^2RJ+m^2Lb+m^2JK_{21})\\&+\frac{d^2r_2(t)}{dt}(m^2LJ)+r_1(t)(m^2bK_{12}+mK_{11})+\frac{dr_2(t)}{dt}(m^2JK_{12})]\\&-[w_L(t)(m^2b_n)+\frac{dw_L(t)}{dt}(m^2J_n)+w_j(t)m=0\end{aligned} \tag{33}$$

Eq. (33) is composed of two parts: the first part corresponds to the evolution of the normal residue (r_{2n}) and the second part represents the evolution related to the uncertainty of the parameters (d_2).

$$\begin{cases}R_2=r_{2n}+d_2\\r_{2n}=r_2(t)(m^2Rb+m^2bK_{22}+mK_{21})+\frac{dr_2(t)}{dt}(m^2RJ+m^2Lb+m^2JK_{21})\\+\frac{d^2r_2(t)}{dt}(m^2LJ)+r_1(t)(m^2bK_{12}+mK_{11})+\frac{dr_2(t)}{dt}(m^2JK_{12})]\\|d_2|=w_L(t)(m^2b_n)+\frac{dw_L(t)}{dt}(m^2J_n)+w_j(t)m\end{cases}$$

3.6.6. Detection and localization of sensor faults

Figure 20 shows the evolution of residues based on the BG model using the BG-DOS and BG-GOS structures, the residue $r_1(t)$ is sensitive to the defects occurring on the Df_1 sensor while the residue $r_2(t)$ is sensitive to faults on the Df_2 sensor.

Table 5(a) and **(b)** represents the binary signatures of the defects for the deduced DOS and GOS structures to perfectly isolate the defects.

4. Conclusion

In this chapter, we proposed a technique for the diagnosis of linear systems by Luenberger observer using the leap graph. We presented the leap graph approach for constructing a full-order observer and proposed a new BG-based observer method. Subsequently, we presented the systems with uncertain parameters modeled by the leap graph approach and we also proposed a new method of diagnosis of systems with uncertain parameters by Luenberger observer.

In the last part of this chapter, we developed and proposed an observer-based diagnostic technique (BG-DOS / BG-GOS) to detect and locate faults.

Author details

Abderrahmène Sallami

Address all correspondence to: abderrahmenesallami@gmail.com

Department of Electrical Engineering, Laboratory ACS, Tunisia

References

[1] Khedher A, Othman KB, Benrejeb M, Maquin D. Adaptive observer for fault estimation in nonlinear systems described by a Takagi-Sugeno model. In: 18th Mediterrranean Conference on Control and Automation, Marrakech, Morocco; 2010. pp. 23-25

[2] Maquin D, Ragot J. Diagnostic Des Systèmes Linéaires. Hermes ed. Paris, France: Hermes; 2000

[3] Sueur C, Dauphin-Tanguy G. Structural Controllability and Observability of linear Systems Represented by Bond Graphs. Journal of the Franklin Institute. 1989;**326**:869-883

[4] Pichardo-Almarza C, Rahmani A, Dauphin-Tanguy G, Delgado M. Observateur Proportionnel-Intégral pour des Systèmes Linéaires Modélisés par Bond Graphs. In: Actes des Journées Doctorales d'Automatique; 2003

[5] Luenberger DG. An introduction to observers. IEEE Transactions on Automatic Control. 1971;**16**:596-602

[6] Rahmani A, Sueur C, Dauphin-Tanguy G. Pole assignment for systems modelled by bond graph. Journal of the Franklin Institute. 1994;**331**(3):299-312

[7] Djeziri MA, Ould Bouamama B, Merzouki R. Modelling and robust FDI of steam generator using uncertain bond graph model. Journal of Process Control. 2009;**19**:149-162

[8] Dauphin-Tanguy G, Sié Kam C. How to Model Parameter Uncertainies in a Bond Graph Framework. Erlangen, Germany: ESS'99; 1999. pp. 121-125

[9] Sié Kam C, Dauphin-Tanguy G. Bond graph models of structured parameter uncertainties. Journal of the Franklin Institute. 2005;**342**:379-399

[10] Redheffer R. On a certain linear fractional transformation. Journal of Mathematics and Physics Banner. 1960;**39**:269-286. l'aide d'éléments caractérisés». 8ème colloque Electronique dePuissance du Futur EPF 2000, Lille; 2000

[11] Alazard D, Cumer C, Apkarian P, Gauvrit M, Fereres G. Robustesse et Commande Optimale. Cépadues ed. 1999. ISBN: 2.85428.516.6

[12] Sié Kam C. Les bond graphs pour la modélisation des systèmes linéaires incertains [Thèse de doctorat]. USTLille1-ECLille; Décembre 2001. N°d'ordre 3065, 2001

[13] Djeziri MA. Diagnostic des systèmes incertains par l'approche bond graph [thèse de doctorat]. École Centrale de Lille; 2007

8

Optimization of Fuzzy Logic Controllers by Particle Swarm Optimization to Increase the Lifetime in Power Electronic Stages

Pedro Ponce, Luis Arturo Soriano,
Arturo Molina and Manuel Garcia

Abstract

In recent years, brushless direct current motor (BLDCM) applications have been increased due to their advantages as low size, mechanical torque, high-speed range, to mention some. The BLDCM control is required to operate at high frequency, high temperature, large voltage, and quick changes of current; as a result of this kind of operation, the power drive lifetime is affected. The power drives commonly utilized insulated gate bipolar transistors (IGBTs) and metal oxide semiconductor field effect transistors (MOSFETs), which present power losses, on-state losses, and switching losses caused by temperature oscillations. Hence, the power losses are related to the command signals generated by the controller. In this sense, the BLDC motor drive controller design, frequently, does not take into account the power losses and the temperature oscillations, which cause the IGBT lifetime decrease or premature fail. In this chapter, a brushless DC motor drive is designed based on a fuzzy controller tuned with the particle swarm optimization algorithm, where the temperature oscillations and speed set points are considered in order to increase IGBT module lifetime. The validation of the proposed fuzzy-PSO controller is carried out by the co-simulation between LabVIEW™ and Multisim™ and finally analysis and conclusion of the work.

Keywords: power electronics lifetime, speed controller, fuzzy logic, PSO, BLDCM

1. Introduction

In this chapter, a design of a BLDCM speed control based and tuned by particle swarm optimization is presented; this proposal control considers two objectives, the first one is the speed set point and the second one is the power electronic lifetime. The BLDC motor covers a wide range of applications in several fields as the robotic systems, aerospace industry, medical industry, automotive industry, and electronic devices [8, 17, 25, 32]. Their structure of control is integrated by BLDC motor stage, power drive stage, and sensor stage. The power drive is integrated by semiconductors as IGBTs and MOSFETs, these semiconductors frequently work under high thermal stress to reach the mechanical reference of speed, and in some cases, after a certain period of operation, they are damaged due to the generation of command signals by the speed controller [7, 9, 12, 23, 34, 35]. There are several methods to predict and evaluate the lifetime of power electronics devices [34], the manufactures develop these studies as an estimation of certain conditions and operation features, but in some applications, these conditions are not always the same and change according to the control objective, which affects the power electronics lifetime used. For these reason, it is important that the controller design not only considers the mechanical references such as position and velocity in the case of BLDCM, as well as the lifetime of semiconductors [10, 22, 23, 24, 33, 34, 35]. Hence, a great effort for predicting and improving the lifetime in semiconductors has been pushing to develop new control algorithms that help to improve the conditions in the semiconductor [16]. In this chapter, the designed controller uses an optimization process based on an objective function that takes into account the temperature of the power electronic stages in order to increase their lifetime, and tracking motor speed reference [1, 10]. Thus, PSO algorithm has been implemented to optimize the controller proposed as in [13, 18, 21, 27, 30], due to its few parameters that need to be tuned and its easy implementation. Finally, the proposed controller in BLDCM is validated through co-simulation, which helps to combine two or more specialized programs utilizing advanced models and avoiding implementation problems between the control systems and power electronics stage [11, 14, 15].

2. Structure and drive mode of BLDC motor

2.1. Types of BLDCM drives

The BLDC motor drive employs a permanent magnet alternating current; the stator windings are commonly star connection, when voltage source is applied between any two terminals, the current flows between two phases, leaving one of the three without energy, and as a result, only one current at a time needs to be controlled. This kind of operation is known as back electromotive force (EMF) and **Figure 1** shows the waveform in the BLDC motor with two phases active.

According to their physical position, the output is related to 60° or 120° phase shift to each other. To drive the windings, a sequence of commutation is necessary; this kind of operation is commonly called "six-step commutation." The trapezoidal waveform is due to the rotor revolving in a counterclockwise direction, for example, to phase A, after the 120° rotation to the top conductor of phase.

The motor position must be measured by three Hall effect sensors (A, B, and C) so a voltage is sent to the specific coils to control the speed and position of the brushless motor [3, 25]. A standard circuit to drive BLDCM in Multisim^{TM} program is shown in **Figure 2**.

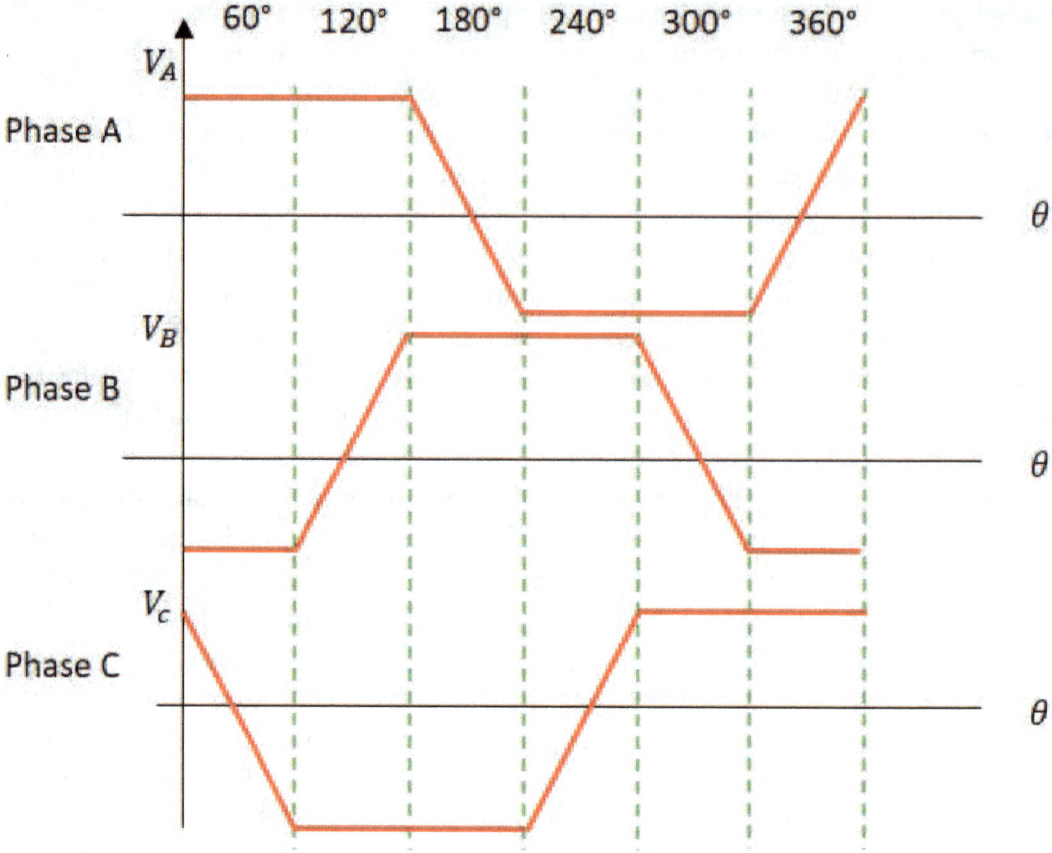

Figure 1. Trapezoidal EMF.

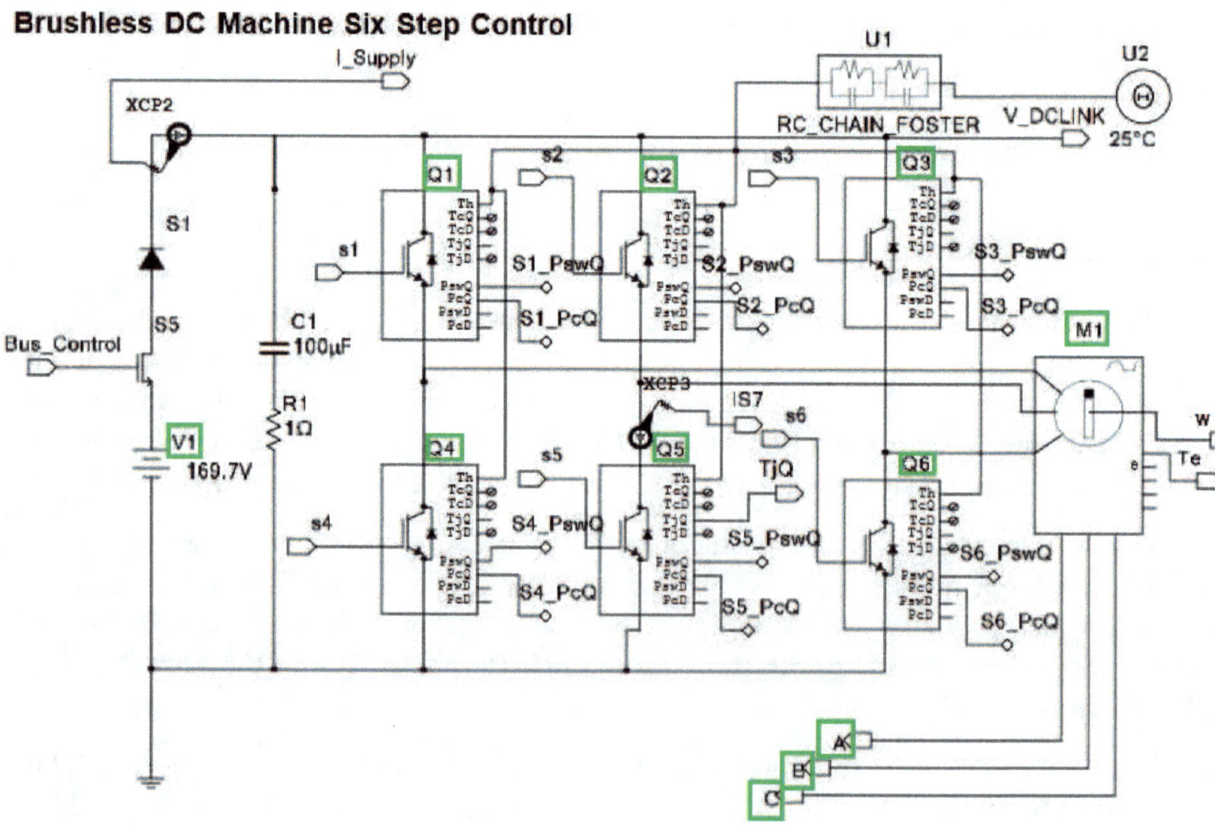

Figure 2. BLDCM drive in Multisim™ for co-simulation.

2.2. States space model

According to [25], the BLDCM can be represented by Eq. (1) where R is the stator resistance per phase, i_{as}, i_{bs}, and, i_{cs} are the stator phase current, and e_{as}, e_{bs}, and e_{cs} are the rotating back-EMF that are produced by the winding flux linkage caused by rotating rotor.

$$\begin{bmatrix} u_{as} \\ u_{bs} \\ u_{cs} \end{bmatrix} = R \begin{bmatrix} 100 \\ 010 \\ 001 \end{bmatrix} \begin{bmatrix} i_{as} \\ i_{bs} \\ i_{cs} \end{bmatrix} + (L-M) \begin{bmatrix} 100 \\ 010 \\ 001 \end{bmatrix} \frac{d}{dt} \begin{bmatrix} i_a \\ i_b \\ i_c \end{bmatrix} + \begin{bmatrix} e_{as} \\ e_{bs} \\ e_{cs} \end{bmatrix} \tag{1}$$

The induced EMF is given by the Eq. (2)

$$\begin{aligned} e_{as} &= \omega\psi_m f_{as}(\theta) \\ e_{bs} &= \omega\psi_m f_{bs}(\theta) \\ e_{cs} &= \omega\psi_m f_{cs}(\theta) \end{aligned} \tag{2}$$

where $f_{as}(\theta)$, $f_{bs} = f_{as}(\theta - 2\pi/3)$ and $f_{cs} = f_{as}(\theta + 2\pi/3)$ are the back-EMF waveform function of phase, respectively. ψ_m is the maximum value of PM; the flux linkage of each winding as $\psi_m = 2NSBm$. On the other hand, the dynamic features of BLCDM are expressed in state equations as shown in Eq. (3):

$$\begin{bmatrix} \dot{i}_{as} \\ \dot{i}_{bs} \\ \dot{i}_{cs} \\ \omega_m \\ \dot{\theta}_r \end{bmatrix} = \begin{bmatrix} -\frac{R_s}{L_1} & 0 & 0 & -\frac{\psi_m}{J} f_{as(\theta_r)} \\ 0 & -\frac{R_s}{L_1} & 0 & -\frac{\psi_m}{J} f_{bs(\theta_r)} \\ 0 & 0 & -\frac{R_s}{L_1} & -\frac{\psi_m}{J} f_{cs(\theta_r)} \\ \frac{\psi_m}{J} f_{as(\theta_r)} & \frac{\psi_m}{J} f_{bs(\theta_r)} & \frac{\psi_m}{J} f_{cs(\theta_r)} & -\frac{B}{J} \end{bmatrix} \begin{bmatrix} i_{as} \\ i_{bs} \\ i_{cs} \\ \omega_m \\ \theta_r \end{bmatrix} + \begin{bmatrix} \frac{1}{L-M} & 0 & 0 & 0 \\ 0 & \frac{1}{L-M} & 0 & 0 \\ 0 & 0 & \frac{1}{L-M} & 0 \\ 0 & 0 & 0 & \frac{1}{J} \end{bmatrix} \begin{bmatrix} v_{as} \\ v_{bs} \\ v_{cs} \\ T_l \end{bmatrix} \tag{3}$$

where windings are symmetrical, the self-inductance will be equal, and so as the mutual inductance, $L_{as} = L_{bs} = L_{cs} = L \text{ and } L_{ab} = L_{bc} = L_{ca} = M$. The electromagnetic torque is given in Eq. (4):

$$T_e = \psi_m \left[f_{as}(\theta_r) i_{as} + f_{bs}(\theta_r) i_{bs} + f_{cs}(\theta_r) i_{cs} \right] (N.m) \tag{4}$$

and, the complete model of the electromechanical system is described by Eq. (5):

$$T_e - T_l = J\,\dot{\omega}_m + B\omega_m \tag{5}$$

where T_l is the load torque, J is the rotor moment of inertia, and B is the viscous coefficient.

3. Power losses

3.1. Semiconductor losses

Power losses can be classified as on-state losses and switching losses. The IGBT power losses in on-state are calculated by Eq. (6):

$$P_{CT} = u_{CEO}{*}I_{Cav} + r_c I_{Crms}^2 \tag{6}$$

and the diode power losses in on-state are calculated by Eq. (7):

$$P_{CD} = u_{DO}{*}I_{Dav} + r_D I_{Drms}^2 \tag{7}$$

These can also be estimated using datasheet information of **Figure 3**. On the other hand, the IGBT energy losses in turn-on are given by Eq. (8):

$$E_{onT} = \int_{ton} (u_{CEO}{*}i_C(t))dt \tag{8}$$

and, the energy losses in turn-off are given by Eq. (9):

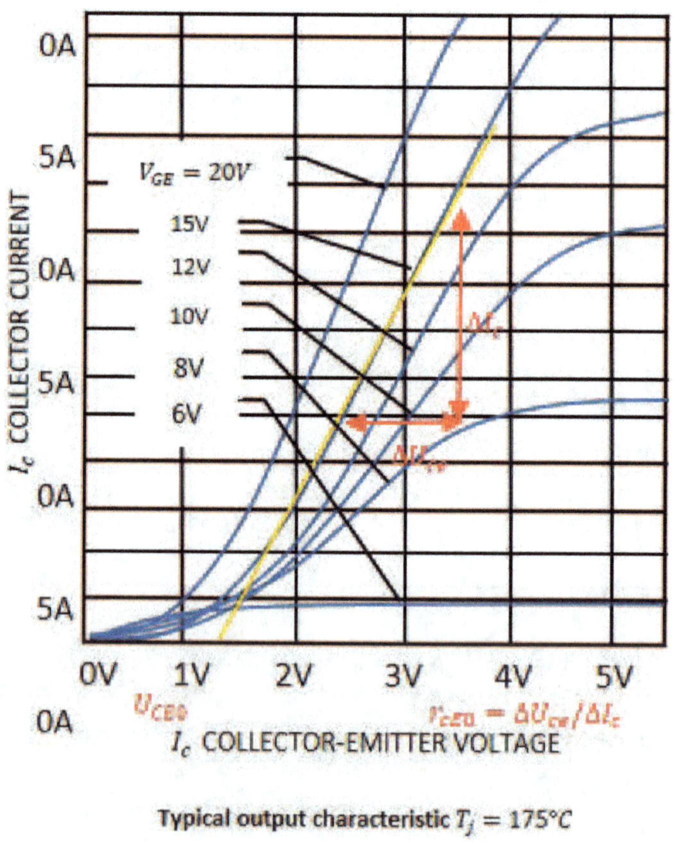

Figure 3. The u_{CEO} and r_c ($r_c = \Delta U_{ce}/\Delta I_c$) values of datasheet diagram.

$$E_{offT} = \int_{toff} (u_{CEO} * i_C(t))dt \tag{9}$$

On the other hand, the diode energy losses in turn-on are mostly presented when the reverse-recovery energy occurs, and they are given by:

$$E_{onD} = \int_{toff} (u_D(t) * i_F(t))dt \tag{10}$$

The switch-off losses in the diode are commonly neglected $E_{offD} \approx 0$. In consequence, the switching losses in the IGBT are calculated in Eq. (11) as:

$$P_{swT} = \left(E_{onT} + E_{offT}\right) * F_{sw} \tag{11}$$

In the case of a diode, they are given as the product of switching energies and the switching frequency F_{sw} as shown in Eq. (12).

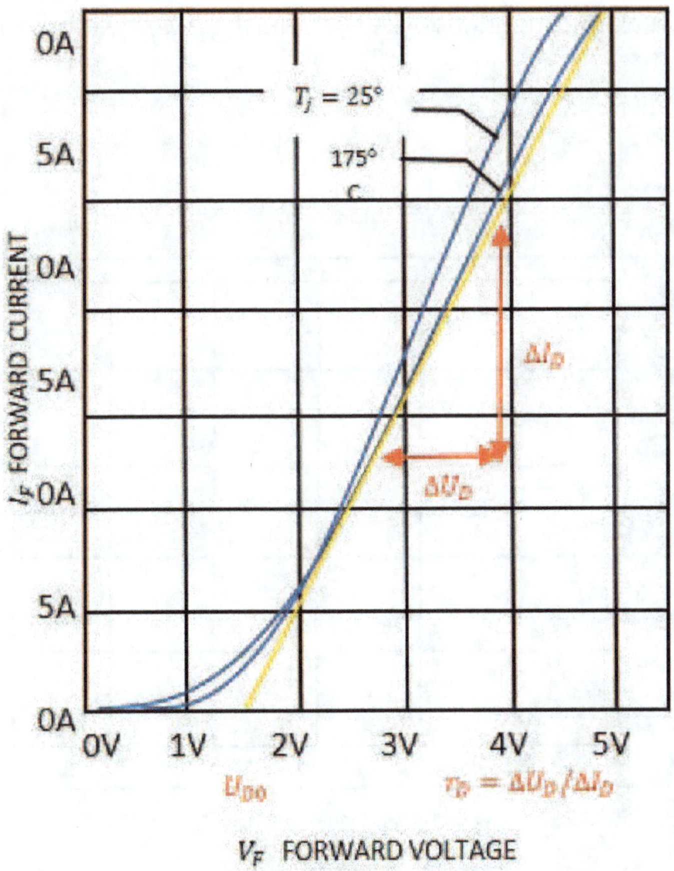

Typical diode forward current as a function of forward voltage

Figure 4. The u_{DO} and r_D ($r_d = \Delta U_{ce}/\Delta I_c$) values of datasheet diagram.

$$P_{swD} = \left(E_{onD} + E_{offD}\right)*(F_{sw}) \approx E_{onD}*(F_{sw}) \tag{12}$$

Finally, the total losses are calculated by Eq. (13).

$$W = P_{CT} + P_{swT} + P_{CD} + P_{swD} \tag{13}$$

Besides, the IGBT and diode switching losses can be estimated from characteristic curve provided by datasheet of manufactures as shown in **Figures 3**, **4** and **5**, respectively [5]. Using the equivalent circuit of **Figure 6(a)**, where W is a module power loss, T_j is the junction temperature IGBT chip, T_f is the heat sink temperature, T_c is the module case temperature, T_a is the ambient temperature, $R_{th(j-c)}$ is the thermal resistance between case and heat sink, $R_{th(c-f)}$ is the contact thermal resistance between case and heat sink, and $R_{th(f-a)}$ is the thermal resistance between heat sink and ambient air. The junction temperature T_j can be calculated using the thermal Eq. (14), according to [1, 10].

$$T_j = W*(Rth(j-c) + Rth(c-f) + Rth(f-a)) + Ta \tag{14}$$

Modeling of commutation and conduction losses as well as the temperature profiles on IGBT junction is done by electrothermic networks considering averages on a modulation period as shown in **Figure 6(b)**.

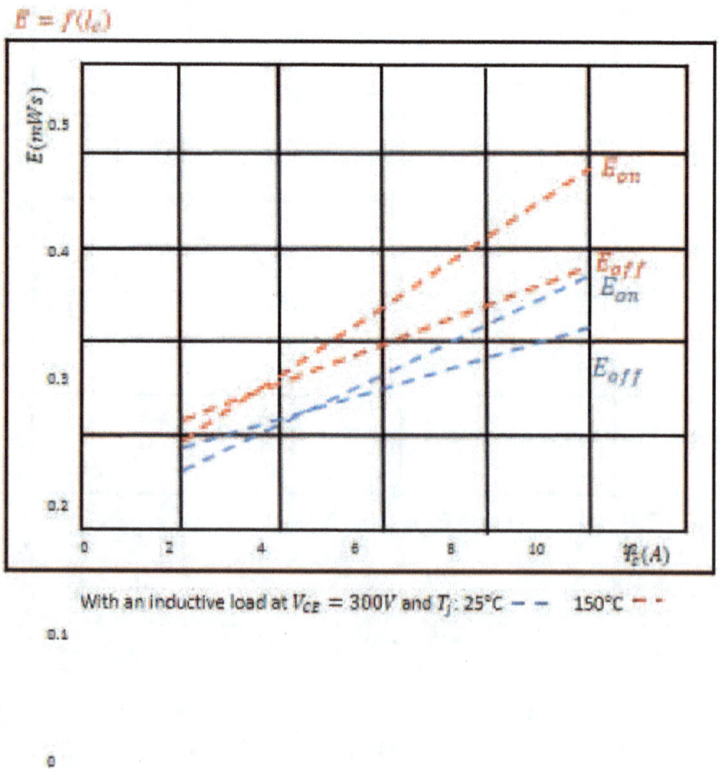

Figure 5. Switching energy losses as a function of collector current.

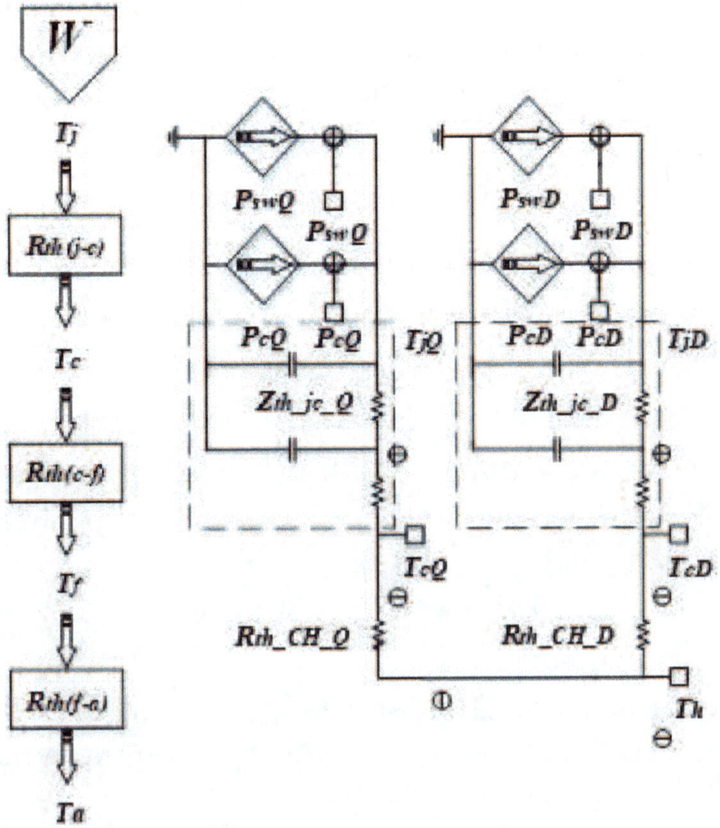

Figure 6. (a) Thermal resistance equivalent circuit, (b) electrothermal model.

4. Power cycle life

The power cycle life can be calculated from the power cycle capability curve that shows the relation between the temperature change ΔT_j and the number cycles. An example of the temperature changes is shown in **Figure 7**.

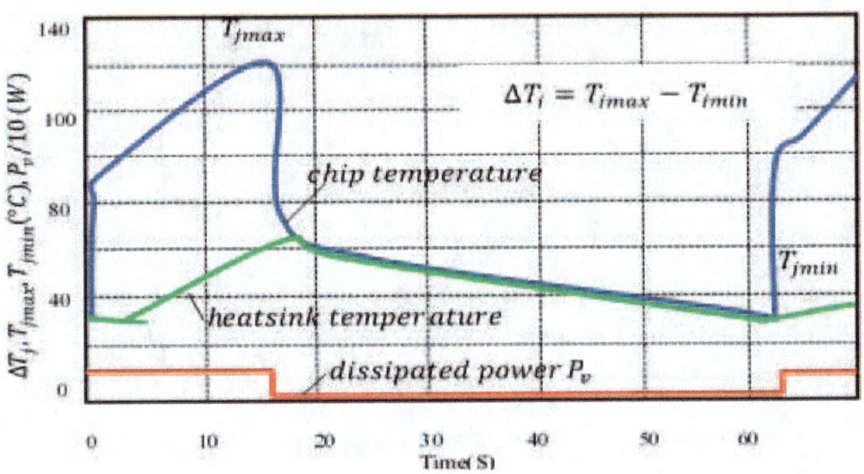

Figure 7. Pattern diagram flow current of ΔT_j power cycle and temperature change.

During the ΔT_j power cycle tests, the junction temperature goes up and down in a short time cycle; as a consequence, the temperature difference between silicon and bonding wire results in thermal stress. The ΔT_j power cycle lifetime is mainly limited by the aluminum bonding wire joints. **Figure 8** shows the power cycle capability curve of the IGBT module to $\Delta T_{jmin} = 25^{\circ}C$ and to $\Delta T_{jmax} = 150^{\circ}C$ according to [5].

On the other hand, the cycles before failure (CBF) can be calculated by Eq. (15), according to [28, 29].

$$CBF = 541162959016419 * \Delta T^{-5.12121} \tag{15}$$

where $\Delta T = P_t Z_{th}$, $P_t = I_{rms}^2 R_{on}$, and $Z_{th} = 2.3354 F_r^{-0.165}$, and the time before failure (TBF) can be calculate in years by Eq. (16).

$$TBF = \frac{CF}{Fr} * 60 * 24 * 265 \ years \tag{16}$$

where F_r is the frequency of the thermal oscillations [5, 29]. **Table 2** shows the parameters of IGBTs, which are considered in Multisim co-simulation; these parameters are obtained from datasheet of IGBT 10-0B066PA00Sb-M992F0 [24].

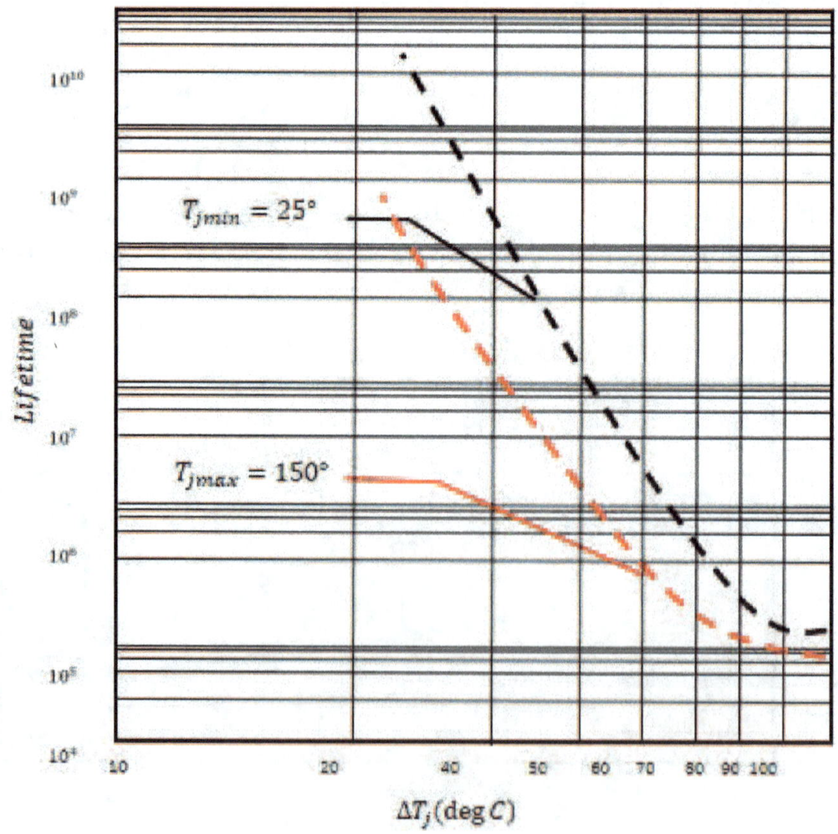

Figure 8. Power cycling lifetime curve.

5. Fuzzy logic controller

The fuzzy logic was proposed in [26] as a class of fuzzy sets with a continuum grade of membership; hence, a useful methodology to design fuzzy logic controllers is based on a linguistic phase plane [19, 20] as shown in **Figure 9**, where the error $e(t)$, change of error $\dot{e}(t)$, and the variation of current i_c are considered as membership functions in the fuzzification stage.

$$e(t) = e_c + e_t \tag{17}$$

$$\dot{e}(t) = \frac{\Delta e(t)}{\Delta t} = e(t) - e(t-1) \tag{18}$$

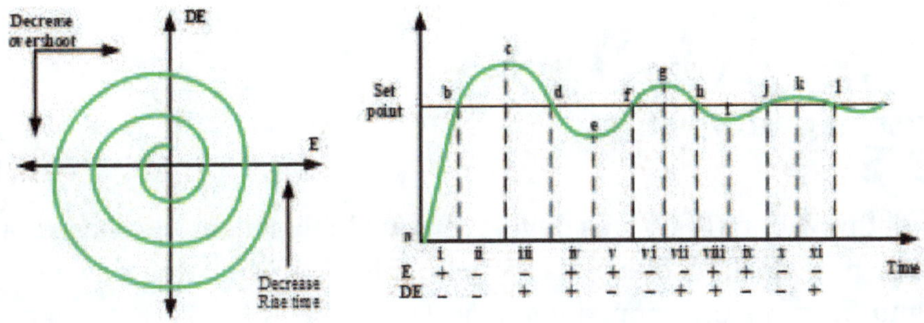

Figure 9. Linguistic phase plane.

Rules	Values			Reference point
	e	$\dot{e}$	i_c	
1	P	Z	P	a,e,i
2	N	N	Z	b,f,j
3	N	Z	N	c,g,k
4	P	P	Z	d,h,l
5	Z	Z	Z	sp
6	P	N	P	i(rt)v
7	N	N	N	ii(os) vi
8	N	P	N	iii,vii
9	P	P	P	iv,viii
10	Z	N	P	ix
11	Z	P	N	xi

Note: The meaning of symbols is as follows, N: negative, Z: zero, P: positive, e: error, $\dot{e}$: error derivative, i_c: output, sp: set point, rt: rise time, os: overshot.

Table 1. Fuzzy control rules for **Figure 12**.

Thus, Eq. (17) is defined as the sum of control error (e_c) and error of temperature (e_t), where e_c is defined as the difference of the desired speed e_d and the real speed e_r; on the other hand, e_t is defined as $e_t = e_{td} - e_{tr}$, where e_{td} is the desired temperature and e_{tr} is the real temperature.

The design of membership functions is according to Mamdani method; the knowledge base can be expressed as if-then statements to design a rules base, which is shown in **Table 1**. At last, the defuzzification stage gets a crisp value from the rule evaluation stage. Hence, the method of the center of gravity is used to compute the crisp value and is given by Eq. (19).

$$z^* = \frac{\int \mu_B(z)\cdot zdz}{\int \mu_B(z)dz} \tag{19}$$

So that, the fuzzy control design is carried out by means of the linguistic plane and the rules presented in **Table 1**. Thus, the controller with small/big rise time as well as small/big overshoot is possible.

6. Particle swarm optimization

Particle swarm optimization (PSO) was developed by [4]; this method is based on the behavior and movement of bird flocks looking for targets; this algorithm was developed to optimize nonlinear and multidimensional functions as in [13, 18, 27, 30, 31]. PSO algorithm needs to initialize the population in a random manner; each particle has a position $x_i(t)$ and velocity $v_i(t)$ with respect to target; then in main loop stage, the $x_i(t)$ and $v_i(t)$ update each iteration; this information is called best local position $P_i(t)$; on the other hand, between all population, there exists a particle that is more closest with respect to the target and it is called global best $g(t)$. Additionally, the position and velocity update are defined by:

$$v_{ij}(k+1) = \omega v_{ij}(k) + r_{1j}C_1\left(P_{ij}(k) - x_{ij}(k)\right) + r_{2j}C_2\left(g_j(k) - x_{ij}(k)\right) \tag{20}$$

$$x_{ij}(k+1) = x_{ij}(k) + v_{ij}(k+1) \tag{21}$$

where

- $i = 1, 2, \ldots, N$, and N is the size of population;
- $j = 1, 2, \ldots, D$, and D is the number of dimensions;
- $k = 1, 2, \ldots, iter$, and $iter$ is the maximum iteration number;
- $x_{ij}(k)$ is the position of particle i, dimension j at iteration k;
- $v_{ij}(k)$ is the velocity of particle i, dimension j at iteration k;
- $P_{ij}(k)$ is the local best position of particle i, dimension j at iteration k;
- $g_j(k)$ is the global best;
- ω is an inertia factor;

- C is an acceleration constant; and
- r_1, r_2 are the independent random numbers, uniformly distributed in $(0, 1)$.

Thus, the primary objective is to find a minimal global value from the cost function to be minimized as it is shown in the next pseudocode [31].

Step 1. Initialization.

For each particle of population N,

1. Initialize the position of each particle.
2. Initialize $P_{ij}(k)$.
3. Initialize $g_j(k)$.
4. Initialize $v_{ij}(k)$.

Step 2. Repeat until the criterion is satisfied.

For each particle of population N,

1. Set random numbers to r_{1j} and r_{2j}.
2. Update its velocity with Eq. (20).
3. Update its position with Eq. (21).
4. If $x_{ij}(k) < P_{ij}(k)$, then
   a. Update the best local position.
   b. If $P_{ij}(k) < g_j(k)$, then update the best global position $\mathrm{g}(k)$.

Step 3. Get the best solution $\mathrm{g}(k)$.

7. Design of fuzzy-PSO controller

The design of controller considers two objectives, the first one reaches the desired motor speed, and the second one is the increase of the semiconductor lifetime. The design of fuzzy-PSO controller to the speed control of BLDCM is based on the diagram shown in **Figure 11**, where μ_d is the desired speed and temperature, μ_t is the temperature sensed, and μ_c is the speed sensed. Furthermore, one of the leading features is the definition of objective function; this is considered as the sum of error of temperature e_t and error of speed e_c, which is given by Eq. (22).

$$f_c = e_c + e_t \tag{22}$$

where e_t is the difference between the desired temperature and the operating temperature; on the other hand, e_c is the difference between the desired speed and the real speed. In optimization

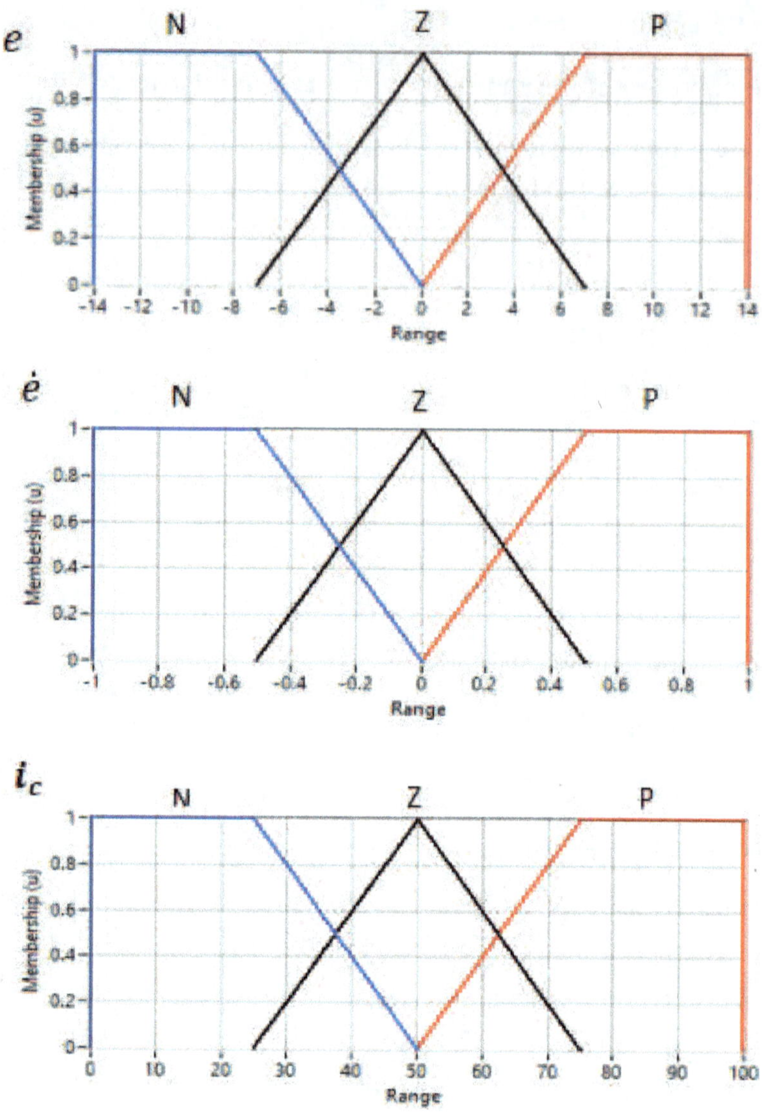

Figure 10. Diagram of fuzzy-PSO control.

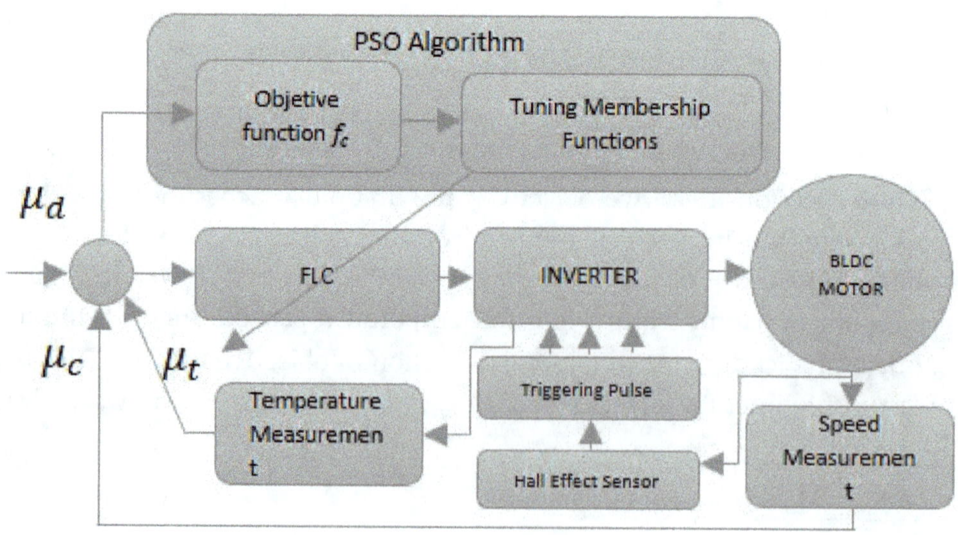

Figure 11. Input-output membership functions where e: error, $\dot{e}$: error derivative, and i_c: output.

IGBT 10-0B066PA00Sb-M992F09		
Parameter	Value	Unit
V_{CE}	600	mH
$R_{CE(ON)}$	0.1	Ω
I_C	8–6	Vs/rad
T_J, junction temperature	80 to 175	Nm/A
R_{J-S} thermal resistance junction to sink	3.50	K/W
BLDC motor		
Stator inductance	0.15	mH
Stator resistance	0.6	Ω
Velocity constant	0.03	Vs/rad
Torque constant	0.03	Nm/A
Number of poles	0.03	
Parameter of IGBT in co-simulation Multisim™		
IGBT control threshold	0.5	V
IGBT on resistance	0.1 meg	Ω
IGBT off resistance	10	Ω
IGBT forward voltage drop	0.7	V
Diode on resistance	1 m	Ω
Diode off resistance	10 meg	Ω
Number of switches in parallel	1	

Table 2. Experiment settings.

process, Eq. (22) is defined as the cost function to be minimized, and each membership functions proposed in **Figure 10** are considered as a decision variable; finally, the mechanical capacity of BLDC and thermal capacity of IGBT shown in **Table 2** are the constraints of the controller optimization.

8. Co-simulation design

The experimental frame is a set of assumptions to obtain the behavior trace and compare it with a real world, which is carried out by co-simulation in order to analyze its dynamic response during a specific period of time [6]. Hence, co-simulation helps to validate how it works under specific assumptions [2, 6]. The problem solution of speed control and lifetime of power electronic components is developed with co-simulation tool of Labview™ and Multisim™, where the BLDCM, the Hall effect sensors, and the six steps are simulated in Multisim™ and the

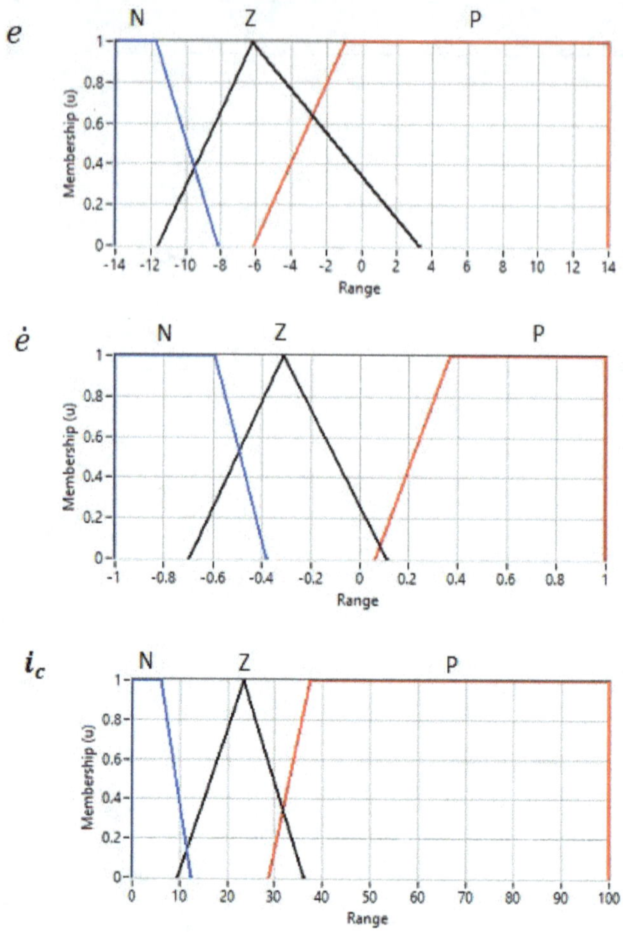

Figure 12. BLDCM speed response.

parameter design of the BLDCM, the six IGBT 10-0B066PA00Sb-M992F09 [24], and the six-steps inverter are shown in **Table 2**.

The proposal control is developed in Labview™ software and its response is shown in **Figure 14**, and the PID and fuzzy control are developed with control design and simulation module and fuzzy logic toolkit [19]; moreover, PSO algorithm is also developed in Labview™, and their parameters are shown in **Table 3**.

Finally, the primary purpose of co-simulation is to analyze the complete behavior of the electric drive with the proposed control.

Parameter	**Values**
Population size	30
Social rate (c_1)	0.005
Cognitive rate (c_2)	0.002
Inertia factor (W)	0.002

Table 3. Parameters of PSO.

9. Discussion and results

The proposed control considers two control objectives, the first one is to track the speed reference and the second one is to keep the desired temperature to increase the lifetime of the IGBTs. Furthermore, the fuzzy-PSO trade-off was tested with PID and fuzzy controllers in co-simulation program, which were evaluated under 25, 30, 35, 40, 60, 80, and 100°C as a temperature desired, 10 and 5 m/s as the reference speed, respectively. As a result, the PID controller response reaches set point of speed in a short time, but it does not present an improvement in the temperature of semiconductors. The fuzzy logic controller improves the response when the desired temperature values are at 40° and 60° and fuzzy-PSO controller presents the best response in order to accomplish the control objective because they reduce the overshoot of current when it reaches the temperature and reference speed. **Figures 13** and **14** show the response of controllers, and **Figure 12** shows the final membership function adaptation of fuzzy-PSO controller.

Table 4 shows the mean square error (MSE) of each controller under different desired temperatures, where the MSE of the fuzzy-PSO response shows good performance. To compare the proposed control response with the traditional estimation, the cycles before failure (*CBF*) given by Eq. (15) was computed to take into account the maximum temperature T_{jmax}, the change of temperature, $\Delta T = T_{jmax} - T_f$, and the IGBT datasheet.

Table 5 shows the calculation of TBF of the PID, fuzzy, and fuzzy-PSO controllers according to Eq. (16), and **Figure 15** shows the behavior of the TBF versus temperature, where the fuzzy-PSO controller reaches a longer lifetime.

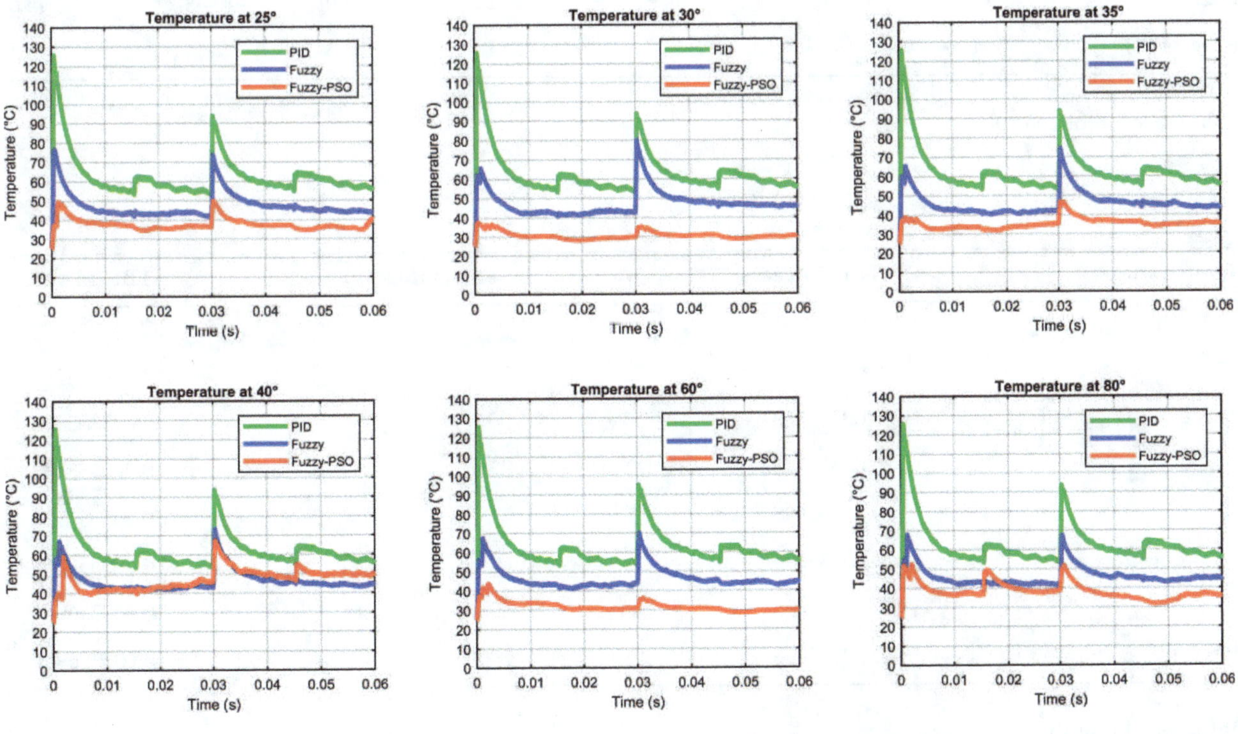

Figure 13. PID, fuzzy, and fuzzy-PSO controllers under different temperatures.

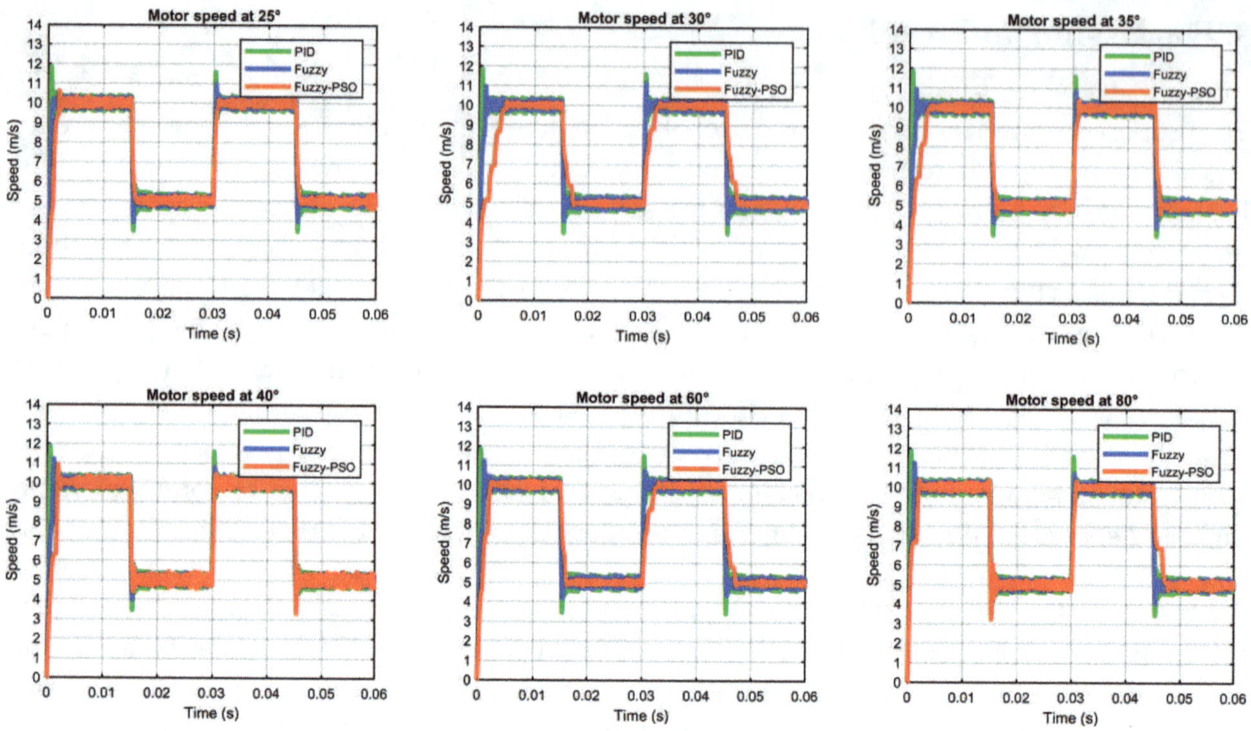

Figure 14. Tuning the input and output membership function parameters by means of the PSO.

MSE/T	25°	30°	35°	40°	60°	80°	100°
PID	61.8	60.8	60.8	60.8	60.8	60.8	60.8
Fuzzy	44.4	44.9	44.5	33.2	39.0	43.7	44.5
Fuzzy-PSO	39.6	37.5	43.7	42.4	52.5	44.0	45.1

Table 4. MSE of controllers.

Temperature	TBF (in years)	Temperature	TBF (in years)
°C	PID	°C	PID
25	0.004105201	25	0.004105201
30	0.005355432	30	0.005355432
35	0.005355432	35	0.005355432
40	0.00535696	40	0.00535696
60	0.00535696	60	0.00535696
80	0.00535696	80	0.00535696
100	0.00535696	100	0.00535696

Table 5. TBF of the controllers.

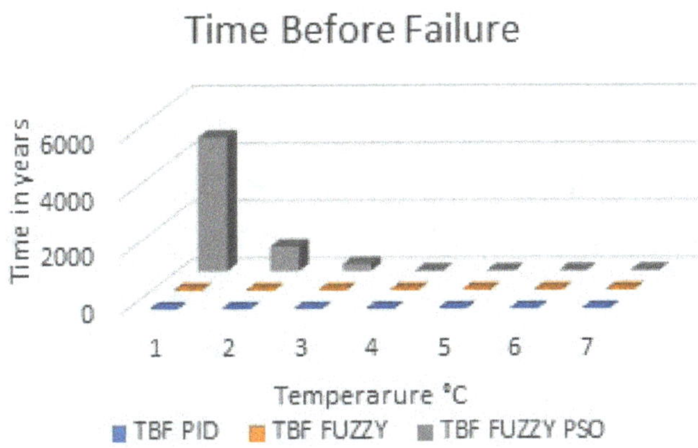

Figure 15. The behavior of TBF versus temperature in years.

10. Conclusions

In this chapter, a control design that takes into account the power electronics lifetime stage and the speed set point for BLDCM is presented. The objective function of optimization is integrated with the error of power stage temperature and the error of BLDCM speed. A voltage source inverter with six IGBT to drive the BLDCM is considered as power stage, where the current temperature in the power electronic stage, reference temperature, current motor speed, and reference motor speed are considered in the controller design. The PID controller, fuzzy logic controller, and fuzzy-PSO controller were designed and validated by NI Labview™ and Multisim™ co-simulation software. As a result, the fuzzy-PSO controller obtains a good response that increases the lifetime and reaches the set point desired due to the time that it takes to reach the desired speed increases but it reduces the overshoot of current during the transition time which produces minimal stress and degradation of the power electronic components.

Acknowledgements

This research is a product of the Project 266632 "Laboratorio Binacional para la Gestión Inteligente de la Sustentabilidad Energética y la Formación Tecnológica" ["Bi-National Laboratory on Smart Sustainable Energy Management and Technology Training"], funded by the CONACYT SENER Fund for Energy Sustainability (Agreement: S0019201401).

Author details

Pedro Ponce[1]*, Luis Arturo Soriano[1], Arturo Molina[1] and Manuel Garcia[2]

*Address all correspondence to: pedro.ponce@itesm.mx

1 Escuela de Ingeniería y Ciencias, Tecnologico de Monterrey, Mexico City, Mexico

2 Escuela Superior de Ingeniería Mecánica y Eléctrica, Instituto Politécnico Nacional, México City, México

References

[1] Bayerer R, Herrmann T, Licht T, Lutz J, Feller M. Model for power cycling lifetime of IGBT modules—Various factors influencing lifetime. In: Integrated Power Systems (CIPS), editor. 5th International Conference on Integrated Power Electronics Systems; March 11–13, 2008; Nuremberg, Germany: VDE; 2008. pp. 1-6

[2] Cellier FE, Kofman, editors. Continuous System Simulation. 1st ed. New York: Springer-Verlag New York, Inc; 2006. 639p

[3] Chau KT, editor. Electric Vehicle Machines and Drives: Design, Analysis and Application. 1st ed. Singapore: John Wiley & Sons, Singapore Pte. Ltd; 2015. 375p. DOI: 10.1002/9781118752555

[4] Eberhart R, Kennedy. A new optimizer using particle swarm theory. In: Micro Machine and Human Science, editor. Proceedings of the Sixth International Symposium on Micro Machine and Human Science; October 4–6, 1995; Nagoya, Japan, Japan: IEEE; 1995. pp. 39-43. DOI: 10.1109/MHS.1995.494215

[5] Fuji Electric. www.fujielectric.com. Fuji IGBT Modules Application Manual [Internet]. 2013 [Updated: January, 2017]. Available from: www.fujielectric.com/products/semiconductor/model/igbt/application/box/doc/pdf/REH984e/REH984e.pdf

[6] Gomes C, Thule C, Broman D, Larsen PG, Vangheluwe H. Co-simulation: State of the art. Cornell University Library. 2017. arXiv preprint arXiv:1702.00686

[7] Gopi Reddy LR, Tolbert LM, Ozpineci B. Power cycle testing of power switches: A literature survey. IEEE Transactions on Power Electronics. 2014;**30**(5):2465-2473. DOI: 10.1109/TPEL.2014.2359015

[8] Shirani M, Aghajani A, Shabani S, Jamali J. A review on recent applications of brushless DC electric machines and their potential in energy saving. Energy Equipment and Systems. 2017;**3**(1):57-71. DOI: 10.22059/EES.2015.13911

[9] Held M, Jacob P, Nicoletti G, Scacco P, Poech MH. Fast power cycling test of IGBT modules in traction application. In: IEEE, editor. International Conference on Power Electronics and Drive Systems; May 26–29, 1997; Singapore, Singapore: IEEE; 1997. pp. 425-430. DOI: 10.1109/PEDS.1997.618742

[10] infineon. www.infineon.com. 1999–2018 [Updated: Aug 12, 2015]. Available from: https://www.infineon.com/dgdl/Infineon-AN2015_10_Thermal_equivalent_circuit_models-AN-v01_00-EN.pdf?fileId=db3a30431a5c32f2011aa65358394dd2

[11] Luo H, Krueger M, Koenings T, Ding SX, Dominic S, Yang X. Real-time optimization of automatic control systems with application to BLDC motor test rig. IEEE Transactions on Industrial Electronics. 2017;**64**(5):4306-4314. DOI: 10.1109/TIE.2016.2577623

[12] Ma K, Vernica I, Blaabjerg F. Advanced design tools for the lifetime of power electronics-study case on motor drive application. In: Power Electronics and Motion Control Conference

(IPEMC-ECCE Asia), editors. 2016 IEEE 8th International; May 22-26, 2016; Hefei, China: IEEE; 2016. pp. 3255-3261. DOI: 10.1109/IPEMC.2016.7512816

[13] Melin P, Olivas F, Castillo O, Valdez F, Soria J, Valdez M. Optimal design of fuzzy classification systems using PSO with dynamic parameter adaptation through fuzzy logic. Expert Systems with Applications. 2013;**40**(8):3196-3206. DOI: 10.1016/j.eswa.2012.12.033

[14] National Instruments. www.ni.com. Introduction to Digital and Analog Co-Simulation between NI LabVIEW and NI Multisim [Internet]. 2013 [Updated: 2018]. Available from: http://www.ni.com/white-paper/13663/en/

[15] Nguyen V, Besanger Y, Tran Q, Nguyen T. On conceptual structuration and coupling methods of co-simulation frameworks in cyber-physical energy system validation. Energies. 2017;**10**(12):1-19. DOI: 10.3390/en10121977

[16] Parker MA, Soraghan C, Giles A. Comparison of power electronics lifetime between vertical- and horizontal-axis wind turbines. IET Renewable Power Generation. 2016;**10**(5):679-686. DOI: 10.1049/iet-rpg.2015.0352

[17] Pillay P, Krishnan R. Application characteristics of permanent magnet synchronous and brushless DC motors for servo drives. IEEE Transactions on Industry Applications. 1991; **27**(5):986-996. DOI: 10.1109/28.90357

[18] Poli R. Analysis of the publications on the applications of particle swarm optimisation. Journal of Artificial Evolution and Applications. 2008;**2008**(ID 685175):10. DOI: 10.1155/2008/685175

[19] Ponce CP, Ramirez FF, editors. Intelligent Control Systems with LabVIEW. 2010th ed. New York: Springer Verlag; 2010. 216p. DOI: 10.1007/987-1-84882-684-7

[20] Prabu MJ, Poongodi P, Premkumar K. Fuzzy supervised online coactive neuro-fuzzy inference system-based rotor position control of brushless DC motor. Institution of Engineering and Technology. 2016;**9**(11):2229-2239. DOI: 10.1049/iet-pel.2015.0919

[21] Premkumar K, Manikandan BV. Speed control of brushless DC motor using bat algorithm optimized adaptive neuro-fuzzy inference system. Applied Soft Computing. 2015;**32**:403-419. DOI: 10.1016/j.asoc.2015.04.014

[22] RENESAa. www.renesas.com. Semiconductor Reliability Handbook [Internet]. Jan.39.2017 [Updated: 2010–2018]. Available from: https://www.renesas.com/en-us/doc/products/others/r51zz0001ej0250.pdf

[23] Soulatiantork P, Alghassi A, Faifer M, Perinpanayagam S. IGBT thermal stress reduction using advance control strategy. In: The Scientific Committee of the 5th International Conference on Through-Life Engineering Services (TESConf 2016); December 2017; Vincent Building Cranfield University, UK. UK: Elsevier; 2017. pp. 274-279. DOI: 10.1016/j.procir.2016.09.040

[24] Vincotech. www.vincotech.com [Internet]. July 23, 2010. [Updated: June 23, 2017]. Available from: https://www.vincotech.com/search.html [Accessed: 10-0B066PA006SB-M992F09]

[25] Xia CL, editor. Permanent Magnet Brushless DC Motor Drives and Controls. 1st ed. Singapore: John Wiley & Sons Singapore Pte. Ltd.; 2012. 304p

[26] Zadeh LA. Fuzzy sets. Information and Control. 1965;**8**(3):338-353. DOI: 10.1016/S0019-9958(65)90241-X

[27] Zhang Y, Wang S, Ji G. A comprehensive survey on particle swarm optimization algorithm and its applications. Mathematical Problems in Engineering. 2015;**2015**:38. DOI: 10.1155/2015/931256

[28] National Instruments. www.zone.ni.com. IGBT Thermal Diode [Internet]. February 2017 [Updated: 2018]. Available from: http://zone.ni.com/reference/en-XX/help/375482B-01/TOC395.htm

[29] National Instruments. http://www.ni.com. Power Electronics Design with NI Multisim [Internet]. Jul 31, 2017 [Updated: Feb 21, 2018]. Available from: http://www.ni.com/white-paper/13714/en/

[30] Amin A, Hegazv OT. Swarm intelligence applications in electric machines. In: Lazinica A, editor. Particle Swarm Optimization. 1st ed. Rijeka: InTech; 2009. pp. 11-39. DOI: 10.5772/56679

[31] Bachache NK, Wen J. Multi objective swarm optimization design fuzzy controller to adjust speed of AC motor drive. In: Tan Y, Shi Y, Mo H, editors. 4th International Conference in Swarm Intelligence ICSI 2013. Harbin, China: Springer, Berlin, Heidelberg; 2013. pp. 522-529. DOI: 10.1007/978-3-642-38703-6_61

[32] Visconti P, Primiceri P. An overview on state-of-art and future application fields of BLDC motors: Design and characterization of a PC-interfaced driving and motion control system. ARPN Journal of Engineering and Applied Sciences. 2017;**12**(17):4913-4926

[33] Lai W, Chen M, Ran L, Xu S, Qin H, Alatise O, et al. Study on the lifetime characteristics of power modules under power cycling conditions. IET Power Electronics. 2016;**9**(5):1045-1052

[34] Abeyasekerab T, Rodriguez P. An overview of the reliability prediction related aspects of high power IGBTs in wind power applications. Microelectronics Reliability. 2011;**51**(9-11):1903-1907

[35] Choi UM, Blaabjerg F, Iannuzzo F. Advanced power cycler with intelligent monitoring strategy of IGBT module under test. Microelectronics Reliability. 2017;**76**:522-526

9

Basics of High-Speed Electrical Machines

Flyur R. Ismagilov, Viacheslav Ye. Vavilov and
Valentina V. Ayguzina

Abstract

The high-speed electrical machines are widely used in different industries, such as machine tools, aerospace engineering, autonomous power engineering, etc. This chapter is devoted to the basics of high-speed electrical machines with high-coercitivity permanent magnets. It is considered in the application areas of high-speed electrical machines and their classifications. In addition, design problems of high-speed electrical machines are shown. To estimate the efficiency, loss calculations are performed. The obtained results can be used in the design of high-speed electrical machines with high-coercitivity permanent magnets and in their future development.

Keywords: high-speed electrical machine, high-coercitivity permanent magnets, amorphous magnetic material, eddy-current losses, windage losses

1. Introduction

The main task of electrical machine (EM) design is to provide maximum power with a minimum energy loss and volume of active and constructive elements. Reduction of the material consumption and the production technology improvement can significantly reduce the cost of electrical machines, and subsequently, the cost of the systems, where they are used (e.g., aircrafts and machine tools), or resources obtained with their help (e.g., electricity and oil). In addition, the development of new industries and the implementation of macro- and microresearch projects lead to an emergence of new tasks to reduce mass-and-size parameters and to increase the efficiency of electrical machines.

The directions that allow the solution of these problems can be determined from the analysis of the mathematical formula proposed by Engelbert Arnold in 1896 [1]:

$$P = \frac{\pi^2}{60 \cdot 10^{10} k_e} \alpha_\delta D^2 l_a B_\delta A n k_{m.f} k_w,$$

where P is EM power, A is a linear current load, D is a rotor diameter, l_a is a length of the EM, B_δ is a magnetic flux density in the air gap, n is a rotational speed, $k_{m.f}$ is a coefficient of the magnetic field shape in the air gap, α_δ is a pole pitch coefficient, and k_w is a winding coefficient.

From Arnold's equation, it can be seen that at the invariable volume of the EM elements, its power can be increased in three ways or their combinations: to increase linear current load, to increase magnetic flux density in an air gap, and to increase rotational speed of EM rotor.

All these three ways have found the niche in different industries. One of the directions is the high-speed and ultra-high-speed EMs, which are widely used in machine tools, aerospace engineering, autonomous power engineering, etc. In this chapter, high-speed and ultra-high-speed EMs are frequency-controlled EMs with high-coercitivity permanent magnets (HCPM). Section 2 presets the classification of high-speed EMs. The main application areas are presented in Section 3. Section 4 is devoted to calculation of losses and efficiency of the high-speed EMs.

2. Classification of high-speed EMs

To develop and generalize the high-speed EM theory, their classification should be considered. In general, high-speed EMs can be classified according to the operation principle, design features, frequency or output voltage, the stator type (slotted or slotless), etc. **Table 1** shows the basic designs of high-speed EMs. The slotless stator designs are not given in **Table 1**. However, the design nos. 1–6 and 8–12 can be made with a slotless stator. The hysteresis motor designs are not presented in **Table 1**, although they are also high speed and are used in gyroscopes.

High-speed EM with electromagnetic excitation and designs nos. 5, 8, 9, 10, 11 are inferior to the high-speed EM with HCPM for reliability, specific parameters, and efficiency. However, they have some development prospects; for example, the design possibility of high-speed EM with rotating rectifiers and a rotational speed of 50,000 rpm is presented in [1, 2].

Axial EMs with HCPM can be promising for power supply systems for rocket and space equipment due to their small axial dimensions. The design and prospect review of axial EMs are presented in [3, 4]. At the same time, axial EMs have been created with rotational speed of 30,000–48,000 rpm.

In addition, the use of HCPM made of NdFeB and SmCo alloys for high-speed damping elements is promising to develop the space systems due to improving their energy characteristics [5]. To develop the alternative energy, the ship and transport systems (railway trains), including the cargo location tracking, the use of single-coordinate and multicoordinate vibrational EMs with HCPM is of interest [6, 7]. To develop the ship power supply systems, energy conversion systems of detonation engines and automotive systems, the use of high-speed linear EM with HCPM as the main generators of the power supply system is topical.

	EM type	Design
1.	High-speed EM with a star-type rotor: 1 is a star-type rotor, 2 is a stator magnetic core, and 3 is a stator winding	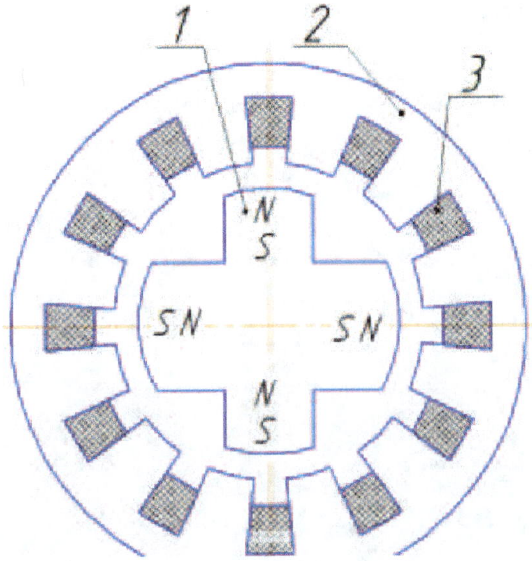
2.	High-speed EM with an assembled star-type rotor: 1 is a cylindrical HCPM, 2 is a stator magnetic core, 3 is a stator winding, 4 is a rotor sleeve, and 5 is a rotor back	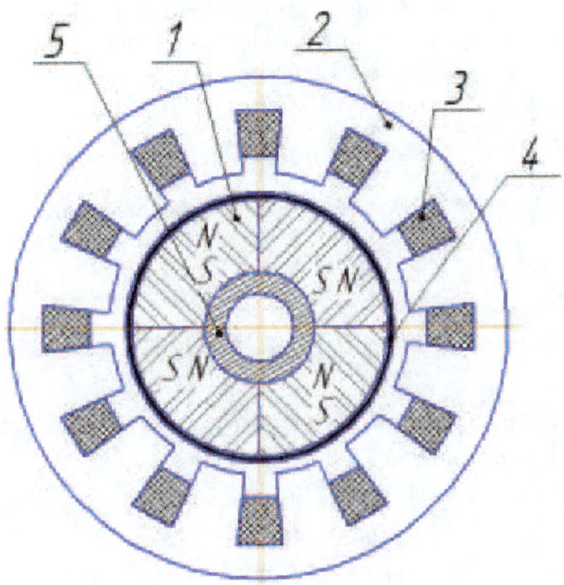
3.	High-speed EM with interior permanent magnets: 1 is the HCPM, 2 is a stator magnetic core, 3 is a stator winding, and 4 is a rotor back	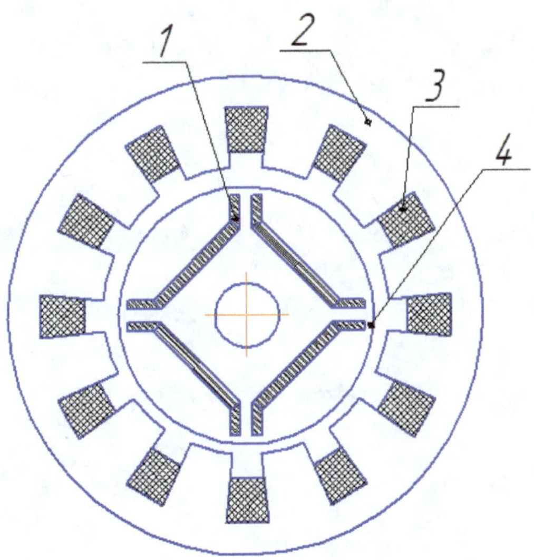

	EM type	Design
4.	High-speed EM with combined rotor magnetic systems (including the Halbach magnetic system): 1 is the combined rotor magnetic system, 2 is the stator magnetic core, and 3 is a stator winding	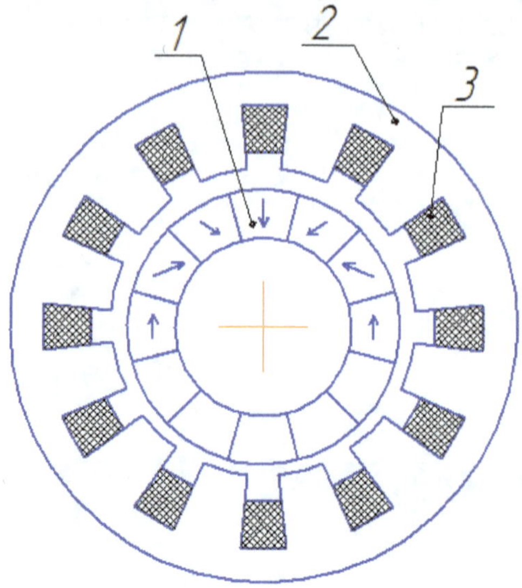
5.	High-speed synchronous EM with rotating rectifiers: 1 is a rotor, 2 is a stator, 3 is a block of rotating rectifiers, and 4 is an excitation system	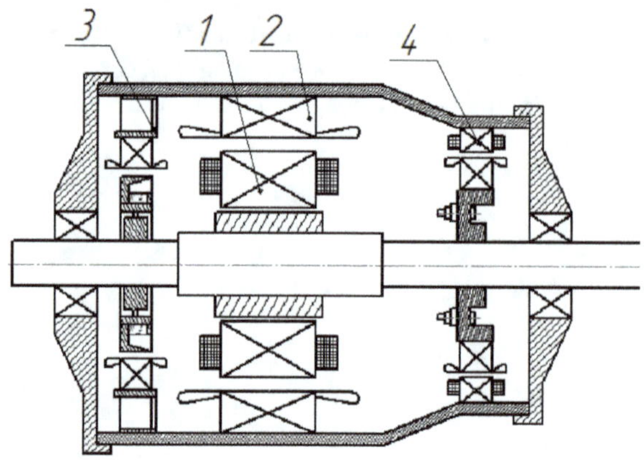
6.	Axial EM with HCPM: 1 is the HCPM, 2 is a stator magnetic core, and 3 is a stator winding	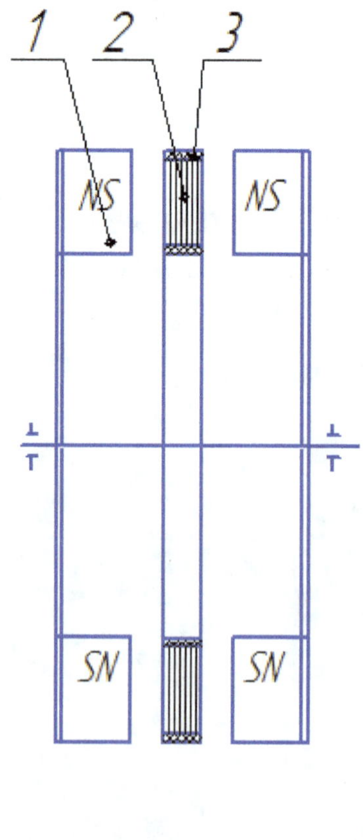

	EM type	Design
7.	High-speed synchronous EM with combined excitation: 1 is a rotor, 2 is a stator magnetic core, and 3 is an additional excitation system	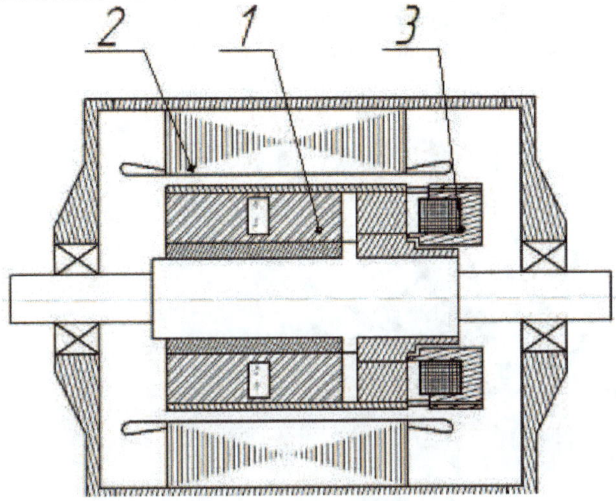
8.	High-speed asynchronous motors with squirrel-cage rotor: 1 is a rotor, 2 is a stator magnetic core, and 3 is rotor bars	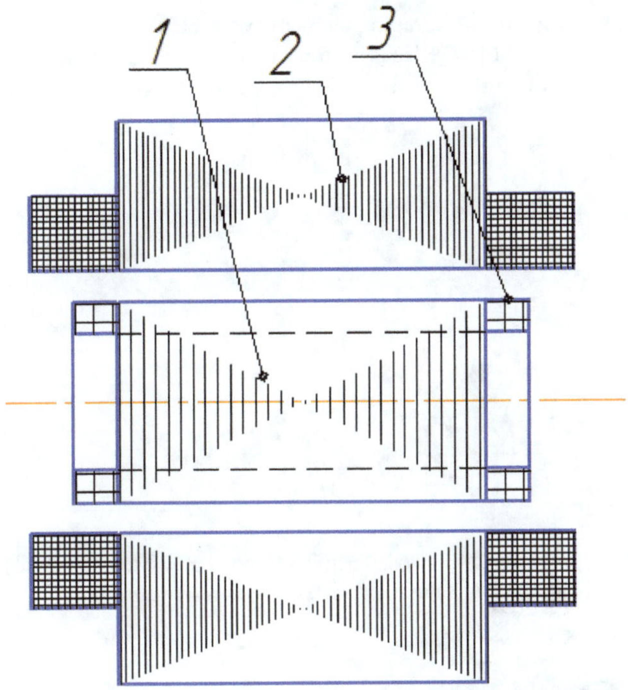
9.	High-speed liquid-metal asynchronous motor: 1 is a annular channel, 2 is a ring winding, and 3 is a magnetic core	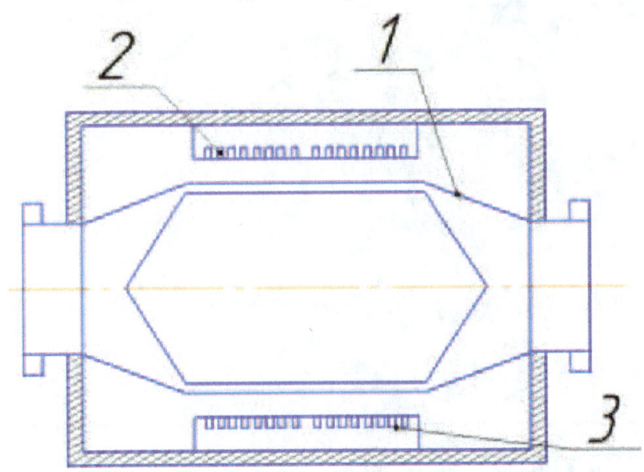

	EM type	Design
10.	High-speed induction EM: 1 is a rotor, 2 is a stator magnetic core, and 3 is a stator winding	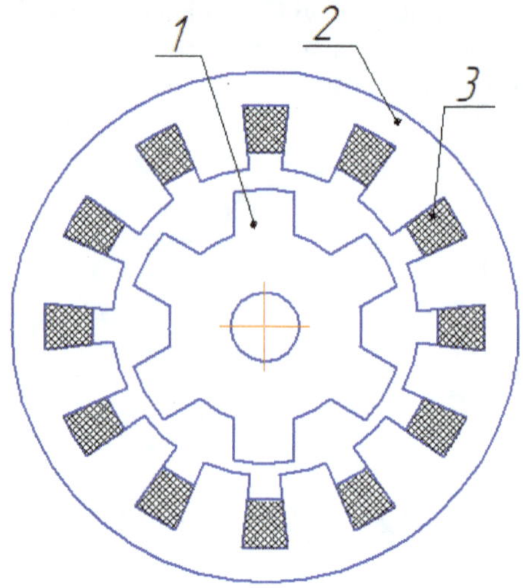
11.	High-speed damping elements with HCPM: 1 is an external HCPM, 2 is an internal HCPM, 3 is a hollow rotor, and 4 is a bearing	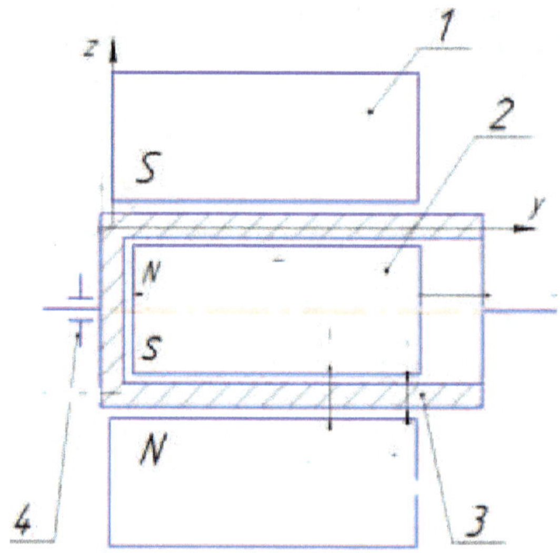
12.	High-frequency oscillating EM with HCPM: 1 is a casing, 2 is a moving part, and 3 is a winding	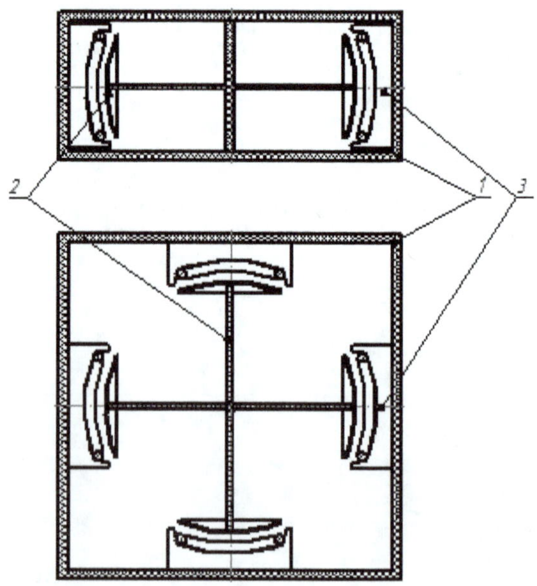

Table 1. Basic designs of high-speed EMs.

The presented classification shows that areas and prospects for the use of high-speed EM with HCPM are quite broad. The improvement of this EM type leads to an expansion of its application areas. Therefore, it seems expedient to consider in more detail the application areas of high-speed EM and to reveal their development prospects.

3. Application areas of high-speed electrical machines and a review of known works

To use the high-speed EMs, it is necessary to have a high-speed drive (for generator mode) or a high rotational speed (for motor mode). Therefore, the main application areas of high-speed EMs are high-speed machine tool, aviation and space power engineering, as well as autonomous power engineering. The initial application of high-speed EMs is associated with the development of rocket and space technology. High-speed induction and hysteresis motors with a rotational speed of 30,000–60,000 rpm have found wide application as a gyro-motor for spinning the gyroscope. In autonomous power engineering, high-speed EMs are developed as a microturbine component. Elliott Group serially produces decentralized power plants with 68,000-rpm magnetoelectric generators [8]. Capstone produces microturbines with high-speed magnetoelectric generators. A distinctive feature of these generators is that the rotor is installed on two air bearings, and the third air bearing is located between the compressor and turbine working wheels [9].

In decentralized energy, the high-speed EMs have found a wide industrial application. Their development is aimed to reduce bearing losses, to increase the working temperature and efficiency. In addition, the high-speed EMs with a power below 250 kW find application in the power supply systems of aircrafts and space vehicles. There are two main variants of high-speed EMs for aircrafts: switched-reluctance EM and EM with HCPM. Comparison of these EM types is presented in different works. It is shown that the high-speed EMs with HCPM have smaller mass-and-size parameters than in switched-reluctance EMs. This determines their preferred use in aircrafts.

The development of such industries as robotics, machine tool, turbomolecular pumps, high-tech medical equipment, etc. and the development of new-generation unmanned aircraft require the creation of ultra-high-speed EMs with a rotational speed from 200,000 to 1,000,000 rpm or more and a power from 50 W to 2 kW [10]. The main advantage of this EM type is a high power density with overall miniature dimensions, which causes wide prospects for their use in various microsystems. In addition, a significant advantage of ultra-high-speed EMs is the uniqueness of the tasks they solve. For example, the application of motor with a 400,000-rpm rotational speed in the machine tool improves the quality of surface treatment [11]. The 500-W microgenerator for an autonomous robotic complex allows abandoning battery cells that have high mass-and-size parameters. **Table 2** shows a comparison of various electricity sources for the electrical supply of robots.

With all the technical advantages of ultra-high-speed EMs with HCPM, this area began to develop relatively recently. The development of this area was facilitated by the emergence of

Electricity sources	Power-to-mass ratio [kW/kg]
Ultra-high-speed microgenerator with microturbine including the fuel mass	0.55
Batteries	0.50
Fuel cells	0.05
Ultra-high-speed microgenerator with flywheel system	0.50

Table 2. Comparison of various electricity sources for the electrical supply of robots.

new electrical materials and the development of microelectronics. Therefore, theoretical studies on this topic are limited and represent a disparate material, which describes individual design solutions for ultra-high-speed EMs for a particular application. It is necessary to consider the main development trends of this EM type and to consider in more detail examples of their practical application.

One of the application areas of ultra-high-speed EMs is autonomous system. Microturbines with ultra-high-speed EMs are planned to be used instead of storage batteries in robotic complexes, unmanned aerial vehicle, cryogenics, etc.

The main developers of ultra-high-speed EMs for microturbine installations are several countries: Japan (IHI Corporation), France (Onera), Switzerland (Power Electronic Systems Laboratory and ETH Zurich), USA, Germany (High-Speed Turbomaschinen GMBH). It seems advisable to consider their development in more detail.

Onera [12] (France, DecaWatt research program) developed a microturbine system with a microgenerator. The purpose of the program is the creation of a microturbine engine with a power of 50–100 W to provide electrical energy for ammunition and equipment for future soldiers. The microgenerator has a power of 55 W and a rotational speed of 840,000 rpm. The generator tests were performed at a rotational speed of 700,000 rpm. It was used in the mechanical high-speed bearings. The pole number of this high-speed EM is equal to 1. The frequency of the output voltage is 14,000 Hz. Ultra-high-speed EMs should be considered as a complex of interconnected systems, since the full efficiency of a high-speed EM and mass-and-size parameters are determined not only by material parameters and geometric dimensions of EM but also by parameters of its control system. In addition, a separate EM with an output frequency of 14 kHz without a power electronics unit is unlikely to find practical application. A similar conclusion was noted in [13, 14]. To achieve the maximum efficiency, Onera used the following technical solutions: amorphous magnetic materials (AMM) as a stator core material to minimize stator losses; a slotless stator design to simplify the manufacturing technology of the stator magnetic core, to reduce the inductance of the stator winding scattering and to minimize the demagnetizing effect of the armature reaction; and a cylindrical permanent magnet (PM) to simplify the rotor manufacturing technology.

According to a similar technology, Power Electronic Systems Laboratory and ETH Zurich developed and implemented two ultra-high-speed PM starter-generators for microturbines [15, 16]; one of which has a power of 500 W and a rotational speed of 500,000 rpm, and the second one has a power of 1 kW and a rotational speed of 500,000 rpm. To implement the ultra-high-speed EM, Power Electronic Systems Laboratory and ETH Zurich used the same technical solutions as Onera. It was a slotless stator, a cylindrical PM and AMM were also used. For the implementation of the starter mode, sensorless control was used.

The same technical solutions are used by High Speed Turbomaschiner, which develops a 150-W high-speed magnetoelectric EM with a rotational speed of 490,000 rpm [17]. The stator magnetic core is slotless and made of AMM.

IHI Corporation developed a microturbine system with a microgenerator [18]. It contains an oil-free gas turbine with a 400-W ultra-high-speed magnetoelectric generator integrated into it. The rotational speed of the microgenerator is 400,000 rpm. Kerosene, propane, light oil, etc. can be used as fuel in the proposed system. Expected applications are power generation for charging portable devices and the use as a power source for robots.

Robot/Mechatronics Research Center [19] is working on the development of a microturbine with a 500-W microgenerator with a rotational speed of 400,000 rpm for unmanned aerial vehicles and robotic complexes. The stator of this generator is made with six slots in which a three-phase winding is laid. The rotor sleeve is made of Inconel 718 alloy. The HCPM type is Sm_2Co_{17}. The disadvantages of this generator are the slotted stator design, which considerably complicates the manufacturing technology of the stator magnetic core made of AMM. If the AMM is not used as a stator core material, the hysteresis losses in the stator magnetic core can significantly reduce the efficiency. For unmanned aerial vehicles and robotic complexes, Stanford University and MDOT Aeronautics develops an ultra-high-speed permanent-magnet EM with a power of 400 W and a rotational speed of 800,000 rpm [20].

For all the above-considered ultra-high-speed EMs for autonomous power supply systems, the shaft and the rotor back are made of nonmagnetic material to increase its mechanical strength. However, this leads to a decrease in its power and in the magnetic flux density in the EM air gap. Therefore, to increase the energy characteristics of ultra-high-speed EMs, it would be more expedient to use a magnetic material, for example, the cobalt alloy Vacodur S Plus, which is close to some titanium brands by its mechanical characteristics [1].

Thus, the ultra-high-speed EMs for autonomous power supply systems are actively developing abroad. They all practically have a common design: a slotless stator made of AMM and a rotor with cylindrical HCPMs.

The publication analysis shows that the research and optimization of the design of all the above-mentioned ultra-high-speed EMs was carried out together with power electronics and control systems [21]. The tendency to reduce number of manufactured parts in mechanical engineering by improving the quality of the treated surface poses the task of creating miniature machines and spindles of small power with ultra-high-speed EMs. In particular, [22] presents a design and prospects of using an ultra-high-speed EM with a 200,000-rpm rotational speed in machine tool to produce electronic components. The stator magnetic core is slotless and made of electrical steel 10SNEX 900 with a sheet thickness of 0.1 mm. The rotor speed is controlled by sensorless algorithms. The rotor sleeve is made of a carbon fiber.

Westwind [23] (United Kingdom) produces PCB spindles with ultra-high-speed EMs with rotational speeds from 85,000 to 370,000 rpm. All spindles are made with foil bearings. EMs with rotational speed of 370,000 rpm are used in a drilling machine to make holes with a

Parameter	Onera	ETH	Robot/mechatronics research center	Electric motor
EM type	PM generator	PM generator	PM generator	PM motor
Power [W]	55	1000	500	—
Rotational speed [rpm]	800,000	300,000–500,000	400,000	200,000
The need of a control system	Yes	Yes	Yes	Yes
Stator design	Slotless	Slotless	Slotted	Slotless
Stator magnetic core material	AMM	AMM	—	Electrical steel 10SNEX 900 with a sheet thickness of 0.1 mm

Table 3. Parameters of ultra-high-speed EMs.

diameter of 75 μm. In addition, ETH Zurich and Power Electronic Systems Laboratory [15] for the first time in the world developed a 100-W electrical motor with a rotational speed of 1,000,000 rpm, which was a breakthrough in rotational speeds of up to 1 million rpm. The electrical motor was made with HCPM and ceramic ball bearings.

One of the important applications of ultra-high-speed EMs is a high-tech medical equipment. Ultra-high-speed EMs are used in blood pumps, dental instruments, surgical operations, etc. For example, Sinotech Motors [24] developed the ultra-high-speed EMs for blood pumps. These motors are slotless. Their control system is integrated, which allows achieving minimum mass-and-size parameters. Sirona has developed the ultra-high-speed EMs for dental instruments [25]. EM data are executed with rotational speeds from 250,000 to 400,000 rpm. In [26], the design of an ultra-high-speed medicine pump based on the ultra-high-speed EM with a rotational speed of 200,000 to 600,000 rpm is described. The stator design of this EM is also slotless. In addition, sensorless control methods are used.

To analyze trends in the high-speed EMs, the EM parameters are listed in **Table 3**. It shows that all ultra-high-speed EMs are operated jointly with power electronics and control systems, which confirms the need to study them not as a separate electrical machine. In addition, all ultra-high-speed EMs have HCPMs, and their stators are slotless.

4. Efficiency of high-speed electrical machines

The EM efficiency is one of the main EM parameters and largely determines the EM design. For example, to increase the EM efficiency, slot skewing or contactless bearings are used, i.e., technical solutions that lead to the complication of the EM design and the production technology. Therefore, for the economic and technical justification of these complications, it is especially important to determine the EM losses and to select the effective EM design.

4.1. Winding losses in the high-speed EM

Total winding losses in the high-speed EM can be present as follow:

$$P_{cu} = P_m + P_{ad}, \tag{1}$$

where P_m is the winding losses and P_{ad} is additional eddy-current losses in winding.

Due to the uneven magnetization of the HCPM and the nonlinearity of loads, the current curve differs from the sinusoidal, i.e., the third, fifth, seventh spatial harmonics have appeared. In this case, the total winding losses are formed as the sum of the losses from these harmonics:

$$P_m = (I_1^2 + I_3^2 + I_5^2 + I_7^2)mr_a, \tag{2}$$

where I_1 is the first harmonic current, I_3 is the third harmonic current, I_5 is the fifth harmonic current, and I_7 is the seventh harmonic current.

To determine the eddy-current losses in winding, the following equation can be used:

$$P_{ad} = \frac{\pi^3 H_{s\,max}^2 \mu_0^2 f^2 d_s^4 \sigma_{cu} l_{cu}}{4}, \tag{3}$$

where $H_{s\,max}$ the maximum value of the magnetic field strength in the slot, f is the current frequency in the EM winding, d_s is the strand diameter, σ_{cu} is the specific electric conductivity of the winding, and l_{cu} is the total strand length.

The maximum value of the eddy-current losses in the winding is in the slotless EM design. For the slotted EMs, they are much smaller, but they also require calculation to determine and select the optimal cooling system. The most convenient calculation way is the computer simulation of the EM magnetic field and determination of the magnetic field crossing the winding in the slots. Analytical calculation methods are known, but they are cumbersome for engineering calculations.

4.2. Stator core losses of the high-speed EM

To determine the stator core losses in the EM, there is no general calculation methodology. In [27, 28], to determine the stator core losses in a wide frequency range, the following equation is proposed:

$$P_{s.c} = k_m P_{s.c50/1} \left(\frac{f}{50}\right)^{\beta} B^2, \tag{4}$$

where $P_{s.c}$ is the stator core losses [W/kg], $P_{s.c50/1}$ is the stator core losses at a 50 Hz frequency and a 1 T magnetic flux density, k_m is a coefficient of the increase in the stator core losses during processing, B is the magnetic flux density in the stator core, and $\beta = 1.3, \ldots, 1.5$.

For a frequency of 400 Hz or more, the following equation [29, 30] is recommended for the stator core loss calculation:

$$P_{s.c} = P_{sz} + P_{sj} = k_m P_{s.c.s} B_j^2 M_j \left(\frac{f}{400}\right)^{1.5} + k_m P_{s.c.s} B_z^2 M_z \left(\frac{f}{400}\right)^{1.5}, \tag{5}$$

where $P_{s.c}$, P_{sz}, P_{sj} are stator core losses, losses in the stator teeth, and losses in the stator back, respectively; k_m is a coefficient that takes into account the increase in the stator core losses during processing; $P_{s.c.s}$ are stator core specific losses at a frequency of 50 Hz and a magnetic flux density of 1 T; B_j, M_j are the magnetic flux density and mass of the stator back, respectively; B_z, M_z are the magnetic flux density and mass of the stator teeth, respectively; f is the current frequency or magnetization reversal frequency of the stator back and teeth.

In [31], the following equation was proposed to determine the stator core losses:

$$P_{s.c} = \left(k_{hyst}B^{\beta}f + k_{eddy}B^{2}f^{2}\right)M_{s}, \tag{6}$$

where $\beta = 1.7–2$; k_{gyst}, k_{vih} are hysteresis and eddy-current factors, respectively, and M_s is a stator mass.

A method for determining the stator core losses is proposed in [32], in which three coefficients characterizing the properties of the stator core material are used: the hysteresis loss coefficient k_{hyst}, the eddy-current loss coefficient k_{eddy}, and excess loss coefficient k_e:

$$P_{s.c} = k_{hyst}B^{2}f + k_{eddy}B^{2}f^{2} + k_{e}(Bf)^{\frac{3}{2}}. \tag{7}$$

These coefficients are often indicated by manufacturers of electrotechnical steels and soft magnetic alloys.

In [22], an empirical equation is presented for calculating the stator core losses of high-speed EM, which is the approximated dependence of the stator core losses on the current frequency of the 10JNEX900 steel:

$$P_{s.c} = \left(7 \cdot 10^{-4}B^{1.75}f^{1.5} + 4.7 \cdot 10^{-4}B^{1.86}f^{1.53}\right). \tag{8}$$

For analytical calculations of the stator core losses in the high-speed EMs in a wide frequency range, Eq. (5) is the most optimal. Coefficients of this equation should be determined based on the experimental studies or data of the specialized literature.

4.3. Influence of heating of EM with HCPM on specific losses in electrotechnical steels and precision soft magnetic alloys

During the operation of the EM with HCPM, their stator core will heat up under the influence of the ambient temperature, and because of the specific losses. An increase in the stator core temperature leads to a change in its magnetic properties. Therefore, it is necessary to evaluate the effect of the EM temperature on the stator core losses.

Measurements of the stator core losses were made under normal climatic conditions at sample temperatures below 23°C, relative air humidity of 25%, and at a temperature below 70°C. For studies at a temperature below 70°C, the sample was heated in a muffle furnace. Then, it was repeatedly measured in a cold state at a temperature below 23°C. Experimental studies have measured specific losses at different temperatures in two annular samples made of AMM (type

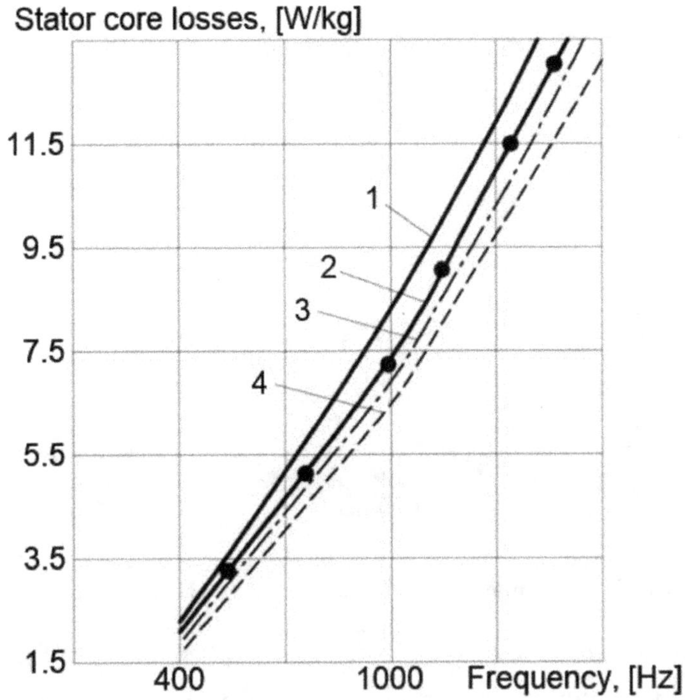

Figure 1. Dependence of the stator core losses on the frequency: curve 1 is for AMM, 5BCDR grade, type T at 20°C; curve 2 is for AMM, 5BCDR grade, type E at 20°C; curve 3 is for AMM, 5BCDR grade, type T at 70°C; and curve 4 is for AMM, 5BCDR grade, type E at 70°C.

E and type T of 5BDSR grade) produced by the Ashinsky metallurgical plant (Russia). The saturation magnetic flux density of these AMMs is 1.3 T. The experimental results at a magnetic flux density of 0.5 T are shown in **Figure 1**. It shows that with the temperature increase of the stator core made of AMM, 5BCDR grade, type T (linear hysteresis loop) to 50°C, the stator core losses decrease by 15%. For temperature increase of the stator core made of AMM, 5BCDR grade, type E (rectangular hysteresis loop) to 50°C, the stator core losses decrease by 14%. Thus, with the temperature increase of the stator core made of AMM, its losses decrease by 15%.

In addition to changing the losses in the EM stator core due to its heating, the losses in the EM stator core during operation will also decrease due to the decrease in the HCPM energy characteristics, and, accordingly, in magnetic flux density of the stator core. The study of this process will be given below. To assess this decrease, a computer model was developed in the Ansoft Maxwell software package, and a computer simulation for the magnetic field distribution along the section of the EM stator core was made. Simulation results for EM at a temperature of 23°C are presented in **Figure 2**. It was found that with an increase of the HCPM to 150°C, the magnetic flux density in the stator core decreases by 6–6.5%. Taking into account the dependence of the stator core losses on magnetic flux density for various materials, this can lead to reducing losses in the stator core by 13–15%. Thus, heating of the stator core leads to a significant loss reduction in the EM.

4.4. Eddy-current losses in the HCPM and rotor sleeve

In the high-speed EMs, spatial harmonics caused by the EM design, the winding type, or the distribution coefficient, as well as the time harmonics caused by an external circuit (for

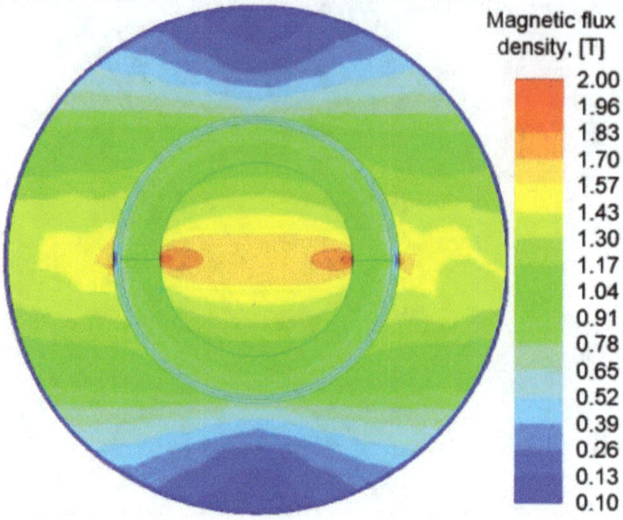

Figure 2. Simulation results for EM at a temperature of 23°C.

example, an inverter or rectifier) will induce significant eddy-current losses in the rotor sleeve or HCPM. This can lead to overheating of the HCPM and to demagnetization of the HCPM. To reduce these losses, HCPMs are usually laminated in the axial direction. To determine these losses, it is necessary to determine separately the losses due to time harmonics (external circuit) and losses due to spatial harmonics (EM design features). Usually, losses caused by time harmonics are greater than the losses caused by spatial harmonics. This statement is valid only for a number of the high-speed EM designs. For example, the EM with a tooth-coil winding has significant spatial harmonics, and losses caused by these harmonics are higher than the losses caused by time harmonics.

In the literature, there is no unambiguous opinion on the loss nature. Some authors argue that the losses in the HCPM are only due to eddy currents. At the same time, several authors argue that these losses are caused by eddy currents and magnetization reversal. In this chapter, the losses in the HCPM and rotor sleeve are considered as created only by eddy currents.

a. Eddy-current losses in the HCPM and rotor sleeve created by time harmonics. As noted above, these losses are formed by an external circuit (rectifier or inverter), which are installed at the EM output for transmission to a standard frequency network. The EM calculation scheme is presented in **Figure 3**. To estimate these losses, expressions obtained by Poliender are convenient to use:

$$P_{\mathrm{HCPM}} = \frac{rlh_{\mathrm{m}}b_{\mathrm{m}}^{2}}{12\rho_{\mathrm{m}}}\left\{\left\{(pa_{\mathrm{m}} + \sin(pa_{\mathrm{m}}))\left(\frac{\mathrm{d}}{\mathrm{d}t}\left(\widehat{B}\cos(p\beta)\right)\right)\right\}^{2} + \left\{(pa_{\mathrm{m}} - \sin(pa_{\mathrm{m}}))\left(\frac{\mathrm{d}}{\mathrm{d}t}\left(\widehat{B}\cos(p\beta)\right)\right)\right\}^{2}\right\} \tag{9}$$

where p is a number of pole pairs, l is the HCPM length, h_{m} is the HCPM height, ρ_{m} is the HCPM resistivity, r is an average air gap radius, a_{m} is a pole angle, b_{m} is a pole arc length, and $\widehat{B}$ is a magnetic flux density created by time harmonics.

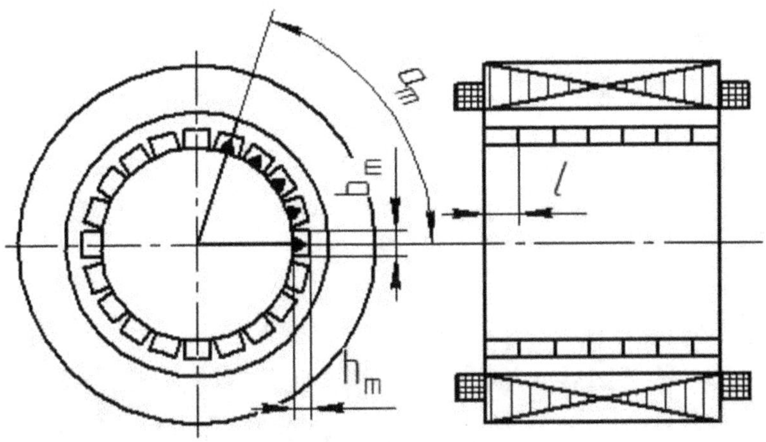

Figure 3. EM calculation scheme.

To determine the magnetic flux density produced by time harmonics, it is necessary to understand the time harmonics magnitude produced by different types of nonlinear load in the EM windings. To solve this problem, a simulation EM model with an external network was developed in the LTSpice IV. It contains a three-phase EM with HCPM, an uncontrolled rectifier, and a DC load (**Figure 4**). The EM was represented by three single-phase sources and three pairs of magnetically coupled inductors forming a three-phase voltage.

Simulation was carried out for a system with a three-phase 115/200 V source (EM with HCPM) with a frequency of 400 Hz, a grounded neutral and a load connected via a single-phase or a three-phase rectifier. In **Figures 5** and **6**, the simulated circuits for the inclusion of three-phase and single-phase nonlinear loads are presented. As a simulation result, the current and voltage curves of the EM and the load are shown in **Figures 7–14**.

To simplify the comparative analysis and to determine the influence of a certain nonlinear load inclusion scheme, the obtained data were summarized in **Table 4**. It shows that when a three-phase rectifier is connected, the phase current of the EM with the HCPM contains spatial harmonics with the numbers 5, 7, 11, 13, 17, 19 in addition to the fundamental harmonic.

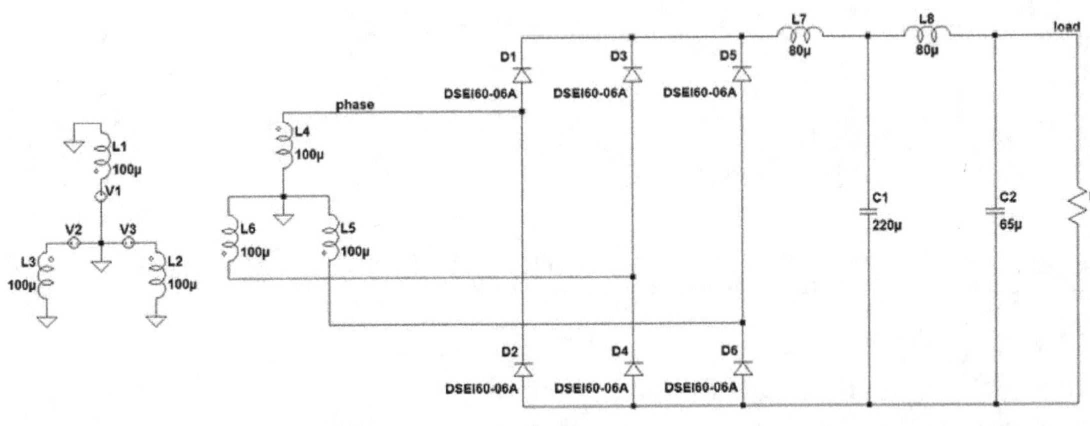

Figure 4. EM simulation model with external network for determining time harmonics and losses generated by them.

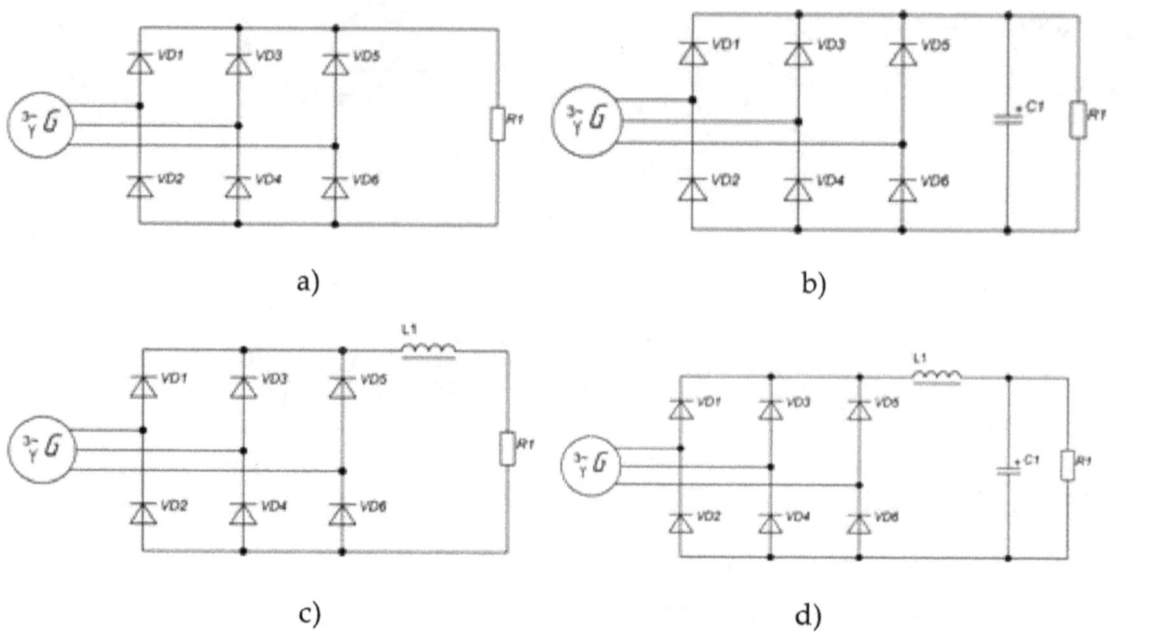

Figure 5. The simulated circuits for the inclusion of three-phase nonlinear loads: (a) active load, (b) active-capacitive load, (c) active-inductive load, and (d) active-inductive-capacitive load.

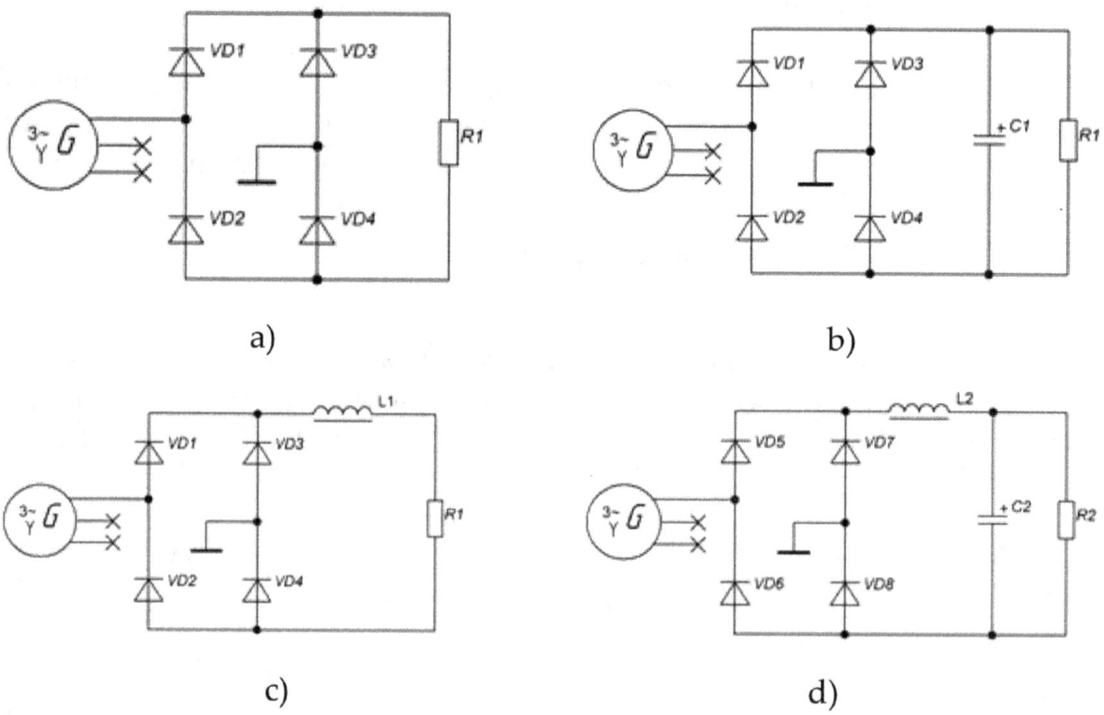

Figure 6. The simulated circuits for the inclusion of single-phase nonlinear loads: (a) active load, (b) active-capacitive load, (c) active-inductive load, and (d) active-inductive-capacitive load.

Harmonics that are multiples of three are absent. For the studied numerical parameters, the worst case scenario is for a three-phase active-capacitive load and a single-phase active-capacitive load. In this case, the harmonic coefficient reaches 95.2%. For an active nonlinear

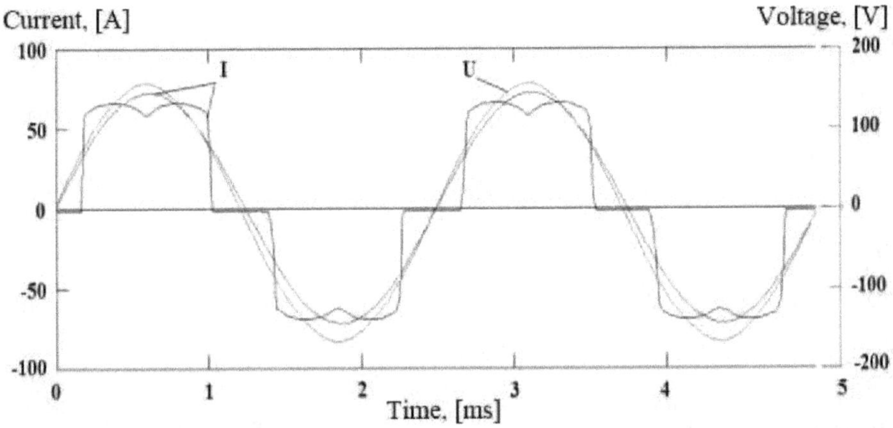

Figure 7. Curves of the voltage, total current, and fundamental current harmonic for the circuit in **Figure 5a**.

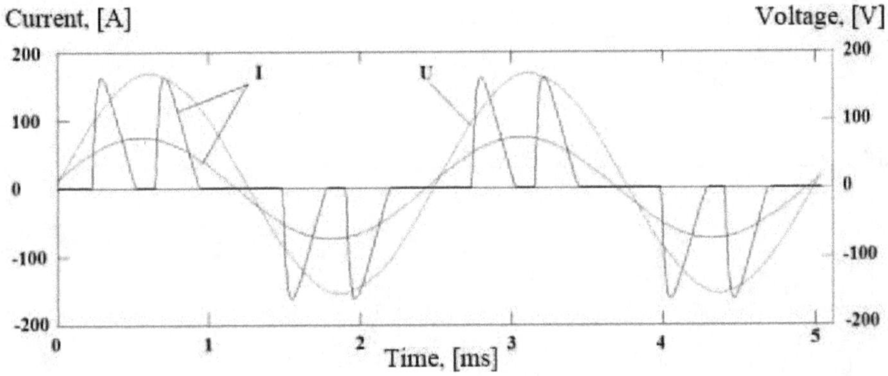

Figure 8. Curves of the voltage, total current, and fundamental current harmonic for the circuit in **Figure 5b**.

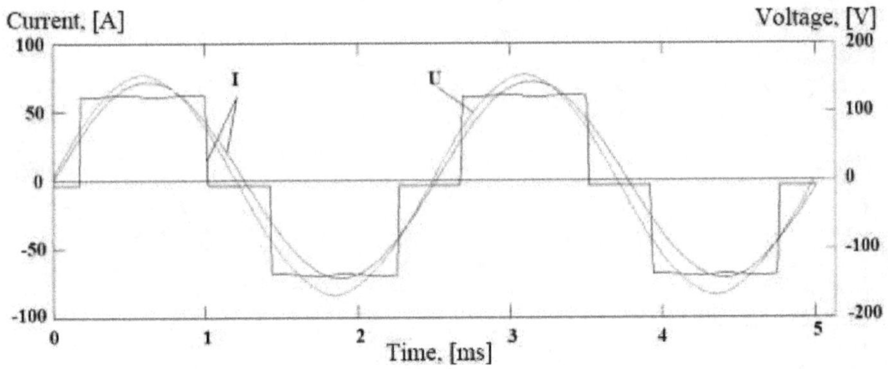

Figure 9. Curves of the voltage, total current, and fundamental current harmonic for the circuit in **Figure 5c**.

three-phase load, the total harmonic distortion (THD) is 29%. For an active-inductive nonlinear load, the THD is 31.1%. It is obvious that the voltage harmonic composition of the EM with nonlinear load characterizes the harmonic composition of the magnetic flux density created by the time harmonics.

To confirm the computer simulation data, an experimental setup was developed. It consists of a DC motor with independent excitation, EM, pulse width modulation (PWM) controller of the

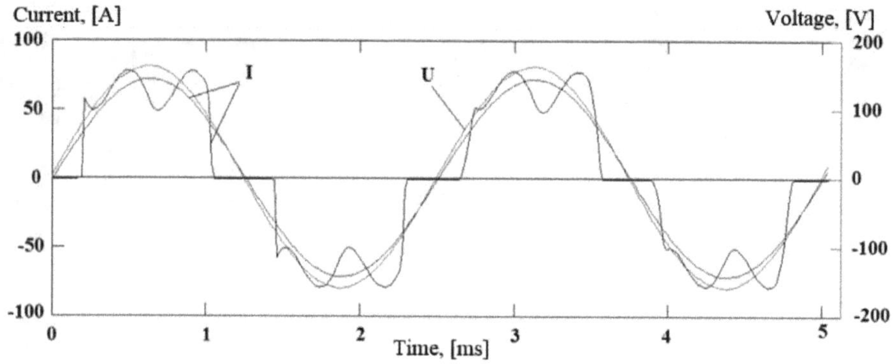

Figure 10. Curves of the voltage, total current, and fundamental current harmonic for the circuit in **Figure 5d**.

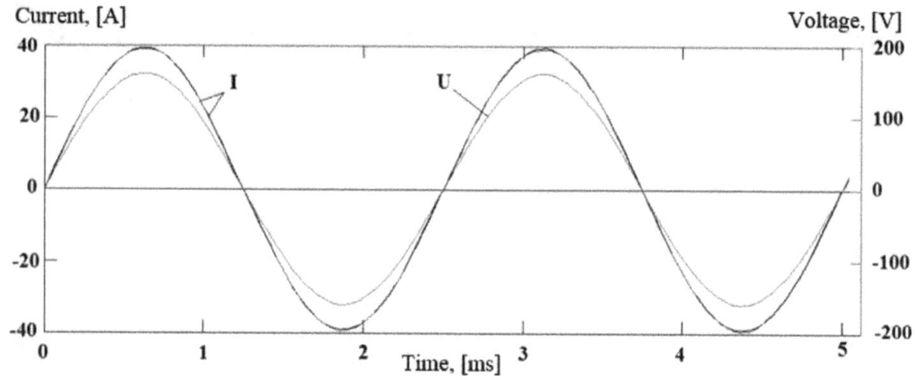

Figure 11. Curves of the voltage, total current, and fundamental current harmonic for the circuit in **Figure 6a**.

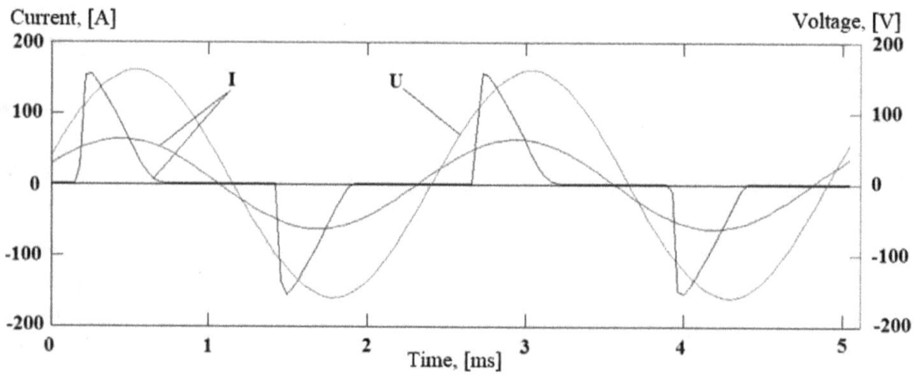

Figure 12. Curves of the voltage, total current, and fundamental current harmonic for the circuit in **Figure 6b**.

rotational speed for the DC motor, and rectifier with an active load. The PWM controller sets the rotational speed of DC motor, which the armature winding is connected to the PWM controller. The excitation winding is supplied from a constant current source with a voltage of 27 V and a current of 2 A. The PWM controller is powered by a 27 V DC source with a current of 15 A. The DC shaft is mechanically connected to the EM shaft. The rotational speed of EM is 8000 rpm, the output voltage is 36 V, and the output current is 3.2 A. The outputs of the EM

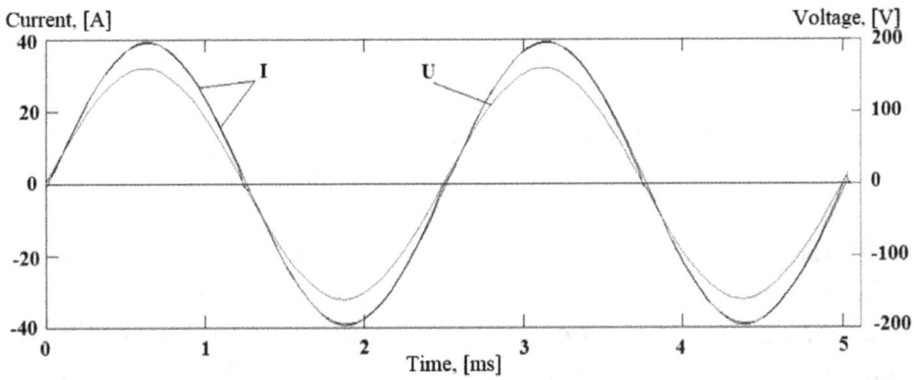

Figure 13. Curves of the voltage, total current, and fundamental current harmonic for the circuit in **Figure 6c**.

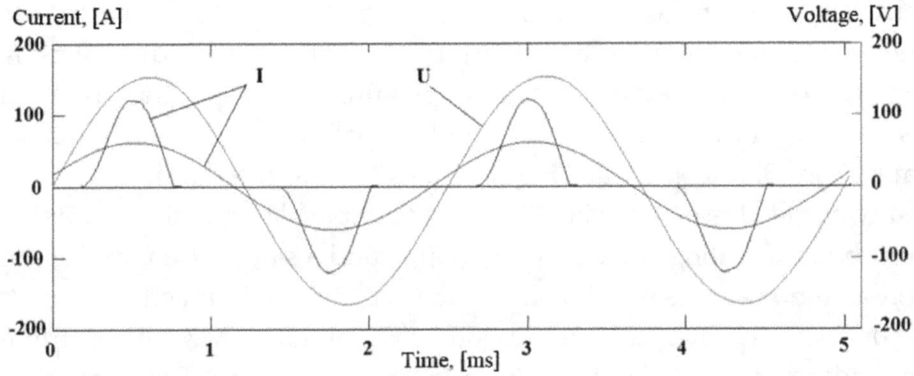

Figure 14. Curves of the voltage, total current, and fundamental current harmonic for the circuit in **Figure 6d**.

Scheme	Total power [kVA]	Power [kW]	cos φ	cos φ(1)	THD [%]
Three-phase active load	18.0	17.3	0.96	1.00	29.0
Three-phase active-capacitive load	24.7	17.5	0.71	0.99	95.2
Three-phase active-inductive load	17.9	17.1	0.95	1.00	31.3
Three-phase active-inductive-capacitive load	18.0	17.1	0.95	1.00	31.7
Single-phase active load	3.1	3.1	1.00	1.00	2.4
Single-phase active-inductive load	3.1	3.1	1.00	1.00	3.6
Single-phase active-inductive-capacitive load	6.2	4.7	0.76	0.97	76.9

Table 4. Comparison of simulation results of connection schemes.

stator winding through a three-phase rectifier are connected to a resistive load of 100 W with the LTS-6NP-type series-connected current transformers.

The experimental results correspond to the simulation data, which indicate the adequacy of the developed simulation model and the obtained theoretical results: the frequency of the fundamental harmonic is 396 Hz, THD is 29.6%, cos φ = 0.95, and cos φ(1) = 1.0.

In addition, it is possible to determine the losses in the HCPM and rotor sleeve by using the obtained data. For example, calculations were made for EM with rotational speeds of 5000 and 32,000 rpm at active-inductive load. It was found that for nonlaminated HCPM, the eddy-current losses are above 280 W at a frequency of 5000 rpm and above 8 W at a frequency of 32,000 rpm.

a. Eddy-current losses in the HCPM and rotor sleeve created by spatial harmonics. These losses are determined by the slot type, the slot size, and the winding type; i.e., it depends on many different factors. Analytically, the definition of these losses is difficult and does not give the necessary accuracy. Therefore, it is more expedient to use the finite element method (FEM) to analyze the losses in the HCPM and rotor sleeve.

To solve this task in Ansys Maxwell software package, it is necessary to specify the specific HCPM conductivity. It is 1,100,000 S/m for SmCo and 625,000 S/m for NdFeB. In the project window, in the excitation section, the eddy-current calculation (Set Eddy Effects) is carried out, and HCPM and the rotor sleeve are selected. In a simulation result, the distribution of eddy-current losses in the HCPM and rotor sleeve on the spatial harmonics will be obtained in the Fields Overlaps-Field-Ohmic_loss tab. **Figures 15–17** show the distribution of the magnetic field and the eddy-current losses in the HCPM caused by spatial harmonics for 250-kW 60,000-rpm EMs with a distributed winding, a tooth-coil winding, and a slotless design. The number of pole pairs for all considered EMs was equal to 4. In addition, the minimum eddy-current losses in the HCPM caused by spatial harmonics have the slotless EM and the EM with a distributed winding. The maximum eddy-current has the EM with a tooth-coil winding. The difference between the losses in EM with distributed and tooth-coil winding is above 500%. Therefore, it is obvious that in the high-speed EMs with a tooth-coil winding, the eddy-current losses in the HCPM caused by spatial harmonics exceed the eddy-current losses in the HCPM and the rotor sleeve caused by time harmonics.

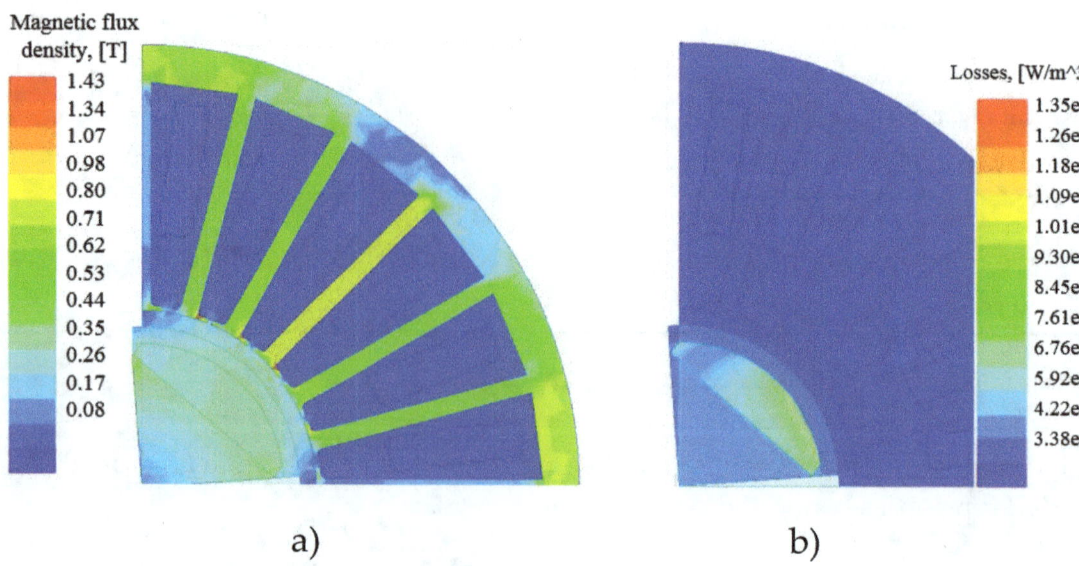

Figure 15. Distribution of the magnetic field (a) and the eddy-current losses (b) in the HCPM of the slotless EM with a distributed winding.

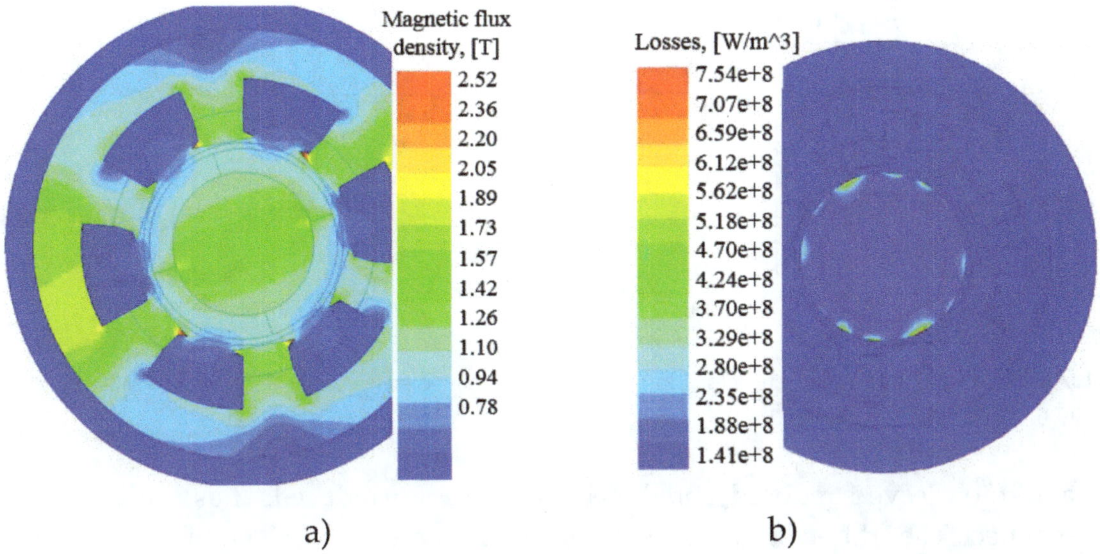

Figure 16. Distribution of the magnetic field and the eddy-current losses in the HCPM of the slotless EM with a tooth-coil winding.

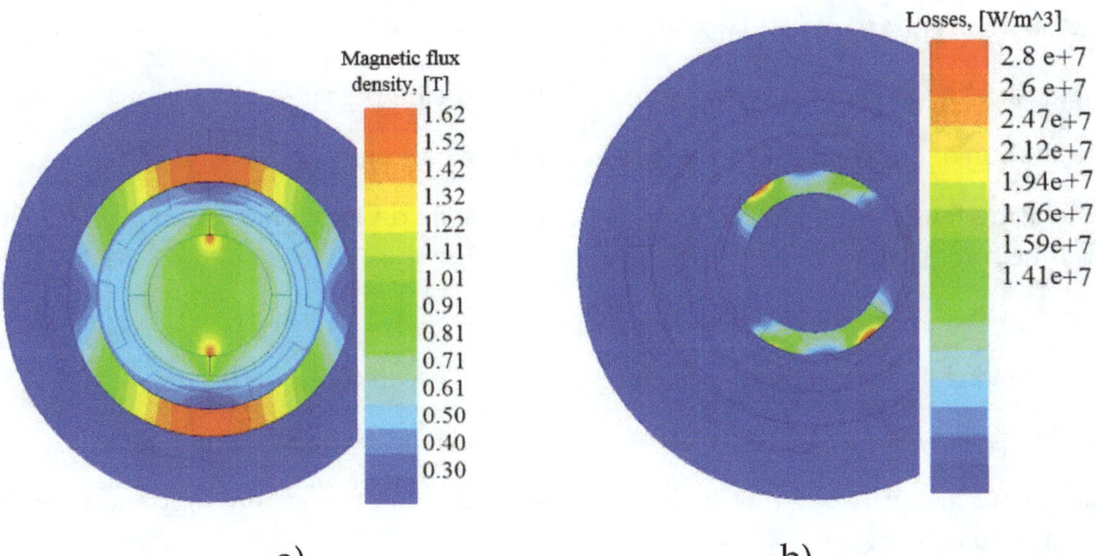

Figure 17. Distribution of the magnetic field (a) and the eddy-current losses (b) in the HCPM of the slotless EM.

In addition, it seems advisable to estimate the influence of the rotor magnetic system (MS) type on the eddy-current losses of HCPM in the high-speed EM.

This task was also solved by computer simulation methods. The results are shown in **Figure 18**. It can be seen that the MS type does not significantly affect the HCPM losses. This is explained by the fact that the HCPM losses are formed by the stator magnetic field. In the case of a constant slot zone, the stator magnetic field remains unchanged. With an increase in the rotational speed, the HCPM losses have a maximum point, after which they begin to decrease. This is because with rotational speed increase, the magnetic field penetration depth in the

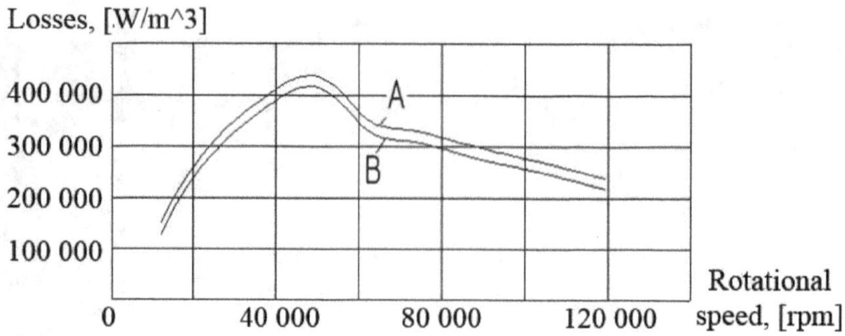

Figure 18. Dependence of eddy-current losses in the HCPM on the frequency and MS type: A—MS with cylindrical HCPM and B—MS with semicircular HCPM.

HCPM and rotor sleeve is reduced, and thereby the losses are reduced. Thus, the MS type has no significant effect. The eddy-current losses in the HCPM can vary considerably due to the change in the load angle. This should be taken into account in the design of the EM with HCPM.

The reduction of the eddy-current losses in the HCPM and rotor sleeve of the high-speed EMs can be achieved by the following ways.

For losses created by time harmonics:

- increasing the number of phases and thereby reducing pulsations,
- lamination of the HCPM and rotor sleeve in the axial direction,
- increasing the air gap and reducing the magnetic field penetration depth into the HCPM and the rotor sleeve.

For losses created by spatial harmonics:

- slot skewing and the selection of a large winding distribution coefficient,
- lamination of the HCPM and rotor sleeve in the axial direction,
- increasing the air gap and reducing the magnetic field penetration depth into the HCPM and the rotor sleeve.

For the EM with a tooth-coil winding, the minimization of the eddy-current losses created by spatial harmonics can be achieved by the correct selection of the number of stator slots and the number of rotor poles.

4.5. Eddy-current losses in bearings of the high-speed EM

For limited axial dimensions of the EM rotor and insignificant bearing removal from the HCPM in the high-speed EM, the bearing balls can be magnetized under the influence of the HCPM stray magnetic field. In a result, eddy currents are induced. This leads to overheating of the bearing caused by its hysteresis losses. With increase in the rotational speed, the overheating increases significantly. Therefore, the study of these processes is important for the high-speed EMs.

The magnitude of the magnetic field flowing through the bearing is determined by the bearing removal from the HCPM. For a numerical evaluation, a computer simulation of the three-dimensional magnetic field of the EM with HCPM was performed. The results of this simulation are presented in **Figure 19**. It shows that the magnetic flux density is 0.32 T at a distance from HCPM. That is quite a sufficient value for the magnetization of the bearing balls and the appearance of eddy-current losses in the bearing.

The object of study is the EM with HCPM. The geometric dimensions of the rotor are shown in **Figure 20**. The shaft is made of steel with a saturation magnetic flux density of 1.7 T. In this case, the effective value of magnetic flux density in the rotor steel is close to saturation (1.6–1.65 T).

The following assumptions are used:

- the magnetic permeability of the environment and air gap is equal to the magnetic permeability of the vacuum; and the magnetic permeability of the magnetic core, shaft, and bearings are equal to infinity;
- the stator winding is represented as a thin electrically conductive layer distributed along the stator core diameter;
- the density of the induced currents along the winding and bearing is constant;
- the mutual influence of temperature and magnetic fields is not taken into account;
- the influence of winding ends on the ball bearing magnetization is not taken into account.

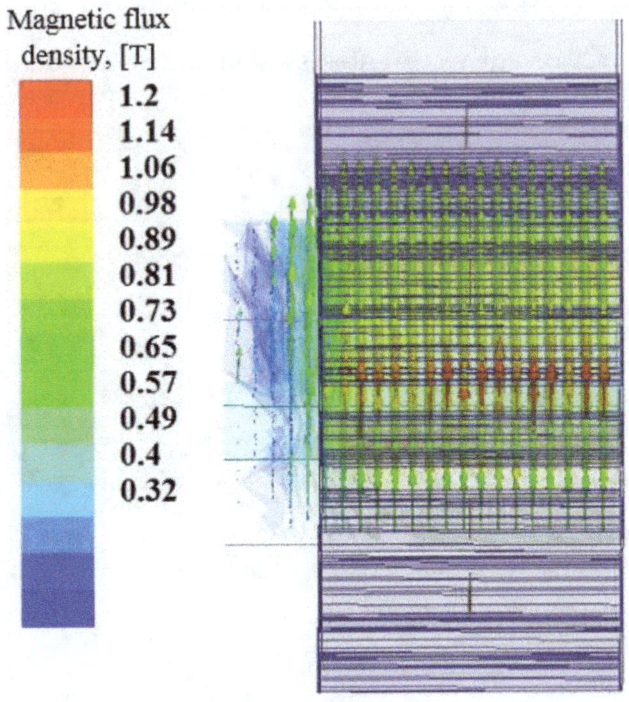

Figure 19. The three-dimensional magnetic field of the EM with HCPM.

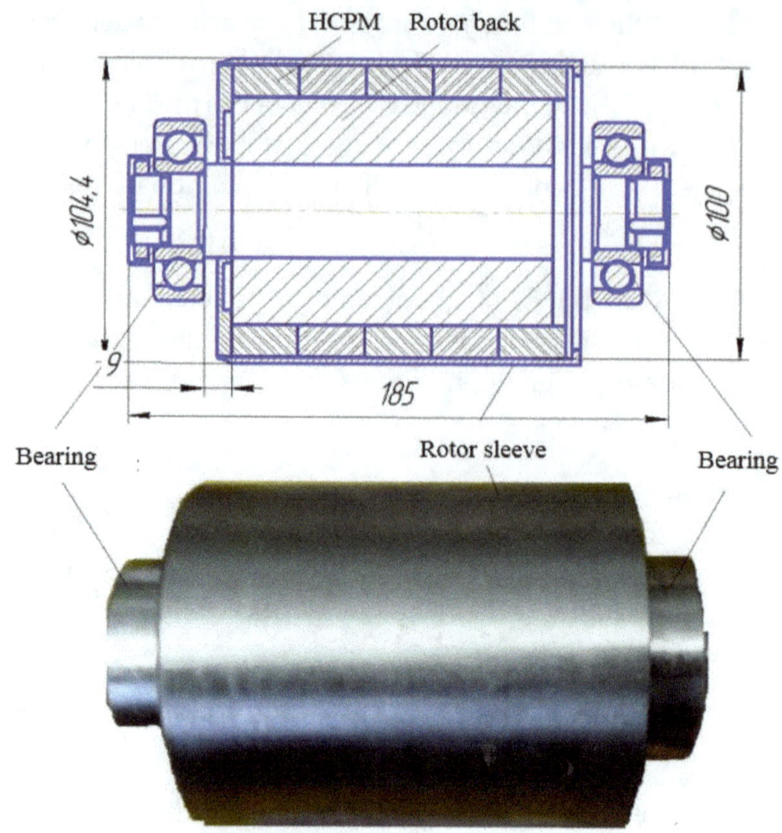

Figure 20. Calculation and experimental models of the research object.

To develop analytical equations for the loss analysis in the bearing, the EM calculation scheme presented in **Figure 21** is considered.

The mathematical analysis of eddy-current losses includes the determination of the three magnetic field components of the EM with the finite-length HCPM. Maxwell's equations are considered here:

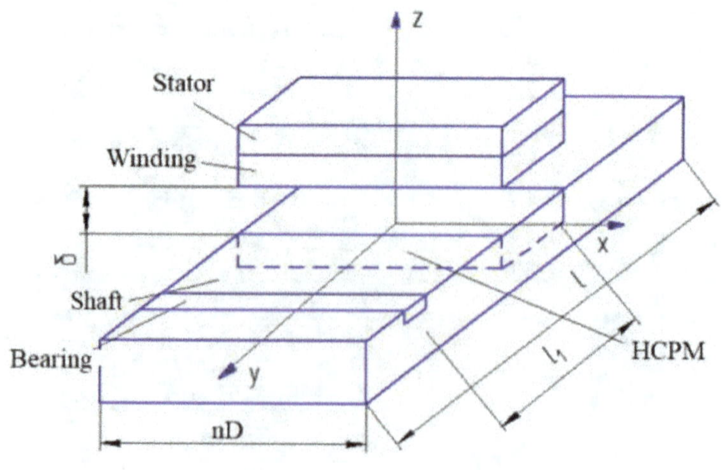

Figure 21. The EM calculation scheme.

$$\mathrm{rot}\vec{H} = \vec{j} + \vec{j}_{ex}, \ \mathrm{rot}\vec{E} = -\frac{\partial \vec{B}}{\partial t}, \ \vec{j} = \sigma\left[\vec{E} + \left(\vec{V} \times \vec{B}\right)\right], \quad \mathrm{div}\,\vec{B} = 0, \mathrm{div}\,\vec{j} = 0, \ \vec{H} = \mu_0\,\vec{B} \quad (10)$$

where $\vec{B}$ is a magnetic flux density vector, $\vec{E}$, $\vec{H}$ are vectors of electric and magnetic fields, respectively, $\vec{V}$ is a rotor velocity vector, σ is an electric conductivity of the stator winding, $\vec{j}$ is an induced current density vector, and $\vec{j}_{ex}$ is an external current density vector.

Since there are no currents in the air gap δ, the magnetic field in the air gap is described by the Laplace's equation in partial derivatives:

$$\Delta\overline{H} = 0, \ \frac{\partial^2 H_z}{\partial z^2} + \frac{\partial^2 H_z}{\partial x^2} + \frac{\partial^2 H_z}{\partial y^2} = 0, \ \frac{\partial^2 H_x}{\partial z^2} + \frac{\partial^2 H_x}{\partial x^2} + \frac{\partial^2 H_x}{\partial y^2} = 0, \ \frac{\partial^2 H_y}{\partial z^2} + \frac{\partial^2 H_y}{\partial x^2} + \frac{\partial^2 H_y}{\partial y^2} = 0. \quad (11)$$

The components H_x, H_y, H_z provide the continuity conditions of the magnetic field lines, while H_z provides the energy conversion.

The magnetic field on the HCPM surface is given as a normal tension component:

$$H_{z0} = H_{z0} \cos q_1 x \cos q_2 y, \quad (12)$$

where $q_1 = \frac{\pi}{\tau}, q_2 = \frac{\pi}{l}, q_3 = \sqrt{{q_1}^2 + {q_3}^2}$, τ is a pole pitch, l is a rotor length, and H_{z0} is a function of coercive force and residual magnetic flux density and is determined by their magnetization.

The mathematical description of HCPM is represented as a dependence of the coercive force, residual magnetic flux density, and magnetization:

$$\vec{B} = \mu\mu_0\left(\vec{M} + \vec{H}\right), \quad (13)$$

where $\vec{M}$ is the HCPM magnetization vector and μ is the HCPM magnetic permeability.

The solution of Laplace's equations is sought in the form:

$$\begin{aligned} H_z &= (C_1 \mathrm{sh} q_3 z + C_2 \mathrm{ch} q_3 z) \cos q_1 x \cos q_2 y, \\ H_x &= (C_3 sh q_3 z + C_4 ch q_3 z) \sin q_1 x \cos q_2 y, \\ H_y &= (C_3 sh q_3 z + C_4 ch q_3 z) \cos q_1 x \sin q_2 y. \end{aligned} \quad (14)$$

The initial conditions determine the integration constants:

1. $z = 0$
2. $z = \delta, yC_1 \mathrm{ch} q_3 \delta + yC_2 \mathrm{sh} q_3 \delta = 0, C_1 = -C_2 \mathrm{th} q_3 \delta;$
3. $\mathrm{div} H = \frac{\partial H_z}{\partial z} + \frac{\partial H_x}{\partial x} + \frac{\partial H_y}{\partial y} = 0$—magnetic field continuity conditions.

Then:

$$H_z = q_3(C_1\text{sh}q_3z + C_2\text{ch}q_3z)\cos q_1x\cos q_2y + q_1(C_3shq_3z + C_4chq_3z)\cos q_1 \\ x\cos q_2y + q_2(C_3\text{sh}q_3z + C_4\text{ch}q_3z)\cos q_1x\cos q_2y. \quad (15)$$

Equal terms can be obtained for the same functions sh and ch:

$$q_3C_1 + q_1C_4 + q_2C_4 = 0, q_3C_2 + q_1C_3 + q_2C_3 = 0. \quad (16)$$

Hence, the following is obtained:

$$C_4 = -C_1\frac{q_3}{q_1+q_2}, C_3 = -C_2\frac{q_3}{q_1+q_2} = -C_2\frac{\sqrt{q_1{}^2+q_2{}^2}}{q_1+q_2}.$$

Taking into account the relations for C_1, C_2, C_3, C_4, the components of the three field strength vectors have the form:

$$H_z = H_{z0m}(\text{ch}q_3z - \text{th}q_3\delta shq_3z)\cos q_1x\cos q_2y,$$

$$H_x = -H_{z0m}\frac{q_3}{q_1+q_2}\text{ch}q_3z(\text{th}q_3\delta - \text{th}q_3z)\sin q_1x\cos q_2y, \quad (17)$$

$$H_y = -H_{z0m}\frac{q_3}{q_1+q_2}\text{ch}q_3z(\text{th}q_3\delta - \text{th}q_3z)\cos q_1x\sin q_2y.$$

It is obvious that the component H_z is the primary magnetic field for the rotating bearing balls at $l = l_1$. This component magnetizes the bearing balls and induces eddy-currents in the bearing. In this case, the bearing is represented as an EM. It is assumed that the pole number of a given EM is equal to the number of bearing balls (**Figure 22**). The air gap in this case tends to zero. It should be noticed that the process of the ball magnetization has a complex character, taking into account possible interactions between neighboring balls, and the pole number of the EM cannot be equal to the number of bearing balls.

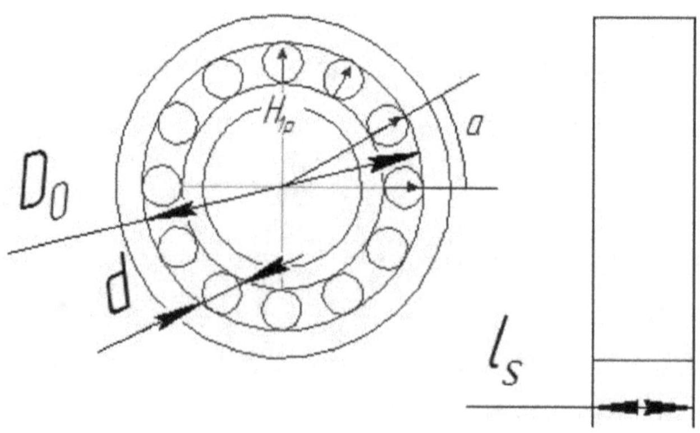

Figure 22. Calculation scheme of the bearing.

The resulting bearing magnetic field can be represented by the sum of two magnetic fields: bearing ball field and bearing cage field: $\vec{H}= \vec{H_1} +\vec{H_2}$ and $\vec{B}= \vec{B_1} +\vec{B_2}$.

The system of Maxwell's equations for the bearing is presented as:

$$\text{rot}\,\vec{H}= j + j_{ex};\ \text{rot}\,\vec{E}= -\frac{\partial B_1}{\partial t} - \frac{\partial B_2}{\partial t};\ \vec{j}=\sigma\left[\vec{E} +\left(\vec{V} \times \vec{B}\right)\right];\ \text{div}\,\vec{B}= 0,\ \text{div}\,\vec{j}= 0;\ \vec{H}= \mu_0\,\vec{B}\,. \tag{18}$$

The field on the bearing ball surface (primary field) is presented as following:

$$H_{1p} = H_{z0m}\left(\text{ch}q_3D_0 - \text{th}q_3\delta sh q_3D_0\right)\cos q_1x\cos q_2l_1. \tag{19}$$

Solving the system of Maxwell's equations concerning the intensity of the secondary field, there is the following:

$$\Delta\vec{H_2} -\mu_0\sigma\left(\text{d}\frac{\vec{H_2}}{\text{d}t} - \text{rot}\left(\vec{V} \times\vec{H_2}\right)\right) = -\mu_0\sigma\left(\text{d}\frac{\vec{H_1}}{\text{d}t} - \text{rot}\left(\vec{V} \times\vec{H_1}\right)\right). \tag{20}$$

The final solution of this system is not given here. The losses in the bearing cage are determined based on that $j_{2y} = -\frac{\partial H_{2z}}{\partial x}$.

Losses in the bearing cage are defined as $P = {j_{2y}}^2/\sigma_s$, where σ_S is the conductivity of the bearing cage material. Since the bearing is represented as an EM, the bearing cage field will create a demagnetizing reaction (armature reaction) that can demagnetize the bearing balls and induce eddy currents in the bearing balls. By analogy of the eddy-current losses in the EM with the HCPM, eddy-current losses can be defined as:

$$P_{sh} = l_s d^2 \sum_{n=1}^{n} \frac{d^2}{12\rho}\left(\frac{\text{d}}{\text{d}t}\left\{B_2\cos\left(p(a-\beta)\right)\right\}\right)^2. \tag{21}$$

where ρ is a specific resistance of the bearing cage material, d is a bearing ball diameter, and p is the number of pole pairs of the EM, which is equivalent to the bearing.

The developed mathematical equations provide a general understanding of the loss occurrence processes in the bearing cage. They are obtained with a variety of assumptions, and the main one is that the number of poles is equal to the number of bearing balls. Therefore, the developed mathematical apparatus can be used for preliminary loss calculations in bearings and the selection of their location on the shaft of the EM with the HCPM.

4.6. Winding losses in the high-speed EMs

In addition to energy losses in the stator windings and stator core, windage losses are important to determine the efficiency of high-speed EMs with slotted and slotless stator design.

Winding loss analysis is considered in different works. In [29], the model proposed takes into account the effect of environmental pressure on the windage losses. Another model proposed in [33] describes the winding losses that are analogous to the model proposed by M. Mack, which has experimental confirmation to use in the EM. In this model, winding losses are determined by the equation:

$$P_f = c_f \pi \rho_{air} \Omega^3 R_r^4 l, \tag{22}$$

is a friction coefficient between rotor end environment, ρ_{air} is an air density, R_r is a rotor radius, Ω is a rotational speed [rad/s], and l is a length of the site on which losses are determined.

This model does not take into account the windage loss dependence on the environment pressure, the air gap temperature, and the stator and rotor slots, for example in the induction EMs. Analysis of Eq. (7) shows that the winding losses depend to the rotational speed. Consequently, the negligence of the air gap temperature, pressure and structural features of the stator, and rotor can cause errors in calculations and can also lead to incorrect selection of the rotor design and to decrease the efficiency of the high-speed EM.

Thus, this model needs refinements. For this, two designs are considered: EM with a smooth rotor and a slotted stator and EM with a slotted rotor and a slotted stator (**Figure 23**).

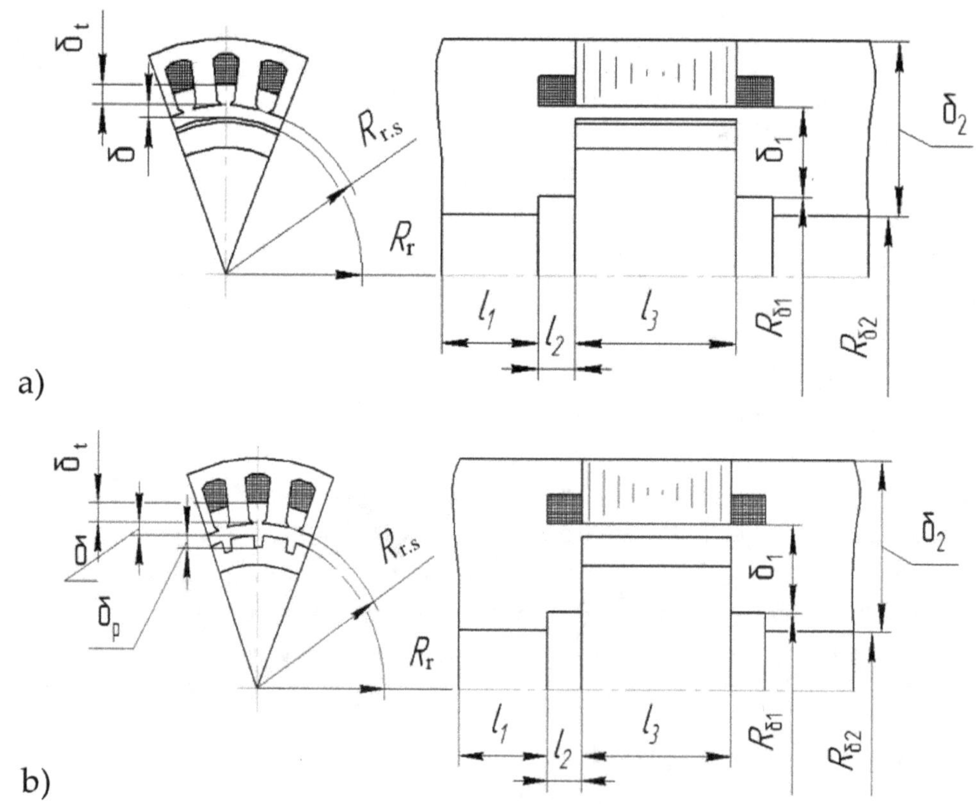

Figure 23. EM designs: (a) smooth rotor and slotted stator and (b) slotted rotor and slotted stator.

The friction coefficient is determined by the Reynolds and Taylor numbers:

$$\mathrm{Re} = \frac{R_r^2\Omega}{\varsigma},\ Ta = \frac{R_r\Omega\delta}{\varsigma}\sqrt{\frac{\delta}{R_r}},\ c_f = \frac{1.8}{\mathrm{Re}}\left(\frac{\delta}{R_2}\right)^{-0.25}\frac{R_{s.r}^2}{R_{s.r}^2 - R_r^2}, \tag{23}$$

where ς is a kinematic air viscosity, $R_{s.r}$ is an initial stator radius, δ is an air gap of the EM, and Ω is the rotational speed.

The model with a smooth rotor and a slotted stator is initially considered. The space between the stator and the rotor is represented in the form of two zones: a tooth and a slot, each of which is characterized by a different air gap.

The friction coefficient for slot zone is considered as following:

$$c_{f1} = \frac{1.8}{\mathrm{Re}}\left(\frac{\delta + \delta_t}{R_2}\right)^{-0.25}\frac{R_{s.r}^2}{[R_{s.r} + \delta_t]^2 - R_r^2}, \tag{24}$$

where δ_t is a distance from the tooth to the winding.

For the tooth zone, the friction coefficient is determined from Eq. (9).

Taking into account that the kinematic air viscosity depends on its temperature, the Reynolds number can be rewritten in a general form:

$$\mathrm{Re} = \frac{R_i^2\Omega}{\varsigma(T, p_a)}, \tag{25}$$

where $\varsigma(T, p_a)$ is a kinematic air viscosity at a certain temperature and pressure, R_i is a rotating part radius of rotor or shaft, and p_a is an air gap pressure.

Taking into account that the air density also depends on the temperature, the windage losses for the slot zone are determined in the following form:

$$P_{slot} = \frac{1.8}{\mathrm{Re}(R_2)}\left(\frac{\delta + \delta_t}{R_2}\right)^{-0.25}\left[\frac{[R_{s.r} + \delta_t]^2}{[R_{s.r} + \delta_t]^2 - R_r^2}\right]\rho_{air}(T, p_a)\Omega^3 R_p^3 l_3 \frac{zb_{\text{п}}}{2}, \tag{26}$$

and, accordingly, for the tooth zone:

$$P_{tooth} = \frac{1.8}{\mathrm{Re}(R_2)}\left(\frac{\delta}{R_2}\right)^{-0.25}\left[\frac{R_{s.r}^2}{R_{s.r}^2 - R_r^2}\right]\rho_{air}(T, p_a)\Omega^3 R_p^3 l_3 \frac{zb_z}{2}, \tag{27}$$

where z is a number of stator teeth, b_z is a tooth width, b_{slot} is a slot width, and $\rho_{air}(T)$ is an air density at a certain temperature.

Taking into account that the air temperature is the same over the entire surface of the air gap, the total windage losses in the design of **Figure 23a** are defined as:

$$\begin{aligned} P_{\sum \text{windage}} = & \frac{1.8}{\mathrm{Re}(R_2)}\left(\frac{\delta}{R_r}\right)^{-0.25}\left[\frac{R_{s.r}^2}{R_{s.r}^2 - R_r^2}\right]\rho_{air}(T, p_a)\Omega^3 R_r^3 l_3 \frac{z b_z}{2} \\ & + \frac{1.8}{\mathrm{Re}(R_2)}\left(\frac{\delta + \delta_t}{R_r}\right)^{-0.25}\left[\frac{[R_{s.r} + \delta_t]^2}{[R_{s.r} + \delta_t]^2 - R_r^2}\right]\rho_{air}(T, p_a)\Omega^3 R_r^3 l_3 \frac{z b_z}{2} \\ & + \frac{3.6}{\mathrm{Re}(R_{\delta 1})}\left(\frac{\delta_1}{R_{\delta 1}}\right)^{-0.25}\left[\frac{(R_{\delta 1} + \delta_1)^2}{2R_{\delta 1}\delta_1^2 + \delta_1^2}\right]\rho_{air}(T, p_a)\Omega^3 R_{\delta 1}^4 l_2 \\ & + \frac{3.6}{\mathrm{Re}(R_{\delta 2})}\left(\frac{\delta_2}{R_{\delta 2}}\right)^{-0.25}\left[\frac{(R_{\delta 2} + \delta_2)^2}{2R_{\delta 2}\delta_2^2 + \delta_2^2}\right]\rho_{air}(T, p_a)\Omega^3 R_{\delta 2}^4 l_1. \end{aligned} \tag{28}$$

As examples, the energy loss calculations of the EM rotors have been performed at various temperatures in the air gap. The calculations were carried out using two EMs produced by Turbec, which are used in microturbine installations T-100: the 100-kW 60,000-rpm EM with a rotor diameter of 60 mm and the 100-W 500,000 rpm micro-EM with a rotor diameter of 5 mm.

Thus, calculations show that the windage losses largely depend on temperature. With a temperature increase in the air gap from 20 to 60°C, the windage losses increase by 6–7%. The known models of windage losses have a significant error because they do not take into account the stator slots, as well as the temperature influence and pressure in the air gap. Minimal windage losses occur when the inner stator surface is smooth, for example, when slots are filled with a compound.

Minimization of winding losses caused by the stator slots can be achieved by pouring the stator teeth with a compound, or by performing closed stator slots, or by inserting a screen over the entire surface of the air gap. At the same time, the screen will be a technological complication of the EM design. And if it is made of an electrically conductive material, eddy currents will be induced in it and will significantly reduce its efficiency.

For the slotted rotor, the winding loss calculation is similar. However, the slotted rotor surface will cause additional winding losses. Therefore, the rotor and stator surfaces should be smooth; that is, the slots should be filled with a compound or other nonelectrically conductive substances with minimal mass parameters.

Minimal winding losses can be achieved by the operation of high-speed EM with HCPM in vacuum, but it should be taken into account that only certain bearing designs can work stably in a vacuum.

4.7. Efficiency of the high-speed EMs

The efficiency of high-speed EM is determined by the well-known formula:

$$\eta = 100\left(1 - \frac{\sum P_{loss}}{P + \sum P_{loss}}\right), \tag{29}$$

where η is the EM efficiency, $\sum P_{loss}$ is the total losses, and P is the EM power.

The efficiency of modern EMs reaches high values from 90 to 95%. Even the well-known ultra-high-speed EMs with a rotational speed of up to 800,000 rpm and a power of 50–500 W have an efficiency of 80–90%. The main losses that significantly reduce the efficiency of high-speed EM are the stator magnetic core losses, which increase significantly with increasing magnetization reversal frequency, windage losses, and bearing losses. To reduce these losses and increase the efficiency of ultra-high-speed EMs, various technical solutions are used. For example, AMMs are used to reduce losses in the stator magnetic core. To reduce the friction losses in the rotor, various antifriction coatings or vacuuming of the internal EM cavity are used. To minimize bearing losses, contactless bearings are used.

4.8. Influence of the EM thermal state on the efficiency

As it was shown in the previous sections, the EM temperature has a significant effect on the losses in various EM active elements. Heating of electrical steel leads to a decrease in its specific losses; increasing of the rotor temperature leads to an increase in the friction losses; and heating the winding leads to an increase in resistance. Thus, the EM efficiency is significantly dependent on temperature. Therefore, it seems reasonable to evaluate the dependence of the EM efficiency on temperature. An analysis of the dependence of the EM efficiency on temperature was made by using the example of the high-speed micro-EM produced by Onera [12]. The parameters of this micro-EM are presented in **Table 5**.

It is important to notice that the main consumers of micro-EMs are systems with constant power consumption such as computer systems, navigation systems of unmanned aircrafts. This type of load will lead to the fact that with the decrease of the HCPM energy characteristics, the magnetic flux density in the air gap decreases under the temperature influence. In the calculation result, it was found that for the indicated numerical parameters, the magnetic flux density in the air gap will decrease by 6% at a temperature of 150°C. Under conditions of constant input mechanical power and constant electric power consumption, this will lead to an

Parameters	Value
Power [W]	55
Rotational speed [rpm]	840,000
Outer stator diameter [mm]	25
Active length [mm]	22
Shaft diameter [mm]	4.5
Efficiency [%]	91
Bearing losses [W]	9.7–9.8
Stator magnetic core losses [W]	0.18
Friction losses [W]	2.63
Winding losses [W]	2.33
Total losses [W]	14.84

Table 5. Parameters of the high-speed micro-EM produced by Onera.

Losses	Temperature of the active parts is below 23°C	Temperature of the active parts is 100–110°C after 10 min of operation without a cooling system
Stator magnetic core losses [W]	0.18	0.13
Friction losses [W]	2.63	3.02
Winding losses [W]	2.33	3.88
Efficiency without the bearing losses [%]	91.4	88.6

Table 6. The efficiency comparison of the micro-EM in the cold state and after 10 min of operation without a cooling system.

increase in the linear current load of EM and the current value by also 6%. Taking into account that the copper resistance at a temperature of 100–110°C will also increase by 38%, the winding losses will increase by 55%.

To evaluate the change in the linear current load and the magnetic flux density, an idealized linear dependence of the current and magnetic flux density in the air gap of the micro-EM was adopted. In real EMs, an increase in the linear current load will lead to an increase in the demagnetizing armature reaction action, to additional winding heating, and, consequently, to an increase in the resistance caused by this heating. Preliminary calculations show that in the cold state and 10 min after the start of operation, the winding losses of the real EM can differ by 60–65%. The efficiency comparison of the micro-EM in the cold and heated states (after 10 min of operation without a cooling system) is presented in **Table 6**. With the increase in the EM heat emissions, losses in the active elements increase and, accordingly, the EM efficiency decreases. For the studied numerical values, the efficiency of the micro-EM decreases by 2.8%. This suggests that in the EM design, the efficiency should be calculated at the operating temperature.

5. Conclusion

In this chapter, basics of high-speed EMs with HCPM are presented. The application areas of high-speed EMs are shown. The EM classification is proposed to generalize the theory of high-speed EMs. Practical recommendations for the selection of active and structural components of high-speed EMs are given. The loss determination in high-speed EMs is presented, including methods for determining losses in the stator magnetic cores at high frequencies.

The obtained results can be used in practice in the EM design and in further research.

Author details

Flyur R. Ismagilov, Viacheslav Ye. Vavilov* and Valentina V. Ayguzina

*Address all correspondence to: s2_88@mail.ru

Ufa State Aviation Technical University, Ufa, Russia

References

[1] Lahne HC, Gerling D. Investigation of high-performance materials in design of a 50000 rpm highspeed induction generator for use in aircraft Applications. In: AST 2015. pp. 1-10

[2] Lahne HC, Gerling D. Comparison of high-speed highpower machines based on the state of the art. In: 41st Annual Conference of the IEEE Industrial Electronics Society (IECON 2015). 2015. pp. 3519-3524

[3] Muljadi E, Butterfield CP. Axial flux, modular, permanent-magnet generator with a toroidal winding for wind turbine applications. Industry Applications Conference. 1998;**1**:174-178

[4] Caricchi F, Crescimbini F, Honorati O, Bianco GL. Performance of coreless-winding axial-flux permanent-magnet generator with power output at 400 Hz, 3000 r/min. IEEE Transactions on Industry Applications. 1998;**34**(6):1263-1269

[5] Gumerova M, Ismagilov F, Khayrullin I, Vavilov V. Electrodynamic brakes for unmanned aerial vehicles. International Review of Aerospace Engineering (I.R.E.AS.E). 2014;**7**(6):202-206

[6] Szabó L, Oprea C, Viorel I-A, Biró KA. A novel permanent magnet tubular linear generator for wave energy converters. In: Electric Machines & Drives Conference, IEEE International, IEMDC '07; Antalya, Turkey. 2007. pp. 516-521

[7] Jung S-Y, Choi S-Y, Jung H-K, Choi Y-S, Choi K-M. Performance evaluation of permanent magnet linear generator for charging the battery of mobile apparatus. In: Electric Machines and Drives Conference, IEEE International, IEMDC 2001; Cambridge, MA, USA. 2001. pp. 516-521

[8] Moore MJ. Micro-turbine Generators. London: Professional Engineering Publishing; 2002. 113 p

[9] Next-generation microturbines [Internet]. Available from: https://www.capstoneturbine.com [Accessed: 2018-03-28]

[10] Hong DK, Woo BC, Lee JY. Ultra high speed motor supported by air foil bearings for air blower cooling fuel cells. IEEE Transactions on Magnetics. 2012;**48**:871-874

[11] Takahashi I, Koganezawa T, Su G. A super high speed PM motor drive system by a quasi-current source inverter. IEEE Transactions on Industrial Electronics. 1994;**30**(3):683-690

[12] Guidez J, Ribaud Y, Dessornes O, Courvoisier T. Micro gas turbine research at Onera. In: International Symposium on Measurement and Control in Robotics; Brussels, Belgium. 2005. pp. 1-7

[13] Borisavljevic A, Polinder H, Ferreira J. On the speed limits of permanent-magnet machines. IEEE Transactions on Industrial Electronics. 2010;**57**(1):220-227

[14] Rahman MA, Chiba A, Fukao T. Super high–speed electrical machines—Summary. IEEE Power Engineering Society General Meeting. 2004;**2**:1272-1275

[15] Zwyssig C, Kolar JW, Round SD. Mega-speed drive systems: Pushing beyond 1 million RPM. IEEE/ASME Transactions on Mechatronics. 2009;**14**(5):564-574

[16] Zwyssig C, Kolar JW, Round SD. Power electronics interface for a 100 W, 500000 rpm gas turbine portable power unit. In: Applied Power Electronics Conference. 2006. pp. 283-289

[17] Krähenbühl D, Zwyssig C, Weser H, Kolar JW. Mesoscale Electric Power Generation from Pressurized Gas Flow [Internet]. Available from: https://www.pes.ee.ethz.ch/uploads/tx_ethpublications/kraehenbuehl_PowerMEMS07.pdf [Accessed: 2018-03-28]

[18] Isomura K, Murayama M, Teramoto S, Hikichi K, Endo Y. Experimental verification of the feasibility of a 100W class micro-scale gas turbine at an impeller diameter of 10 mm. Journal of Micromechanics and Microengineering. 2006;**16**:254-261

[19] Park CH, Choi SK, Ham SY. Design and experiment of 400,000 rpm high speed rotor and bearings for 500W class micro gas turbine generator. In: International Conference on Micro and Nanotechnology for Power Generation and Energy Conversion Applications (PowerMEMS). 2011

[20] Kang S. Fabrication of Functional Mesoscopic Ceramic Parts for Micro Gas Turbine Engines. Sangkyun Kang: Stanford University; 2001. 137 p

[21] Tüysüz A, Zwyssig C. A novel motor topology for high-speed micro-machining applications. IEEE Transactions on Industrial Electronics. 2014;**61**(6):2960-2968

[22] Borisavljevic A. Limits, Modeling and Design of High-Speed Permanent Magnet Machines. Wormann Print Service; 2011. p. 209

[23] Westwind [Internet]. Available from: http://www.westwind-airbearings.com/pcb/index.html [Accessed: 2018-03-28]

[24] Sinotech [Internet]. Available from: http://www.sinotech.com/ [Accessed: 2018-03-28]

[25] Sirona [Internet]. Available from: http://www.sirona.com [Accessed: 2018-03-28]

[26] Casey MV, Krähenbuhl D, Zwyssig C. The design of ultra-high-speed miniature centrifugal compressors. In: European Conference on Turbomachinery Fluid Dynamics and Thermodynamics ETC 10. 2013. pp. 1-13

[27] Kopylov IP. Electrical Machines (in Russian). Moscow: Energoatomizdat; 1986. 360 p

[28] Goldberg OD. Design of Electrical Machines (in Russian). Moscow: Vysshaya Shkola; 1984. 431 p

[29] Ledovsky AN. Electrical Machines with High-coercive Permanent Magnets (in Russian). Moscow: Energoatomizdat; 1985. 169 p

[30] Balagurov VA. Electric Generators with Permanent Magnets (in Russian). Moscow: Energoatomizdat; 1988. 279 p

[31] Bertinov AI. Special Electrical Machines. Sources and Converters of Energy (in Russian). Moscow: Energoatomizdat; 1982. 552 p

[32] Co H, Zheng L, Acharya D. Losses in high speed permanent magnet machines used in microturbine applications. Journal of Engineering for Gas Turbines and Power. 2009; **131**(2):1-6

[33] Sharov VS. High-frequency and Superhigh-frequency Electrical Machines (in Russian). Moscow: Energy; 1973. 248 p

10

Electric Machines for Smart Grids and Electric Vehicles Applications

Adel El-Shahat

1. Introduction

The rule of electromechanical vitality change was first shown by Michael Faraday in 1821. The first electric machine, really a DC one, was imagined by William Sturgeon in 1832. Nikola Tesla considered the turning attractive field in 1882 and utilized it to develop the first AC machine, really an enlistment engine, in 1883. Now the AC machines, including synchronous generators and enlistment generators, have been broadly acknowledged to function admirably for nonrenewable energy source control age. As of now, all electric machines for sustainable power source collecting are reached out from nonrenewable energy source control plants. Conventional generators are moderately ineffectual to change over sustainable power sources into power, especially inefficient for tackling wind power or wave one [1].

In September 1821, Faraday constructed a first electric motor utilizing a compartment of mercury with a permanent magnet in its inside. At that point, Faraday spent the following decade, on and off, attempting to comprehend the material science behind electromagnetism. At that time, he depicted various fizzled analyses in which he attempted to show what is called "electromagnetic induction." From that point onward, Faraday's superseding inspiration was to "change over magnetism into electricity," and it was "Arago's disk" that was the platform for his delightful analyses. His analysis demonstrated that a current could be actuated all the time when there was relative movement between the conductor and the attractive field. We presently know this impact as "electromagnetic induction." He next found an undeniably compelling method for changing the magnetic field: by moving two wires, one associated with a battery and the other to a galvanometer. Faraday's induction ring was, essentially, the simple first electrical transformer. It gets by right up till the present time and is in plain view in the Royal Institution's historical center as shown in **Figure 1** [2].

Figure 1. Faraday's induction ring (1831) [2].

The American researcher, Joseph Henry (1797–1878), from numerous points of view, reflected that of Michael Faraday, was additionally working autonomously on electromagnetism on the opposite side of the Atlantic—in spite of the fact that enthusiasm for the subject was absolutely circling over the Atlantic by the 1830s. Henry in certainty beat Faraday to the disclosure of inductance by a couple of months in 1831, yet it was Faraday who distributed first and, in spite of the defers that so baffled him, is in this way credited with the revelation. The French instrument producer, Hippolyte Pixii (1808–1835), constructed an unrefined electric generator as ahead of schedule as 1832, constructed specifically in light of Faraday's thoughts of enlistment. A recommendation by Ampère and others prompted the presentation of the commutator—a revolving switch that inverts the association with the outer circuit when the present turns around, giving a beating DC current rather than an AC one. A business, substantial-scale use of Faraday's disclosure was made by the electroplaters of Birmingham as right on time as 1844. The pace of innovation at that point got in the 1850s as plans for evermore groundbreaking generators known as "magnetoelectric machines" were produced fully expecting the business uses of the electric machine [2].

By the mid-1860s, some researchers and innovators were creating handy outlines for the dynamoelectric machine. These gadgets were utilizing self-controlling electromagnetic field curls rather than lasting magnets to empower far more noteworthy power age out of the blue. By the 1880s, the alleged "war of the streams" was going all out between those, for example, Thomas Edison, who favored DC current for control age and those, drove by George Westinghouse and Nikola Tesla, who trusted that AC current was the path forward. The last two would, in the end, unequivocally win that intense war. The advancement of AC control transmission, utilizing transformers (whose causes lie in Faraday's straightforward enlistment ring) to transmit power at high voltage and with low misfortune, permitted focal influence stations to wind up financially down to earth. Today, the alternator overwhelms huge-scale power production and depends on a liquid, normally steam, that goes about as a transitional vitality transporter, to drive the turbines and produce power.

2. Permanent magnet (PM) machine overview

PM synchronous machines have been a point of enthusiasm for the last quarter century. Previously, AC synchronous machines were utilized for the most part of generator applications. Their utilization as a motor was restricted because of the trouble of controlling the frequency of their supply voltages. The presentation of intensity gadgets PWM inverters has enabled the motor drive to have finish control over the extent and recurrence of machine stage to stage voltages. Another factor that helped the improvement of PM synchronous machines is the development of mechanical creation of permanent magnets. The permanent magnet composed to be delivered on a modern scale was the Alnico in the mid-twentieth century. The rotor of line-begin PMSM is made of lasting magnet installed inside a squirrel-confine winding. Enlistment of current in the squirrel confine produces torque at zero or higher paces; a similar way torque is created in acceptance engines. PMSM with line-starting can create torque at zero speed and keep running as an enlistment engine, until the point that the synchronous speed is reached [3, 4]. Once the synchronous speed is achieved the rotor is synchronized with the power source. After this, the motor keeps running as a synchronous motor.

However, high-speed PM machines permit a lessened framework weight, higher working proficiency, diminished upkeep costs, and a littler envelope than a customary arrangement in a similar power rating. Notwithstanding with the higher-power thickness and frequency likewise comes higher-power misfortune thickness. Exceptional consideration should then be paid to the decision of overlay material, loop development, and cooling component for what might somehow be a normal stator and lodging plan. On account of a rapid PM motor, temperature affectability of the magnet material is an extra factor. Hence, samarium cobalt is frequently the decision to acknowledge higher-temperature plans [3, 4]. Permanent magnetic machines can be generally divided into subgroups as shown in **Figure 2**. In addition, **Figure 3** illustrates schematic cross section of a four-pole motor with surface-mounted permanent magnets and an interior PM motor.

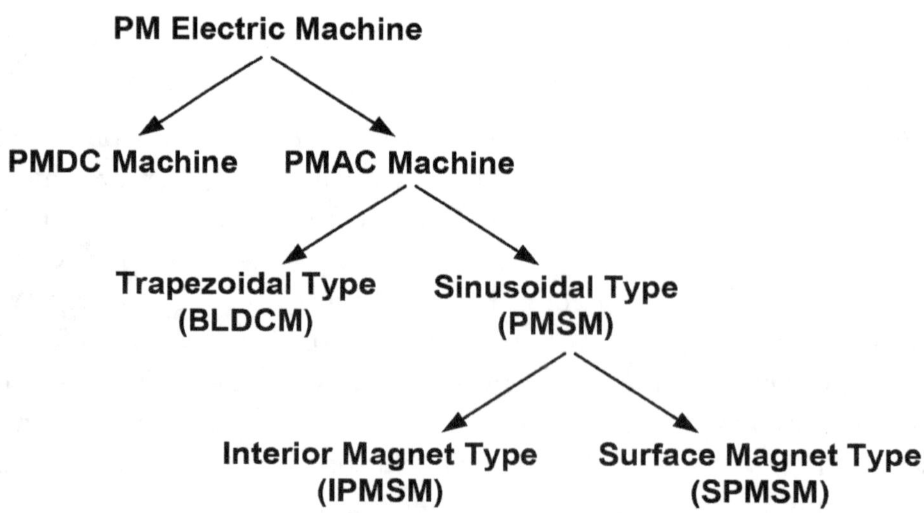

Figure 2. Permanent magnetic machine subgroups.

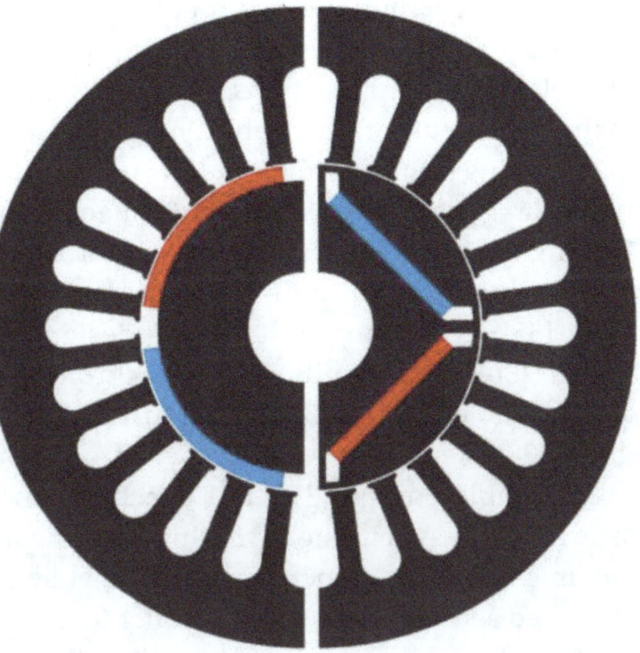

Figure 3. Schematic cross section of a motor with surface-mounted permanent magnets (PM; left) and of an interior PM motor (right) [3].

3. Editor's contributions to this field

This book has recent advances of modeling, control, optimization, analysis, and design of electric machines especially for smart grid applications and electric vehicles. However, the editor himself has some contributions for different types of electric machines in many applications as the following: electric machines (EM) in nano-power grids [5]; EM in smart home systems [6]; performance improvement for induction motor [7]; PM synchronous machine new aspects for modeling, control, and design [8]; sizing high-speed micro-generators for smart grid systems [9]; high-speed PM generator for optimized sizing based on particle swarm for smart grids [10]; EM in small-scale hydropower generator electrical system modeling [11]; for ANN robot energy modeling [12]; small-scale wind power dispatchable energy source modeling [13]; PMSM sensorless speed control drive [14]; ANN interior PM synchronous machine performance improvement unit [15]; PMSM performance improvement [16]; sizing a high-speed PM generator for green energy applications [17]; 400 kW six analytical high-speed generator designs for smart grid systems [18]; nonlinear global sizing of high-speed PM synchronous generator for renewable energy applications [19]; micro-generator design for smart grid system (comparative study) [20]; PM synchronous motor drive system for automotive applications [21]; high-speed PM synchronous machine [22]; PM synchronous motor dynamic modeling with genetic algorithm performance improvement [23]; spacecraft flywheel high-speed PM synchronous motor design (classical and genetic) [24]; high fundamental frequency PM synchronous motor design neural regression function [25]; PM synchronous motor control strategies with their neural network regression functions [26]; high-speed PM generator optimized sizing based on particle swarm optimization for smart grids [27]; energy modeling of differential drive robots [28]; high-speed micro-turbine modeling and control for micro-grid applications [29]; optimized sizing of high-speed PM generator for renewable energy applications [30]; PM synchronous motor genetic algorithm performance improvement for renewable energy applications [31]; high-speed synchronous motor basic sizing neural function for renewable energy applications [32]; PM synchronous motor genetic algorithm performance improvement for green energy applications [33]; high-speed PM synchronous motor basic sizing neural regression function for renewable energy applications [34]; permanent magnet synchronous motor dynamic modeling [35]; generating basic sizing design regression neural function for HSPMSM in aircraft [36]; neural unit for PM synchronous machine performance improvement used for renewable energy [37]; PM synchronous machine stabilization control for electric vehicle [38]; neural unit for PM synchronous machine performance improvement used for renewable energy [39]; PM synchronous machine stabilization control for electric vehicle [40]; PM synchronous machine stabilization control for aircraft and spacecraft [41]; dynamic modeling of permanent magnet synchronous motor using MATLAB-Simulink [42]; speed sensorless neural controller for induction motor efficiency optimization [43]; neural model of three-phase induction motor [44]; and high-speed switched reluctance motor generator for automotive turbocharger applications [45].

Author details

Adel El-Shahat

Address all correspondence to: aahmed@georgiasouthern.edu

Electrical and Computer Engineering Department, Georgia Southern University, Georgia, USA

References

[1] Chau K, Li W, Lee C. Challenges and opportunities of electric machines for renewable energy (Invited Paper). Progress in Electromagnetics Research, B. 2012;**42**:45-74. DOI: 10.2528/PIERB12052001

[2] Al-Khalili J. The birth of the electric machines: A commentary on Faraday (1832) 'experimental researches in electricity'. Philosophical Transactions of the Royal Society A. 2015;**373**:20140208

[3] Krishnan R. Electric Motor Drives: Modeling, Analysis & Control. Prentice Hall. USA: Pearson Publisher. 2006

[4] Aydin M. Axial flux mounted permanent magnet disk motors for smooth torque traction drive application [PhD]. Electrical and Computer Engineering, University of Wisconsin; 2004

[5] El-Shahat A. Nano-Grids (Future of Power Systems Grids Notes). Germany: Scholar Press Publishing. 2017. ISBN: 978-3-330-65035-0

[6] El-Shahat A. Smart Homes Systems Technology. Germany: Scholar Press Publishing. 2015. ISBN 978-3-639-51681-4

[7] El-Shahat A. Induction Motor Performance Improvement. Germany: LAP Lambert Academic Publishing. 2013. ISBN 978-3-659-35255-3

[8] El-Shahat A. PM Synchronous Machine New Aspects; Modeling, Control, and Design. Germany : LAP Lambert Academic Publishing. 2012. ISBN 978-3-659-25359-1

[9] El-Shahat A, Keyhani A. Chapter 6: Sizing high speed micro generators for smart grid systems. Smart power Grids 2011 Book, Series: Power Systems. Germany: Springer-Verlag Berlin Heidelberg;2012. pp. 177-234. ISBN 978-3-642-21577-3

[10] El-Shahat A, Haddad R, Kalaani Y. High-speed PM generator for optimized sizing based on particle swarm for smart grids. International Journal of Modern Engineering. 2016; **17**(1):5-18

[11] Saggus C, Coley R, Floyd Z, El-Shahat A. Small scale hydropower generator electrical system modelling based on real-measurements. European Journal of Advances in Engineering and Technology. 2016;**3**(8):1-12

[12] Rios-Gutierrez F, El-Shahat A, Wahab M. ANN robot energy modeling. IOSR Journal of Electrical and Electronics Engineering. 2016;**11**(4 Ver. III):66-81

[13] Cannon J, Moore D, Eason S, El-Shahat A. Small scale-wind power dispatchable energy source modeling. International Journal of Scientific Engineering and Applied Science. 2016;**2**(1):34-39. ISSN: 2395-3470

[14] Kalaani Y, Haddad R, El-Shahat A. PMSM sensorless speed control drive. Journal of Engineering Research and Technology. 2014;**1**(4):132-143

[15] El-Shahat A, Abdul-Kader F, El-Shewy H. ANN interior PM synchronous machine performance improvement unit. Journal of Automation & Systems Engineering. 2013;**7** (4, P3):164-175. ISSN 1112-8542

[16] El-Shahat A, Abdul-Kader F, El-Shewy H. PMSM performance improvement–Review paper. Journal of Automation & Systems Engineering. 2013;**7**(2, P3):78-93. ISSN 1112-8542

[17] El-Shahat A, Keyhani A, El Shewy H. Sizing a high speed PM generator for green energy applications. Journal of Electrical Systems. 2010;**6**(4):501-506. ISSN1112-5209

[18] El-Shahat A, Keyhani A, El-Shewy H. 400 kW six analytical high speed generator designs for smart grid systems. International Journal of Energy and Power Engineering. 2010;**3**(4): 210-227

[19] El-Shahat A, Keyhani A, El-Shewy H. Non-linear global sizing of high speed PM synchronous generator for renewable energy applications. International Journal of Research and Reviews in Applied Sciences. August 2010;**4**(3):83-101. ISSN: 2076-734X, EISSN: 2076-7366, Special Issue on Advances in Nonlinear Sciences

[20] El Shahat A, Keyhani A, El Shewy H. Micro-generator design for smart grid system (Comparative Study). International Journal on Smart Sensing and Intelligent Systems. 2010;**3**(2):176-216. ISSN 1178-5608

[21] El-Shahat A, El-Shewy H. PM Synchronous motor drive system for automotive applications. Journal of Electrical Systems(JES). ISSN 1112-5209. June 2010;**6**(2):1-16

[22] El-Shahat A, El-Shewy H. High speed PM synchronous machine–Review paper. Journal of Arab Research Institute for Science & Engineering. 2010(April-June). pp. 7-21. ISSN 1994-3253; Second Quarter

[23] El-Shahat A, El-Shewy H. PM synchronous motor dynamic modeling with genetic algorithm performance improvement. International Journal of Engineering, Science and Technology. 2010;**2**(2):93-106. ISSN 2141-2839 (Online); ISSN 2141-2820 (Print)

[24] El-Shahat A, Keyhani A, El-Shewy H. Spacecraft flywheel high speed PM synchronous motor design (classical & genetic). Journal of Theoretical and Applied Information Technology. 2010;**13**(1):83-100. E-ISSN 1817-3195; ISSN 1992-8645

[25] El-Shahat A, El-Shewy H. High fundamental frequency PM synchronous motor design neural regression function. Journal of Electrical Engineering. March 2010;**10**:1-10. ISSN 1582-4594; Article 10.1.14, Edition 1

[26] El Shahat A, El Shewy H. PM synchronous motor control strategies with their neural network regression functions. Journal of Electrical Systems (JES). 2009;**5**(4):1-16. ISSN 1112-5209

[27] El-Shahat A, Haddad R, Kalaani Y. High speed PM-generator optimized sizing-based on particle swarm optimization for smart grids. 2016 IAJC/ISAM 5th, International Conference in Orlando, FL; 6-8 November 2016, Orlando, Florida

[28] Wahab M, Rios-Gutierrez F, El-Shahat A. Energy modeling of differential drive robots. IEEE SoutheastCon 2015 Conference; 9-12 April 2015; in Fort Lauderdale, Florida

[29] El-Shahat A. High speed micro-turbine modeling & control for micro grids applications. 18th Annual EUEC 2015–USA's Largest Energy, Utility & Environment Conference San Diego Convention Center; 16-18 February 2015; San Diego, CA

[30] El-Shahat A, Keyhani A, El-Shewy H. Optimized sizing of high speed PM generator for renewable energy applications. Fourteenth International Middle East Power Systems Conference; 19-21 December 2010; (MEPCON'10); Cairo, Egypt

[31] El-Shahat A, El-Shewy H. PM synchronous motor genetic algorithm performance improvement for renewable energy applications. MDGEN11, The International Conference on Millennium Development Goals (MDG): Role of ICT and Other Techno logies; 27-29 December 2009; in Chennai, India

[32] El-Shahat A, El-Shewy H. High speed synchronous motor basic sizing neural function for renewable energy applications. MDGEN05, The International Conference on Millennium Development Goals (MDG): Role of ICT and Other Technologies; 27-29 December 2009; in Chennai, India

[33] El-Shahat A, El-Shewy H. PM synchronous motor genetic algorithm performance improve ment for green energy applications. Paper ID: X792, 2nd International Conference on Computer and Electrical Engineering (ICCEE 2009); 28-30 December 2009; Dubai, UAE

[34] El-Shahat A, El-Shewy H. High speed PM synchronous motor basic sizing neural regression function for renewable energy applications. Paper ID: X304, 2nd International Conference on Computer and Electrical Engineering (ICCEE 2009); 28-30 December 2009; Dubai, UAE

[35] El-Shahat A, El-Shewy H. Permanent magnet synchronous motor dynamic modeling. Paper ID: X305, 2nd International Conference on Computer and Electrical Engineering (ICCEE 2009); 28-30 December 2009; Dubai, UAE

[36] El-Shahat A. Generating basic sizing design regression neural function for HSPMSM in aircraft. EP-127, 13th International Conference on Aerospace Science & Aviation Technology; 26-28 May 2009; ASAT 2009 – Military Technical College; Cairo, Egypt

[37] El-Shahat A, El-Shewy H. Neural unit for PM synchronous machine performance improvement used for renewable energy. Ref: 93, The Third Ain Shams University International Conference on Environmental Engineering (Ascee-3); 14-16 April 2009; Cairo, Egypt

[38] El-Shewy H, El-Shahat A. PM synchronous machine stabilization control for electric vehicle. Ref: 118, The Third Ain Shams University International Conference on Environmental Engineering (Ascee-3); 14-16 April 2009; Cairo, Egypt

[39] El-Shahat A, El-Shewy H. Neural unit for PM synchronous machine performance improvement used for renewable energy. Paper Ref: 910, Global Conference on Renewable and Energy Efficiency for Desert Regions (GCREEDER2009); Amman, Jordan

[40] El-Shewy H, El-Shahat A. PM synchronous machine stabilization control for electric vehicle. Paper Ref: 911, Global Conference on Renewable and Energy Efficiency for Desert Regions (GCREEDER2009); Amman, Jordan

[41] El-Shewy H, El-Shahat A. PM synchronous machine stabilization control for aircraft and spacecraft. CT-128, 13th International Conference on Aerospace Science & Aviation Technology; 26-28 May 2009; ASAT 2009 – Military Technical College; Cairo, Egypt

[42] El Shewy HM, Abd Al Kader FE, El Kholy M, El-Shahat A. Dynamic modeling of permanent magnet synchronous motor using MATLAB–Simulink. EE108, 6th International Conference on Electrical Engineering ICEENG 6; 27-29 May 2008, Military Technical College; Egypt

[43] Tawfik A, El Shahat A. Speed sensorless neural controller for induction motor efficiency optimization. 1st International Conference on Advanced Control Circuits and Systems (ACCS'05), ACCS Catalog No: 080610 M 05; 6-10 March 2005; Cairo, Egypt. ISBN: 0-146-6310-7933-2

[44] El-Shewy H, Tawfik A, El-Shahat A. Neural model of 3 phase induction motor. 1st International Conference on Advanced Control Circuits and Systems (ACCS'05), ACCS Catalog No: 080610 M 05; 6-10 March 2005; Cairo, Egypt. ISBN: 0-145-6310-7933-2

11

The Design of Motor Drive for Brushless DC Motor

Tze-Yee Ho

Abstract

The electric machines have been applied to the industrial and commercial areas for several decades. The high performance of control and fast dynamic response of electric machines can be achieved via motor drive because of high development of semiconductors such as the microcontroller and power transistors. Brushless DC motor (BLDC) and permanent magnet AC synchronous motors drives have many advantages such as high efficiency, small size, less rotor losses, and fast torque response over the induction motor drives. Therefore, this chapter will focus on the design and implementation of a motor drive as well as its realization of control algorithm. At the beginning of chapter, the fundamental of energy conversion for electric machines is introduced. The dynamic model for brushless dc motor is derived and explained in detail in section two. The 120° conduction and 180° conduction of switching device for an inverter are also described and illustrated. The control design of a speed-controlled motor drive is described in section three. The hardware and software design of a motor drive based on a microcontroller is given in section four. The experimental results are also presented and discussed. Finally, a conclusion is made in section five.

Keywords: brushless DC motor, motor drive, inverter, 120° conduction

1. Introduction

The electrical machine is the bridge of energy conversion between the electrical system and mechanical system, as shown in **Figure 1**. The electrical machine can be an electric motor. For an electric motor, the input is electrical energy represented by voltage and current, and the output is mechanical energy delivered to the mechanical load, which is represented by speed and torque. On the contrary, the mechanical energy can be converted to electric energy and transferred to electrical system by electrical machine, which is called as electric generator.

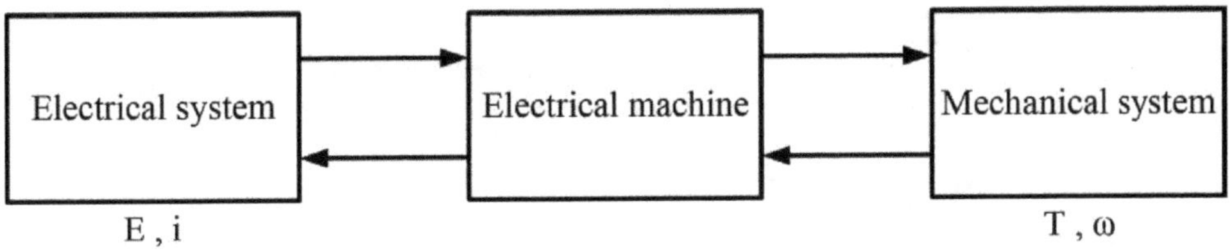

Figure 1. The structure of energy conversion.

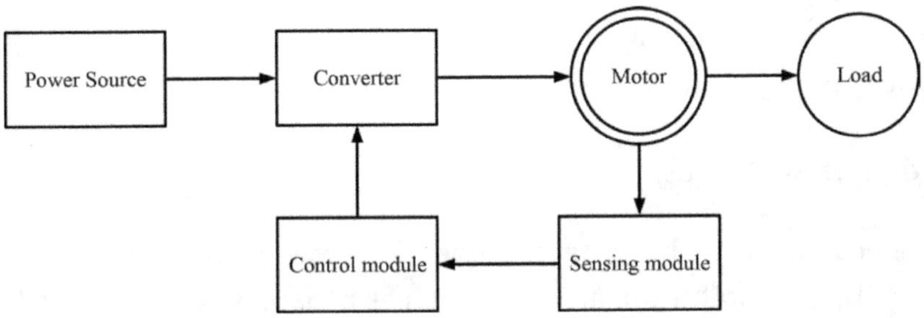

Figure 2. The block diagram of an electrical drive.

A simplified block diagram of an electrical drive is shown in **Figure 2**. It mainly consists of power source, converter, motor, load, sensing module, and control module. Power source can be supplied by AC line supply or DC battery. Load can be washing machine, range hood, pumps, air conditioners, fans, machine tools, or electric drills. The motor has the characteristics required by the load and the capability to transfer the energy from the source to the load. Most commonly used electrical motors are DC motors, brushless DC motors, induction motors, permanent magnet synchronous motors, stepper motors, and switched reluctance motors. The converter converts the electrical energy of the input source in the form suitable to the motor. Depending on the energy from required by the input power source and motors, AC to DC converters (rectifiers or controlled rectifiers), DC to DC converters (step-up converters or step-down converters), inverters, and cycloconverters are commonly used in the electrical drives design [1, 2]. Sensing module is employed to detect rotor position, rotor speed, motor current, or motor terminal voltages. It is required for the closed-loop control. The control module is the core controller of the overall electrical drive. Due to the high development of power semiconductors, the control module such as microcontrollers, DSPs, ARMs, or FPGA are widely adopted according to the complexity functions for the motor drive requirements.

1.1. Torque equation

A motor generally drives a load through some transmission systems as shown in **Figure 3**. The energy keeps balance, which performs the operation of transmission. Hence, torque developed by motor is counter balanced by a load torque T_L and a dynamic torque. The energy transferred from the motor to load is maintained constant. It can be expressed by the following torque equation [3]:

$$T_e = J\frac{d\omega_m}{dt} + B\omega_m + T_L \tag{1}$$

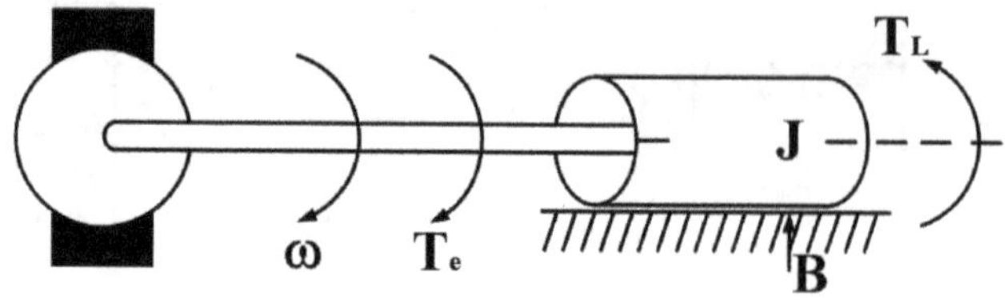

Figure 3. Motor load system.

T_e is instantaneous value of developed motor torque, N m. ω_m is instantaneous angular velocity of motor shaft, rad/s. J is inertia of motor-load system, kg m^2 and T_L is the instantaneous value of load torque, N m.

1.2. The conduction of switching device

In recent decade, because of the high developments of power electronics and advanced semiconductor devices, the adjustable speed drive such as brushless DC motor (BLDC) drives, has been well developed and widely employed to the industrial and commercial applications. A typical motor drive consisting of an AC/DC converter, LC DC link, and an inverter is shown in **Figure 4**. The AC/DC conversion of electric power converts AC input power source to the DC link voltage. The DC to AC conversion is known as inverter, which converts the DC power to AC power required by motor and is normally accomplished by pulse width modulation technique [4]. The inverter may receive DC power from a battery but in most applications it is fed by a rectifier. The switching devices (Q1–Q6) can be BJTs, MOSFETs, IGBTs, GTOs, or thyristors with external commutation circuits and accompanied with six diodes [5].

Assume the inverter feeds a balanced three-phase wye connected resistor load. Two modes of control signals can be applied to the switching transistors: 120° conduction mode and 180° conduction mode.

1.2.1. 120° conduction mode

Each switching device is turned on for one-third of a cycle, or 120°. Consequently, only two switching devices can conduct simultaneously at any instant. It is most suitable for a delta connected load because it has a six-step output voltage waveform across any of its phases. The gating signals of each switching device are shown in **Figure 5(a)** and the corresponding three-phase line voltages are shown in **Figure 5(b)**. The output voltage waveforms of an ideal

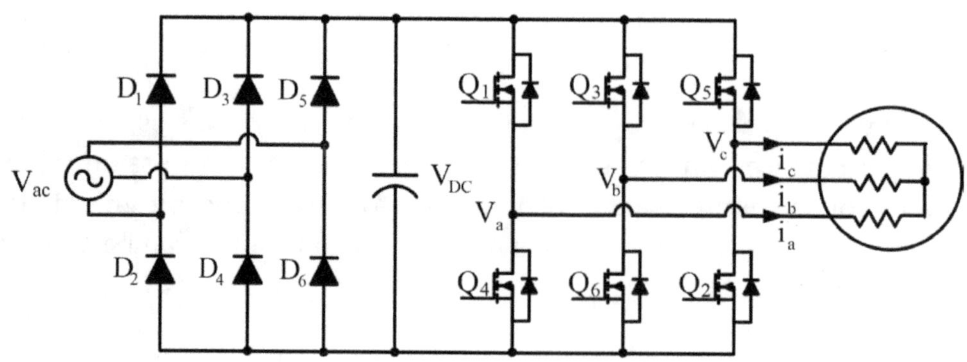

Figure 4. A typical three phase bridge inverter.

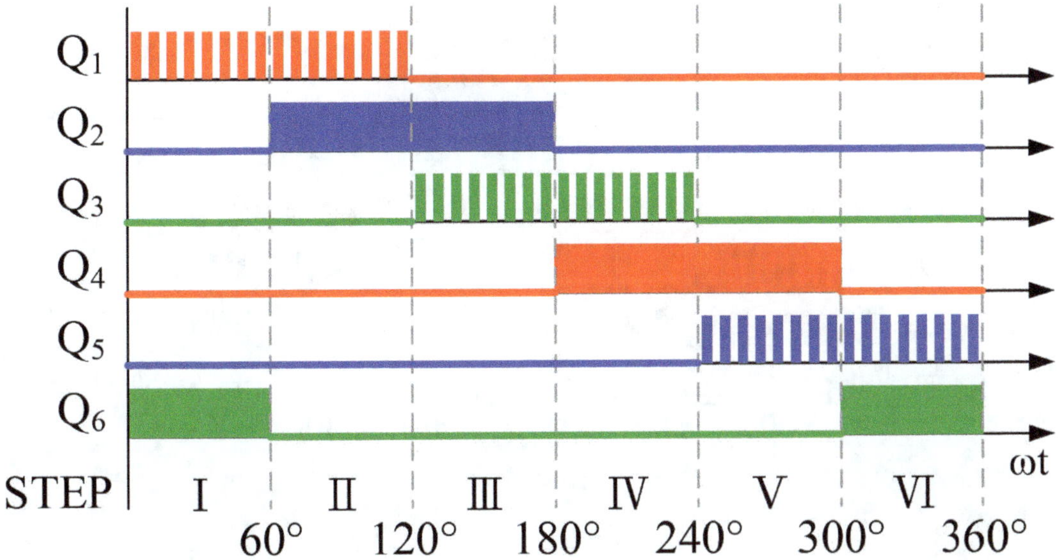

(a)

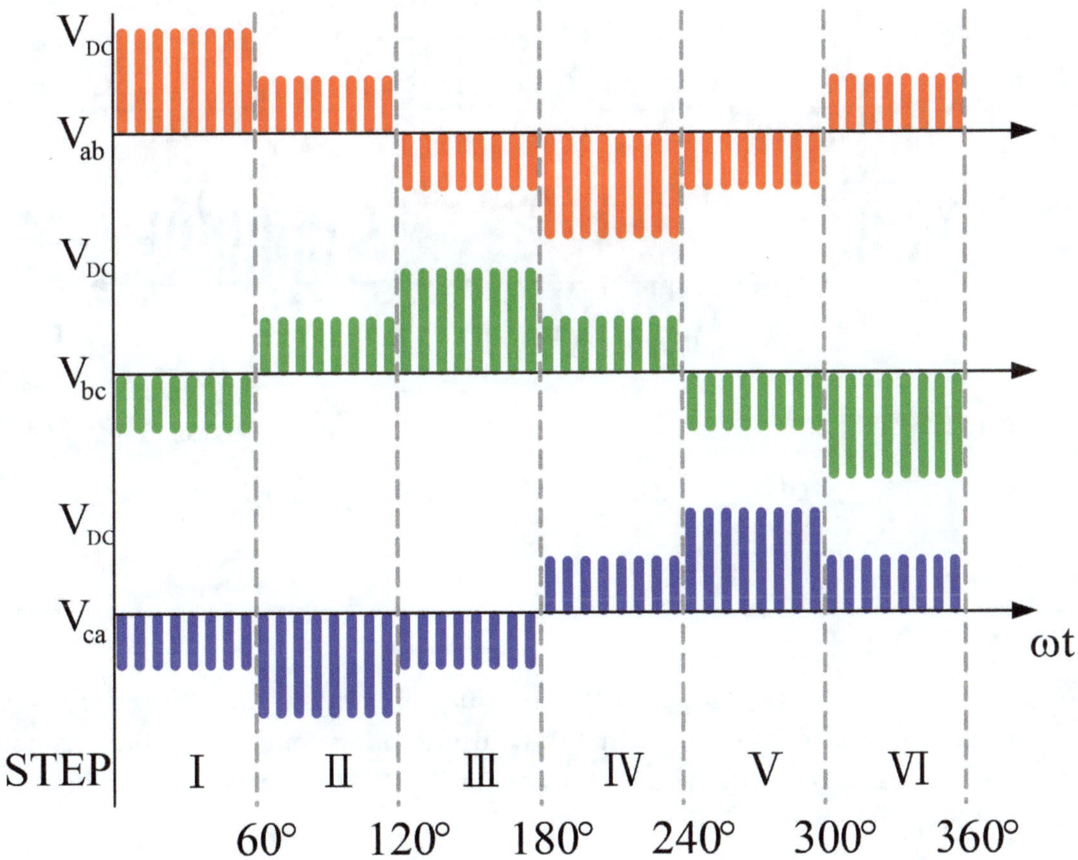

(b)

Figure 5. (a) Gating signals of 120° conduction mode and (b) the line-to-line voltages for 120° conduction mode.

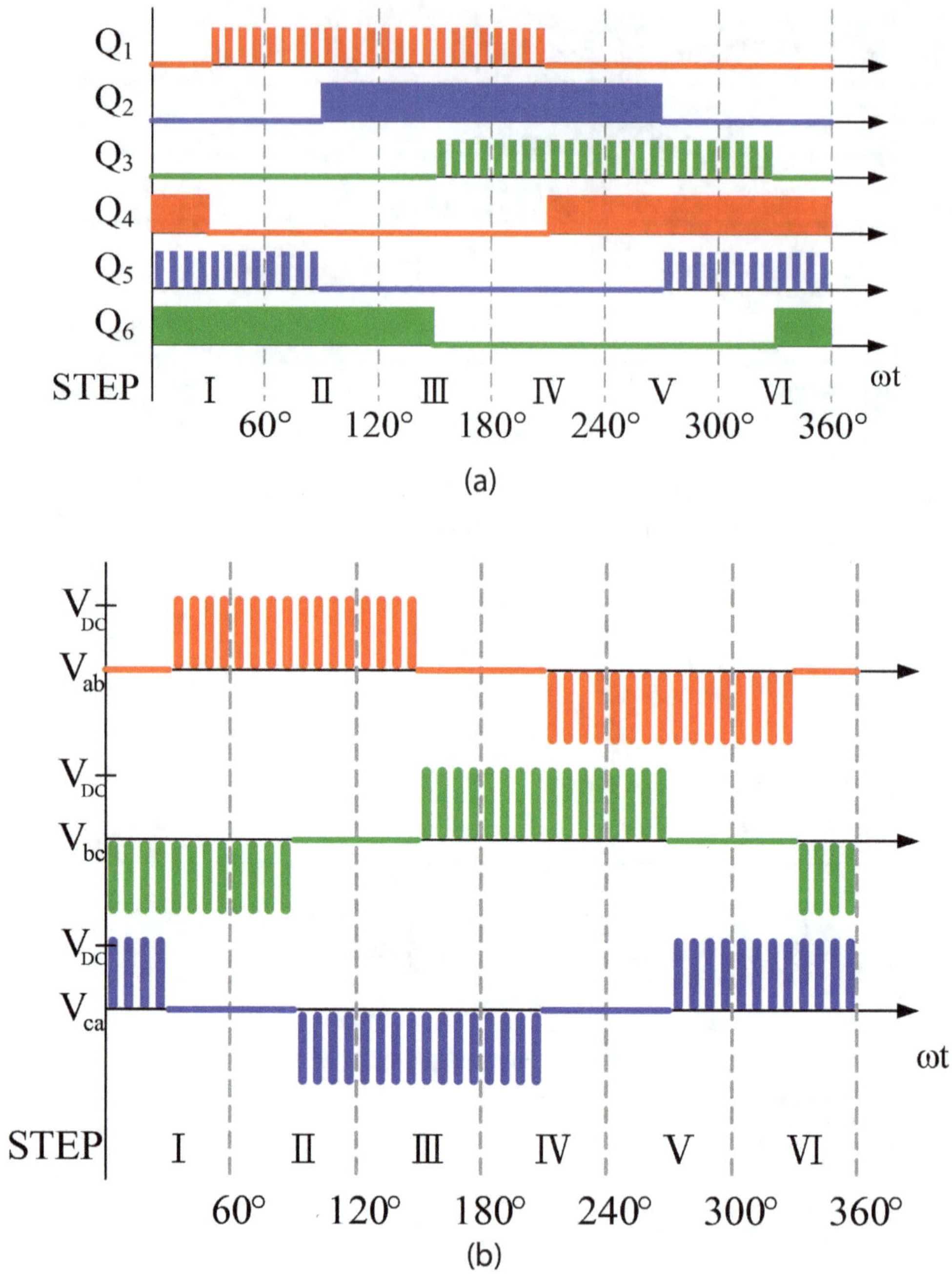

Figure 6. (a) Gating signals of 180° conduction mode and (b) the line-to-line voltages for 180° conduction mode.

inverter should be sinusoidal. However, the output voltages are usually square waveform or quasi square waveform. This means that the output voltage inherently contains harmonic contents. The rms value of fundamental phase voltage is 0.39 V_{DC} and can be found from the calculation of Fourier series for line voltages [5, 6].

1.2.2. 180° conduction mode

For 180° conduction mode, each switching device conducts is turned on and off at 180° interval. At any instant of time, three switching devices always remain turn-on state. The gating signals for each switching device are phase displacement of 120° as shown in **Figure 6(a)**. Consequently, a three-phase output voltage can be obtained with respect to the corresponding switching sequence pattern, as shown in **Figure 6(b)**. The rms value of fundamental phase voltage is 0.45 V_{DC}. Compared to the 120° conduction mode, it has much higher output voltage under any operating condition. In other words, it has better voltage utilization of switching devices [5–7].

2. The dynamic model of a BLDC motor

The equivalent circuit of a BLDC motor is shown in **Figure 7**. The stator phase voltage equations (V_{an}, V_{bn}, and V_{cn}) related to the stator phase currents (i_a, i_b, and i_c) and back electromotive force (e_a, e_b, and e_c) for a BLDC motor, are expressed by (1) [5, 6].

$$\begin{aligned} V_{an} &= R_a i_a + L_{aa}\frac{di_a}{dt} + L_{ab}\frac{di_b}{dt} + L_{ac}\frac{di_c}{dt} + e_a \\ V_{bn} &= R_b i_b + L_{ba}\frac{di_a}{dt} + L_{bb}\frac{di_b}{dt} + L_{bc}\frac{di_c}{dt} + e_b \\ V_{cn} &= R_c i_c + L_{ca}\frac{di_a}{dt} + L_{cb}\frac{di_b}{dt} + L_{cc}\frac{di_c}{dt} + e_c \end{aligned} \tag{2}$$

where R_a, R_b, and R_c, represent the phase resistance for each phase, L_{aa}, L_{bb}, and L_{cc} represent the self-inductance for each phase, L_{ab}, L_{bc}, and L_{ca} represent the mutual inductance between either of two phases, and e_a, e_b, e_c, represent the back EMF for each phase. If a three-phase balanced system is considered, that is, $i_a + i_b + i_c = 0$, the stator voltages in (1) can be rearranged to matrix form as

$$\begin{bmatrix} V_{an} \\ V_{bn} \\ V_{cn} \end{bmatrix} = \begin{bmatrix} R_s & 0 & 0 \\ 0 & R_s & 0 \\ 0 & 0 & R_s \end{bmatrix} \begin{bmatrix} i_a \\ i_b \\ i_c \end{bmatrix} + \frac{d}{dt} \begin{bmatrix} L_s & 0 & 0 \\ 0 & L_s & 0 \\ 0 & 0 & L_s \end{bmatrix} \begin{bmatrix} i_a \\ i_b \\ i_c \end{bmatrix} + \begin{bmatrix} e_a \\ e_b \\ e_c \end{bmatrix} \tag{3}$$

In steady state, the air gap power in terms of the electromagnetic torque and speed is expressed by

$$e_a i_a + e_b i_b + e_c i_c = T_e \omega_m \tag{4}$$

The electromagnetic torque is then given by

$$T_e = \frac{e_a i_a + e_b i_b + e_c i_c}{\omega_m} \tag{5}$$

Assume the flux linkage is continuous. The induced emfs are derivatives of the flux linkage. By proper control of the motor, the electromagnetic torque can be expressed by

$$T_e = K_t i_a \tag{6}$$

The electromagnetic torque expressed in (6) indicates that it can be controlled by the stator current with the torque constant, K_t. This is very similar to the control of separately DC motor drive. The load model can be expressed in terms of the motor speed, ω_m, a moment of inertia, J,

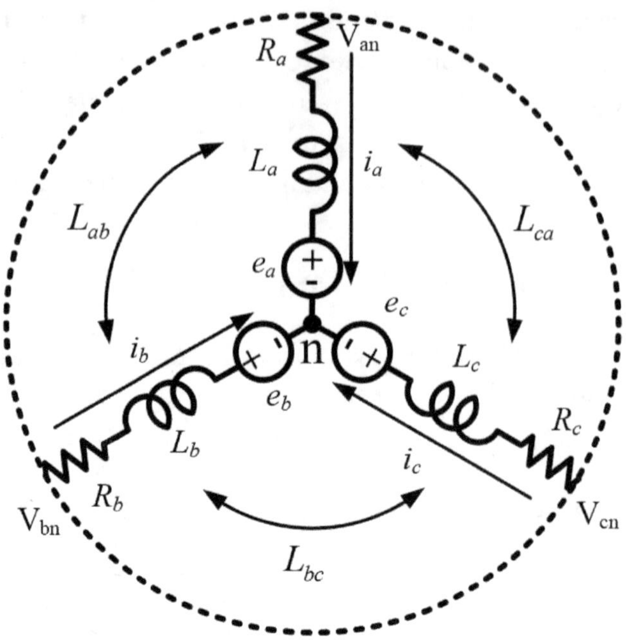

Figure 7. The equivalent circuit of a BLDC motor.

in kg m²/s² with a viscous friction B, in N m/rad/s. The electromagnetic torque, T_e, in N m then drives the load torque, T_L, in N m as represented in (1) and repeated here.

$$J\frac{d\omega_m}{dt} + B\omega_m + T_L = T_e \tag{7}$$

3. The design of a speed controller

For a speed-controlled motor drive, the time constants of the current loop are usually much smaller than that of the load and the speed controller. In such situation, the control system with simplified current loop for a motor drive is shown in **Figure 8**. If the 3-dB frequency of current control loop is properly selected, the current controller, $G_i(s)$ can be approximated to [8].

$$G_i(s) \cong \frac{1}{1+sT_c} \tag{8}$$

Assume that T_c is very small in the operating frequency interval, $G_i(s)$ is approximate to be unity. The PI control is usually employed to the speed controlled motor drive system and has proved to achieve good performance. The transfer function of speed controller, $G_s(s)$, can then be expressed by [9].

$$G_s(s) = K_{ps} + \frac{K_{is}}{s} \tag{9}$$

where K_{ps} is the proportional gain and K_{is} is the integrator gain. As we observe the speed control loop, the open-loop transfer function of the speed response has three poles, which can be obtained from the integrator time delay constant, current loop time delay, and the load time constant, respectively.

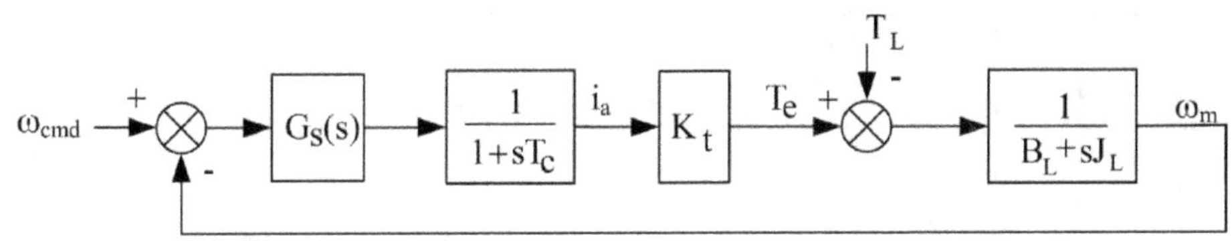

Figure 8. The current loop for motor drive.

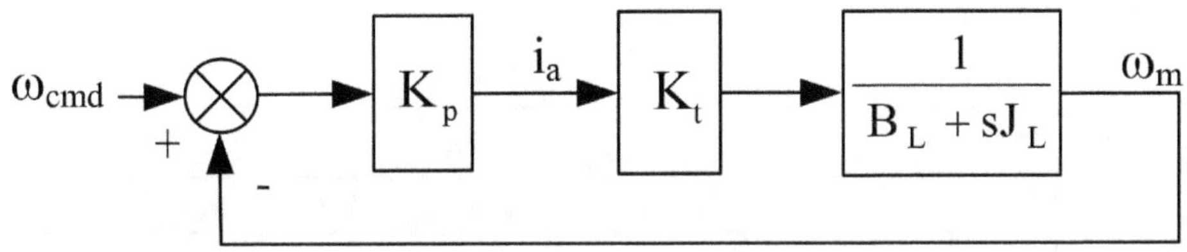

Figure 9. The overall control system of speed-controlled motor drive.

For a speed-controlled drive system, the 3-dB frequency generated from the current loop time delay is always designed to be higher than the rotor frequency. In addition, the 3-dB frequency is obtained from the integrator time delay of PI controller is usually chosen to be smaller than the rotor frequency. From the approximation of current loop crossover frequency and the corner frequency of PI controller, the complete speed-controlled drive system is redrawn in **Figure 9**. The system is inherently stable since the maximum phase angle of the complete control system is −90°. Adjusting the proportional gain of speed controller, K_{ps}, can be applied to meet the requirements of the overshoot for the overall control system [8, 9].

4. The implementation of a BLDC motor drive

The system structure with control algorithm of motor drive based on a microcontroller is shown in **Figure 10**. It consists of a microcontroller, protection circuit, optical coupling isolation, inverter, current sensor, Hall effect sensor, and communication interface. For the design and implementation of an electric bicycle, the AC power source and rectifier are replaced by DC battery. The microcontroller dsPIC 30FXX series manufactured by Microchip technology incorporate is the core controller of the electric bicycle. It is a 16-bit CPU with the capability of digital signal processing. Moreover, it supports many powerful modules such as built-in PWM module, addressable encoder interface module, and input capture module; these make the design friendly and thus shorten the development schedule. The three-phase bridge inverter comprises six switching power MOSFETs. The photocoupler is used for electrical isolation between the microcontroller system and bus voltage. The motor currents from the DC bus are sensed through the current detection circuit. The magnet pole and rotor position are detected by the Hall effect sensor. The speed and rotor position can be calculated and precisely controlled, accordingly. The 120° conduction of pulse width modulation technique for MOSFETs switching, is employed to drive the thee-phase inverter.

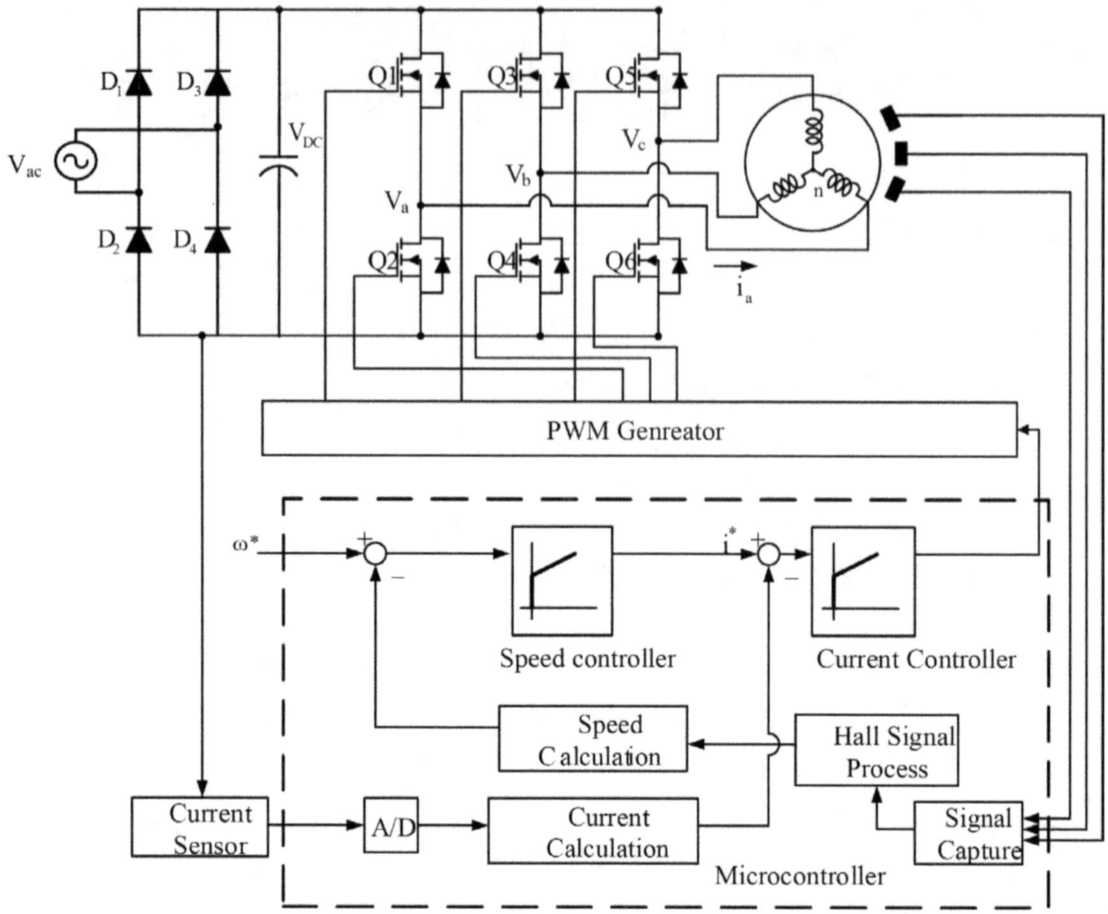

Figure 10. The system structure of a motor drive.

Figure 11 is the main program of a motor drive for electric bicycle design. The initializations for I/O configuration, Timer 1, Timer 2, ADC, and PWM settings are firstly processed in the main program. Most of the functions for electric bicycle are programmed in the microcontroller firmware, which includes the circuit protection mechanism, PWM generation, motor currents calculation, rotor position and speed calculation, and rotor pole position [8, 10].

The speed and currents are usually implemented in the PWM interrupt routine, as shown in **Figure 12**. The sensing current is firstly calculated and fed to the current controller. Since PWM frequency is 20 kHz, the current controller is updated for every 50 μs. After the calculation of current controller, the speed calculation is then performed by speed controller. The speed is constrained by the limiter in the operating speed range [11].

In order to obtain the proper control of the BLDC motor, the timing between rotor position detected by Hall effect sensor and the corresponding back emf, must always be kept in a fixed sequence. **Figure 13** shows the timing sequence between the rotor position (Ch1–Ch3) and the corresponding back emf (Ch4). The motor parameters for the experimental test are $K_t = 0.0761021$ N m/A, $J_L = 0.000245$ kg m^2, and $B_L = 0.00731$ N m s/rad. The magnitude frequency response of Bode plot with different K_{ps} is shown in **Figure 14**. Observing the **Figure 14**, it can be found that the 3-dB rotor frequency is around 800 rad/s. **Figure 15** shows the step response of the speed from zero to

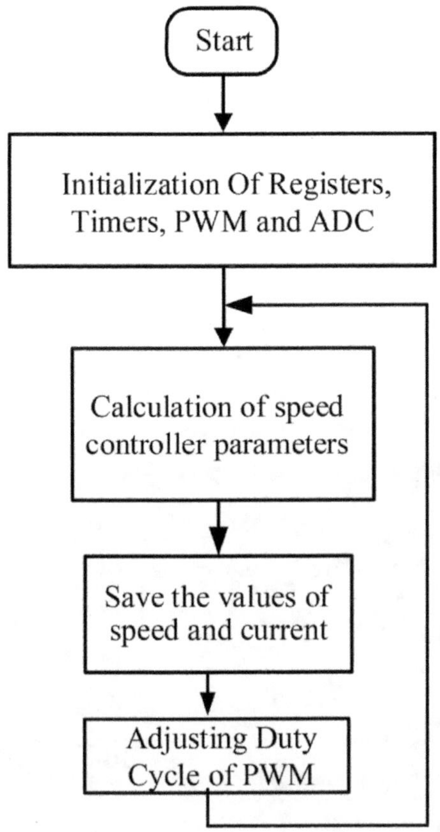

Figure 11. Main program of a motor drive for electric bicycle.

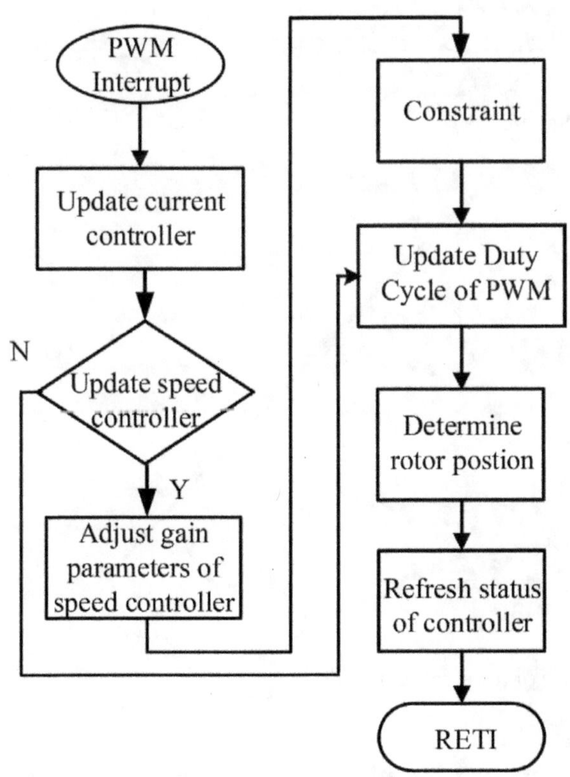

Figure 12. The PWM interrupt routine.

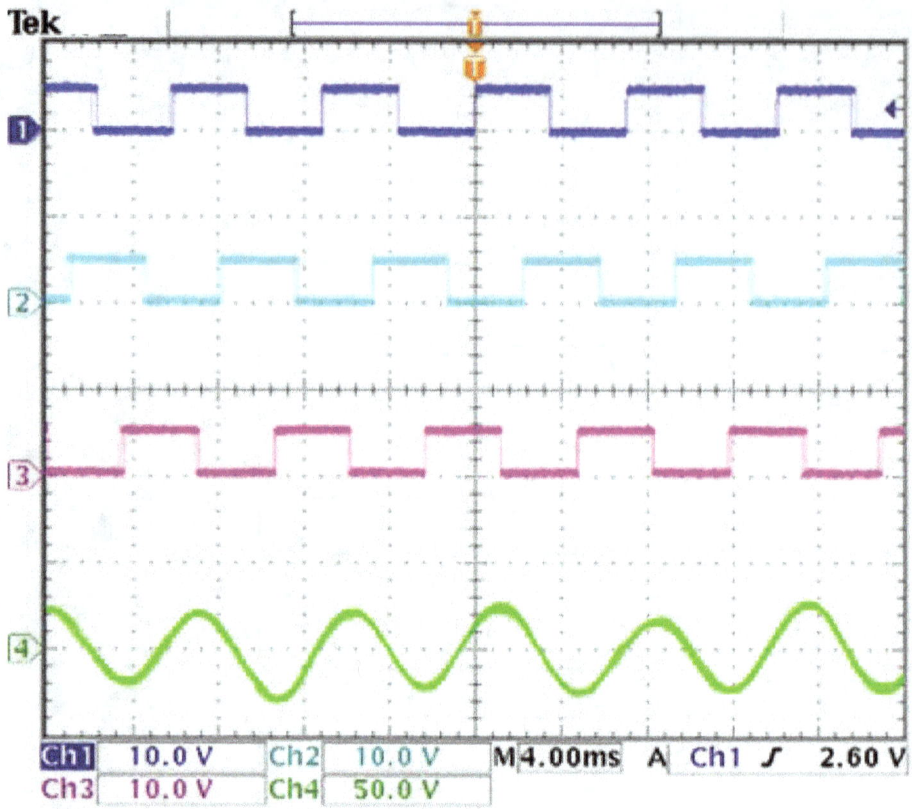

Figure 13. The timing sequence of Hall effect sensing signals and back emf.

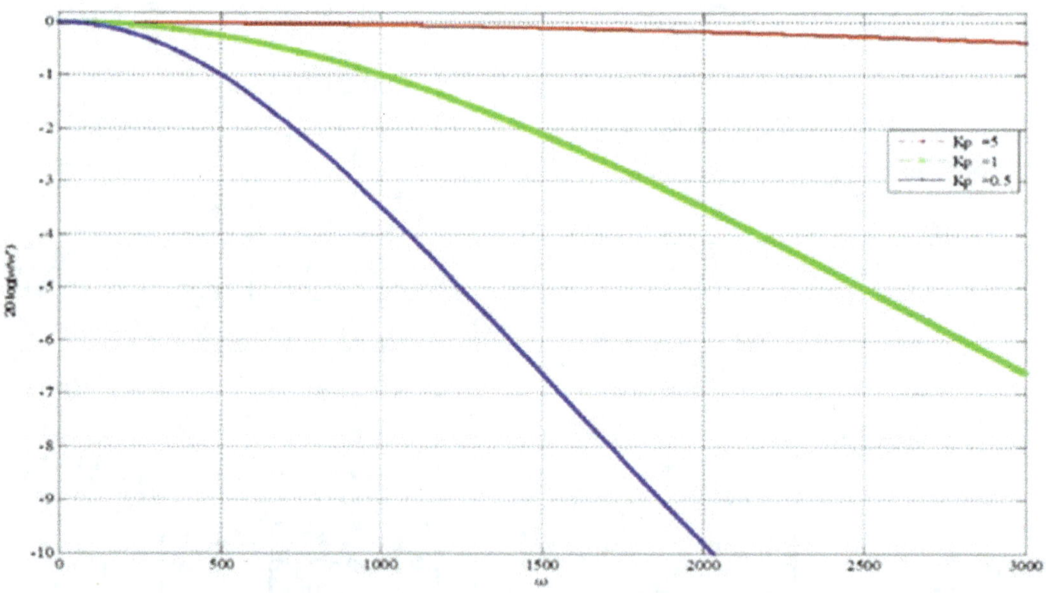

Figure 14. The magnitude response of rotor frequency [8].

200 rpm under the load of 10 kg cm. It can be observed that the motor drive can reach the command speed within the 1.5 s. The phase currents under the load of 10 kg cm is shown in **Figure 16**, where Ch1, Ch2, and math represent phase currents A, B, and C, respectively.

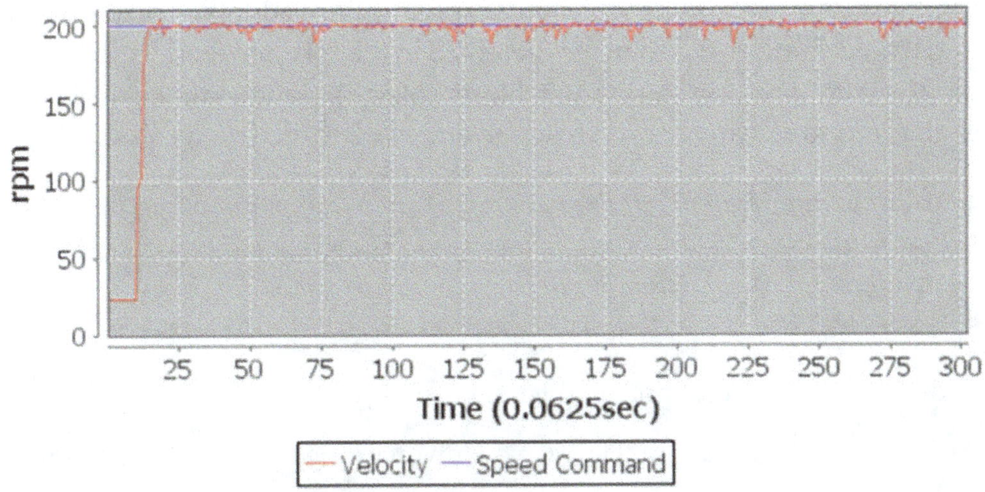

Figure 15. The step response of speed under the load of 10 kg cm.

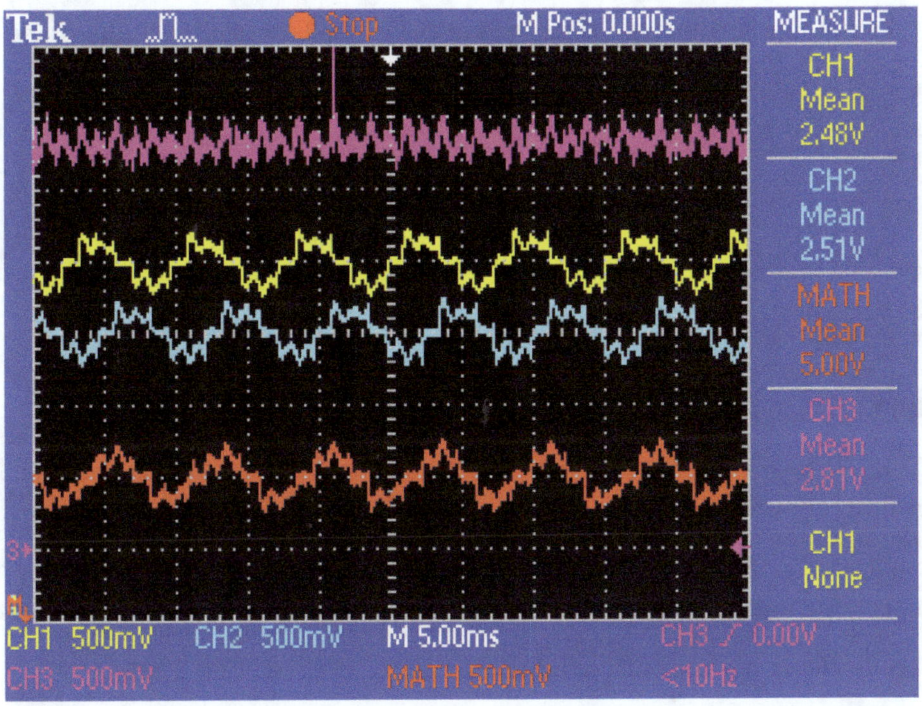

Figure 16. The phase currents under the load of 10 kg cm.

5. Conclusions

In summary, the speed control of motor drive for an electric bicycle is described in this chapter. The hardware structure including a microcontroller, protection circuit, optical coupling isolation, three-phase inverter, current sensor, and communication interface is well designed. Further, the system software and firmware of microcontroller are programmed and described in detail.

Based on the proposed simple method to find the rotor corner frequency, the range of proportional gain of speed controller can be determined so as to avoid the system oscillation. This result can not only speed up the design and implementation of speed controlled motor drive, but also reduce the development time. Further, the contents presented in this chapter about basic knowledge and fundamentals of a motor drive system are suitable for college graduate students. Moreover, I hope it can help engineers and researchers when design and implement a motor drive.

Author details

Tze-Yee Ho

Address all correspondence to: tyho@mail.fcu.edu.tw

Department of Electrical Engineering, Feng Chia University, Seatwen, Taichung, Taiwan, R.O.C.

References

[1] Sen PC. Principles of Electric Machines and Power Electronics. New York: John Wiley & Sons; 1997

[2] Dubey GK. Fundamentals of Electrical Drives. 2nd ed. Kanpur: Alpha Science Int. Ltd.; 2001

[3] Ho Tze-Yee, Chen Yuan-Joan, Chen Po-Hung. The design and implementation of motor drive for foot rehabilitation. Journal of Computers and Electrical Engineering. 2016;**56**: 795-806

[4] Murphy JMD, Turnbull FG. Power Electronic Control of AC Motors. New York: Pergamon Press; 1988

[5] Rashid MH. Power Electronics: Circuits, Devices, and Applications. 3rd ed. New Jersey: Pearson Education Inc.; 2004

[6] Ahmad M. High Performance AC Drives: Modelling Analysis and Control. Berlin Heidelberg: Springer; 2010

[7] Krishnan R. Electric Motor Drives. New Jersey: Prentice Hall; 2001

[8] Tze-Yee H, Mu-Song C, Wei-Chieh C, Chiang C-H. The design and implementation of motor drive for an electric bicycle. Lectures Notes in Computer Science 7928. Advances in Swarm Intelligence Proceeding Part II. 2013. pp. 488-495

[9] Crnosija P., Krishnan R., Bjazic T. Optimization of PM brushless DC motor drive speed controller using modification of Ziegler-Nichols methods based on Bodé plots. In: IEEE International Power Electronics and Motion Control Conf. 2006. pp 343-348

[10] Krishnan R. Selection criteria for servo motor drives. IEEE Transactions on Industry Applications. 1987;**IA-23**(2):270-275

[11] Ho T-Y, Mu-Song C, Lung-Hsian Y, Jia-Shen L, Po-Hung C. The design of a high power factor brushless DC motor drive. International Journal of Advancements in Computing Technology. 2012;**4**(18):P141-P149. ISSN: 2005-8039

12

Mechanical Design of Rotors with Surface Mounted Permanent Magnets

Simon Barrans and Levi Mallin

Abstract

Electric machines with permanent magnet rotors are becoming increasingly popular due to the high power density that they offer relative to other configurations. Where the speed of rotation is high, the magnets are typically mounted on the surface of the rotor and retained by an outer sleeve. In the literature, a variety of analytical models have been proposed to aid the mechanical design process. Many of these models contain inherent assumptions about the stress and strain field in the rotor which may not always be apparent. In this article the range of rotor stress models are presented and explained and the limitations due to the inherent assumptions are investigated. This will allow the designer of such rotors to assess mechanical performance without introducing unforeseen errors.

Keywords: plane stress, plane strain, generalized plane strain, surface mounted permanent magnets, rotor stress, rotor failure, rotor sleeve

1. Introduction

High speed electric machines are increasingly being used in place of mechanical drive systems. As [1–3] observe, they are more efficient and reliable than their mechanical equivalents and have high power densities. Gerada et al. [4] observe that this makes them an attractive choice across a number of areas including flywheel energy storage and high speed spindles. A particularly demanding application is in electrically assisted turbochargers [5].

In order to achieve the increasingly high speeds being demanded by a number of applications, synchronous machines with permanent magnet rotors are typically used. These machines can be broadly categorized as having interior permanent magnets (IPM's) embedded within the iron core of the rotor or surface-mounted permanent magnets (SPM) attached to the outer surface of the rotor. When considering the mechanical design of these rotors, SPM rotors allow higher rotational speeds than IPM rotors and will be the focus of this article.

The materials used for the permanent magnets include samarium cobalt (SmCo5 or Sm2-Co17) and neodymium (Nd2Fe14B). These materials are formed from sintering or bonding and are typically very weak in tension. If the magnets were simply glued to the outer surface of the rotor, the tensile stresses induced by rotation would cause them to fail. They are therefore typically retained onto the rotor shaft or core with a sleeve, as shown in **Figure 1**. The interference fit between the sleeve and the magnets places the magnets into compression. A sufficiently large interference will ensure that this initial compressive state will prevent tensile stresses being generated under rotation. However, a large interference can induce large circumferential forces in the sleeve. Increasing the sleeve thickness increases the effective air gap between the rotor and the stator which reduces electromagnetic efficiency. The mechanical design of SPM rotors is therefore challenging and has been approached in a number of ways using both analytical and numerical approaches: The analytical approaches treat the sleeve as either a thin walled or thick walled cylinder. Where a thick walled cylinder approach is used, it may be approximated as being in plane stress, plane strain or generalized plane strain. All of these approaches incorporate some degree of approximation which may limit their applicability to a particular rotor design. This article will explore these approximations and limitations.

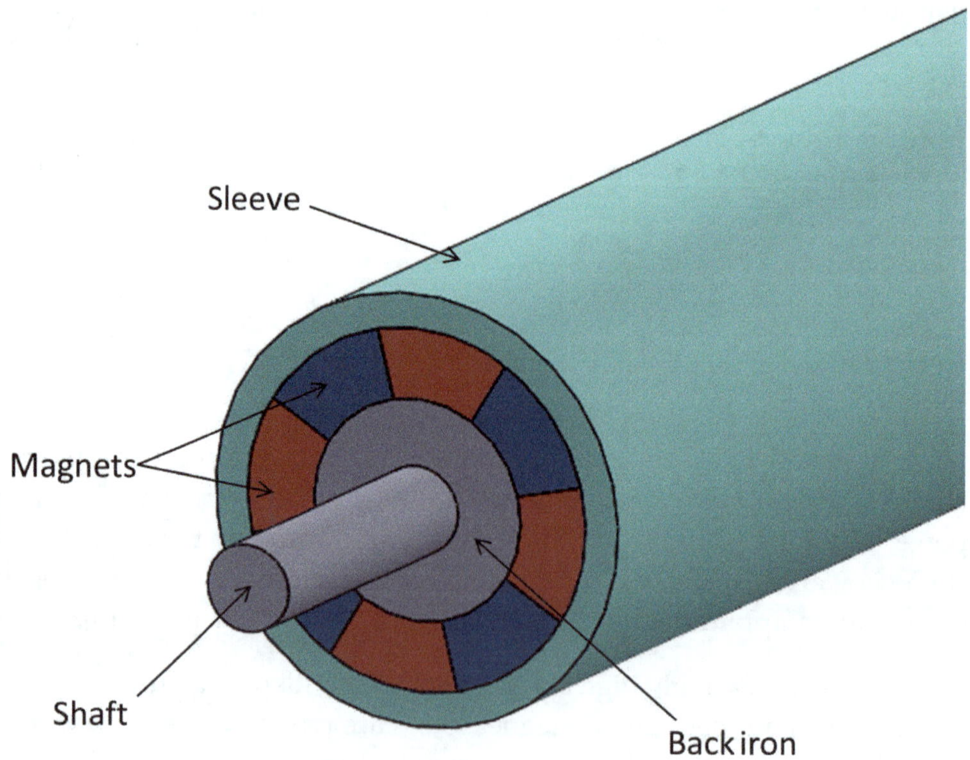

Figure 1. SPM rotor assembly.

2. Classical analytical methods

2.1. Thin shell approach

One of the simplest approaches to SPM rotor mechanical design is to assume that the sleeve acts as a thin walled cylinder, wall thickness t and radius R, as proposed in [6] and illustrated in **Figure 2**. When the rotor is stationary, the radial interference, δ, between the sleeve and the magnets generates a circumferential strain, ε_θ and stress, σ_θ, in the rotor and an interference pressure, S, between the sleeve and the magnets, given respectively by Eqs. (1)–(3).

$$\varepsilon_\theta = \frac{\delta}{R} \tag{1}$$

$$\sigma_\theta = E\varepsilon_\theta = E\frac{\delta}{R} \tag{2}$$

$$S = \frac{\sigma_\theta t}{R} = \frac{E\delta t}{R^2} \tag{3}$$

where E is the Young's modulus of the sleeve material. It is important to note here that all the deformation required to assemble the sleeve onto the magnets is assumed to take place in the sleeve, i.e. the magnets, back iron (if present) and shaft are assumed to be rigid.

The 'thin walled' cylinder approximation used here assumes that there is no variation in circumferential stress and strain through the thickness of the sleeve. Hence, in Eq. (1)–(3) only a single value for these factors is generated. The rather misleadingly termed 'thick walled' cylinder approach correctly allows for variation in stress and strain through the sleeve wall. The 'thin walled' model is therefore an approximation to the 'thick walled' model and for any

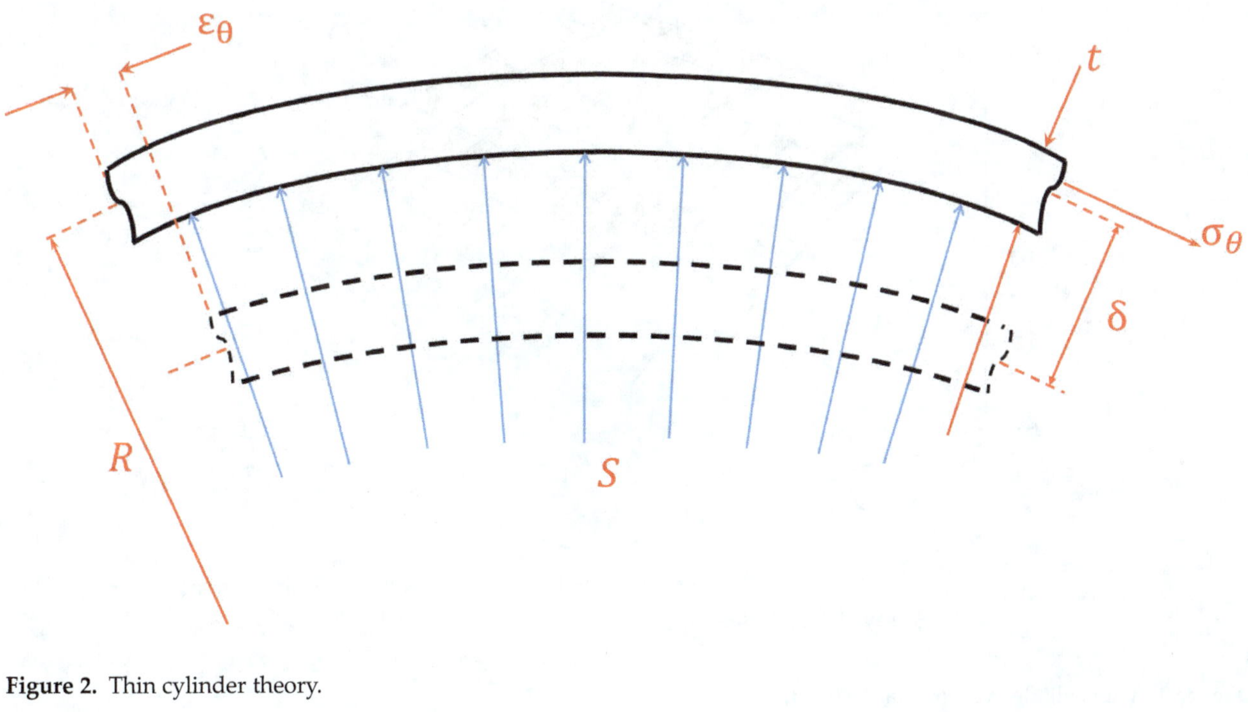

Figure 2. Thin cylinder theory.

sleeve thickness, will introduce an error in the calculation of interference pressure. The various forms of the thick walled cylinder approach will be discussed in the following sections. Taking either the plane stress approach or the generalized plane strain approach gives the relationship between interference pressure and interference as:

$$S = \frac{\delta E\left(R_o^2 - R_C^2\right)}{R_C\left\{R_O^2(1+\nu) + R_C^2(1-\nu)\right\}} \tag{4}$$

where R_O and R_C are the outer and inner radii of the sleeve respectively and ν is the Poisson's ratio of the sleeve material and the magnets, back iron and shaft are assumed to be rigid.

Assuming that the radius of the 'thin walled' sleeve, R, is the mean of the outer and inner radii:

$$R_C = R - \frac{t}{2} \text{ and } R_O = R + \frac{t}{2} \tag{5}$$

Taking typical properties for an Inconel sleeve ($E = 205$GPa, $\nu = 0.294$) and $\delta = 0.1$ mm, the error in the shrinkage pressure introduced by the 'thin walled' cylinder approximation is as shown in **Figure 3**. For small sleeve thicknesses, the errors are small. However, as sleeve thickness increases and sleeve diameter reduces, the thin cylinder approach significantly underestimates the circumferential stress.

Having established the interference pressure acting between the sleeve and the magnets when the rotor is stationary, the effect at maximum rotor speed must be determined. Various approaches can be taken with different criteria being used to assess the mechanical integrity of the rotor. The method proposed in [6] assumes that the interference pressure between the magnets and the back iron is the same as that between the sleeve and the magnets and then

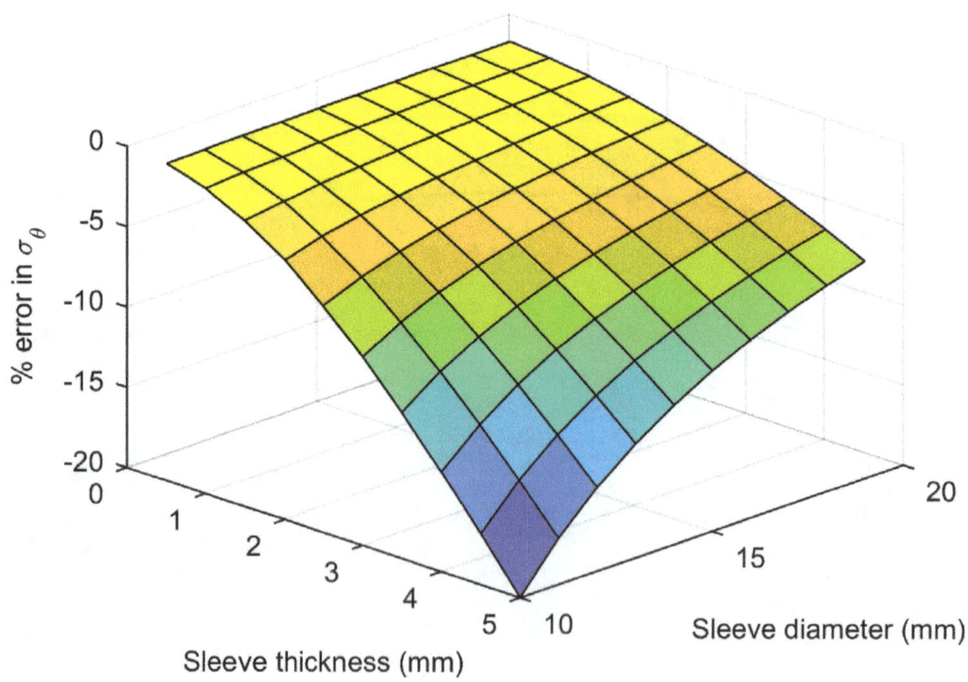

Figure 3. Thin cylinder approximation error.

determines the centripetal force required to reduce this pressure to zero (indicating that the magnets may then slip on the back iron). This approach introduces two further approximations: Firstly, to maintain force equilibrium in the radial direction, the inner surface area of the magnet must be identical to the outer surface area. Secondly, the magnets must have a sufficiently small circumferential dimension (i.e. be segmented) for no significant circumferential stress and no bending to be generated during rotation. The impact of the first effect can be readily determined and indeed can be corrected for. The impact of the second effect is more difficult to determine. If the permanent magnet is a single cylinder, it will resist circumferential expansion during rotation and there will be a smaller reduction in the interference pressure between the magnets and the back iron. However, if the magnet is formed from segments which abut in the circumferential direction, the centripetal forces will tend to separate them. This will lead to a substantial increase in the stresses in the sleeve at the point of separation and the generation of tensile stresses due to bending in the magnets. Whilst the magnet segments remain in full contact in the circumferential direction, their behavior will be identical to that of a single cylinder. This article will be restricted to rotors with magnets which are either single cylinders or are formed from segments which abut and remain in full contact during rotation.

2.2. Thick walled cylinder or disk analysis

All 'thick walled' approaches to the mechanical analysis of rotors are founded on the fundamental principles of force equilibrium, compatibility and the constitutive relationship between stress and strain within the material. These relationships are initially applied to a two dimensional

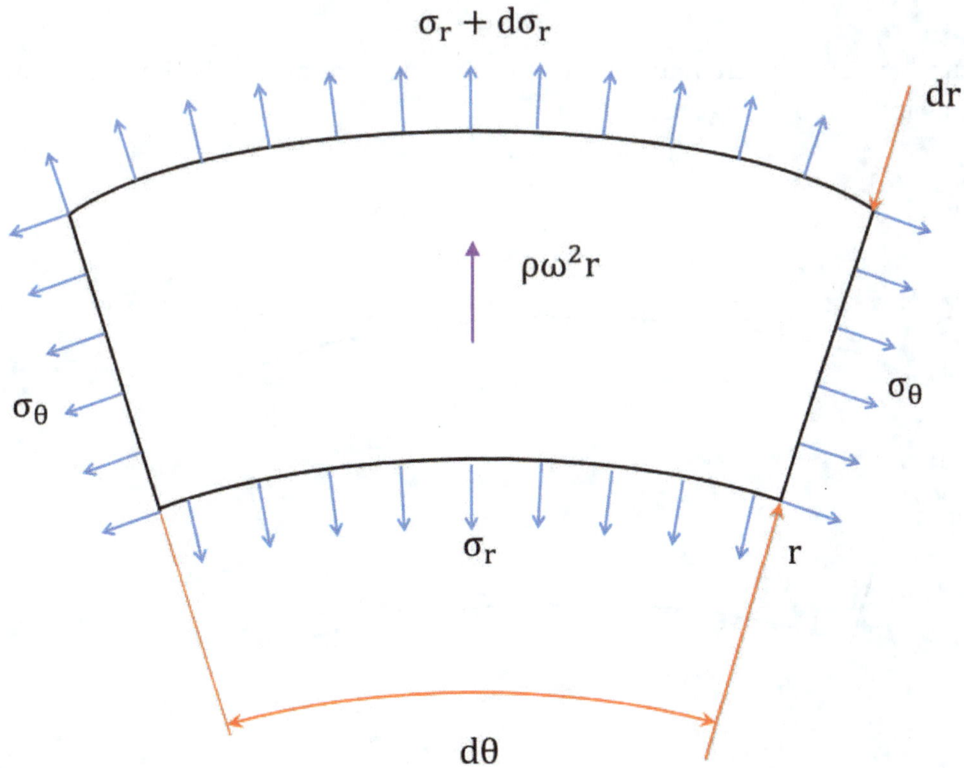

Figure 4. Internal stresses in the cylinder/disc.

cross section of the rotor, perpendicular to the axis and are presented in many text books (see for example [7]). This theoretical material is repeated here to allow the reader to appreciate the differences between the different analysis approaches.

Figure 4 shows a segment, angle $d\theta$ and thickness dr, of an annular ring, radius r, taken from a cylinder or disk which may be rotating at an angular velocity, ω, and may be subjected to either internal or external pressures. The disk material has a density, ρ. It is assumed that the radial stress, σ_r, and circumferential stress vary with radius.

Considering equilibrium of forces on this segment in the radial direction, the following relationship is obtained:

$$\sigma_\theta - \sigma_r - r\frac{d\sigma_r}{dr} = \rho\omega^2 r^2 \tag{6}$$

The constitutive relationship can then be expressed as Eqs. (7)–(9) where ε_r is the radial strain, ε_z the axial strain, α the coefficient of thermal expansion of the material and ΔT the change in temperature imposed on the material.

$$\varepsilon_r = \frac{ds}{dr} = \frac{1}{E}(\sigma_r - \nu\sigma_\theta - \nu\sigma_z) + \Delta T\alpha \tag{7}$$

$$\varepsilon_\theta = \frac{s}{r} = \frac{1}{E}(\sigma_\theta - \nu\sigma_r - \nu\sigma_z) + \Delta T\alpha \tag{8}$$

$$\varepsilon_z = \frac{1}{E}(\sigma_z - \nu\sigma_r - \nu\sigma_\theta) + \Delta T\alpha \tag{9}$$

In Eqs. (7) and (8), s is the change in radius of the segment and ds is the change in the radial dimension, as indicated in **Figure 5**.

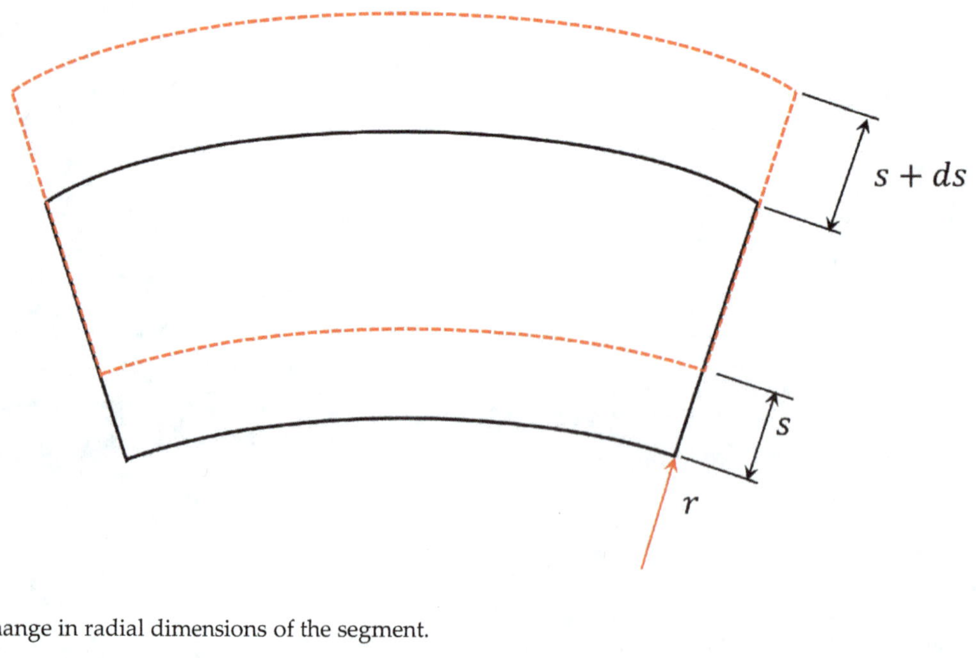

Figure 5. Change in radial dimensions of the segment.

To maintain compatibility:

$$\frac{d(r\varepsilon_\theta)}{dr} = \frac{ds}{dr} = \varepsilon_r \tag{10}$$

or, with reference to Eqs. 7 and 8:

$$(\sigma_r - \sigma_\theta)(1+\nu) = r\left(\frac{d\sigma_\theta}{dr} - \nu\left(\frac{d\sigma_r}{dr} + \frac{d\sigma_z}{dr}\right)\right) + E\alpha r\frac{d\Delta T}{dr} \tag{11}$$

As Li et al. [8] determined, the thermal barrier resisting heat transfer from the outer surface of the rotor to the air gap is much larger than the thermal resistance within the rotor components. Hence, temperatures are essentially uniform with respect to radius and vary only slightly along the axis. Assuming that $d\Delta T/dr = 0$ and combining Eq. 11 with the equilibrium equation (6), gives:

$$\frac{d\sigma_r}{dr} + \frac{d\sigma_\theta}{dr} + \frac{d\sigma_z}{dr} = -(1+\nu)\rho r\omega^2 \tag{12}$$

It is worth emphasizing that Eq. 12 is common to all the analytical approaches presented. However, further development of the analysis requires additional information and the solution paths diverge.

2.2.1. Plane stress approach

An approach previously used in some work [9–12] has been to assume that the axial cross section is in a state of plane stress, where $\sigma_z = 0$. Introducing this condition into Eq. (12) gives:

$$\frac{d\sigma_r}{dr} + \frac{d\sigma_\theta}{dr} = -(1+\nu)\rho r\omega^2 \tag{13}$$

which can be integrated to give:

$$\sigma_\theta + \sigma_r = -\frac{(1+\nu)\rho\omega^2 r^2}{2} + 2A \tag{14}$$

where A is a constant of integration. Reintroducing the equilibrium equation, 6, and further integrating leads to:

$$\sigma_r = A - \frac{B}{r^2} - \frac{(3+\nu)}{8}\rho\omega^2 r^2 \tag{15}$$

$$\sigma_\theta = A + \frac{B}{r^2} - \frac{1+3\nu}{8}\rho\omega^2 r^2 \tag{16}$$

with B being a further constant of integration.

In [10–12] the plane stress assumption is justified by the fact that there is no significant axial stress in the rotor as there is nothing to constrain axial extension or compression. However, as will be demonstrated later, the lack of an external force does not prevent an axial stress being generated in the rotor under rotation. The only valid justification for the plane stress assumption is that the rotor is in the form of a disk which is of insufficient thickness for axial stress to develop.

Where the rotor is modeled as a pair of interfering cylinders, there are two sources of stress: stresses due to the interference and stresses due to rotation. Assuming linear behavior, these stresses can be determined separately and then simply summed to determine the total stress. There are also two load cases to consider:

- Stationary rotor with stress due only to interference.
- Rotating rotor with stresses generated by both rotation and interference.

For the second load case it is important to account for the change in interference due to rotation [13].

Considering **Figure 6**, the two interfering cylinders must achieve a common radius, R_c, and an interference or shrinkage pressure, S, is generated. Treating the two cylinders separately, the boundary conditions are:

$$\begin{aligned} \text{Inner cylinder : At } r &= R_i; \sigma_{rii} = 0 \\ r &= R_c; \sigma_{ric} = -S \\ \text{Outer cylinder : At } r &= R_o; \sigma_{roo} = 0 \\ r &= R_c; \sigma_{roc} = -S \end{aligned}$$

where σ_{rii} is the radial stress in the inner cylinder at the inner radius, σ_{ric} is the radial stress in the inner cylinder at the common radius, σ_{roo} is the radial stress in the outer cylinder at the outer radius and σ_{roc} is the radial stress in the outer cylinder at the common radius.

Substituting these boundary conditions into Eq. (15) allows the constants of integration to be found and Eqs. (15) and (16) become:

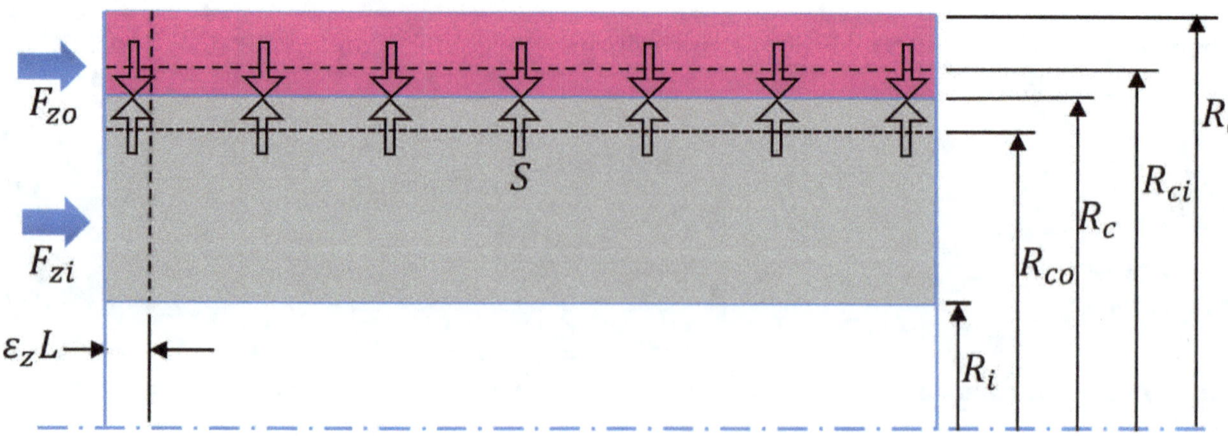

Figure 6. Interference condition.

$$\text{Inner cylinder}: \sigma_{ri} = \frac{SR_c^2}{R_i^2 - R_c^2}\left(1 - \frac{R_i^2}{r^2}\right) \tag{17}$$

$$\sigma_{\theta i} = \frac{SR_c^2}{R_i^2 - R_c^2}\left(1 + \frac{R_i^2}{r^2}\right) \tag{18}$$

$$\text{Outer cylinder}: \sigma_{ro} = \frac{SR_c^2}{R_o^2 - R_c^2}\left(1 - \frac{R_o^2}{r^2}\right) \tag{19}$$

$$\sigma_{\theta o} = \frac{SR_c^2}{R_o^2 - R_c^2}\left(1 + \frac{R_o^2}{r^2}\right) \tag{20}$$

The compatibility requirement that the two cylinders establish a common radius can be expressed as:

$$R_{ci}(1 + \varepsilon_{\theta ics}) = R_{co}(1 + \varepsilon_{\theta ocs}) \tag{21}$$

where $\varepsilon_{\theta ic}$ is the circumferential strain in the inner cylinder at the common radius, $\varepsilon_{\theta oc}$ is the circumferential strain in the outer cylinder at the common radius, R_{ci} is the outer radius of the inner cylinder prior to shrinkage and R_{co} is the inner radius of the outer cylinder prior to shrinkage.

Noting that $\sigma_z = 0$, Eqs. (17)–(20) can be used with Eq. (8) to determine the strain terms in Eq. (21) in terms of the dimensions of the cylinders and the shrinkage pressure. The shrinkage pressure can then be determined from:

$$S = \frac{R_{ci} - R_{co} + \Delta T[R_{ci}\alpha_i - R_{co}\alpha_o]}{\frac{R_{co}\left\{R_o^2(1+v_o)+R_c^2(1-v_o)\right\}}{E_o\left(R_o^2-R_c^2\right)} + \frac{R_{ci}\left\{R_c^2(1-v_i)+R_i^2(1+v_i)\right\}}{E_i\left(R_c^2-R_i^2\right)}} \tag{22}$$

where the subscripts i and o on the material properties E, v and α indicate different properties for the inner and outer cylinders respectively.

Having determined the shrinkage pressure, the stresses due to shrinkage can be determined from Eqs. (17)–(20).

To determine the stresses due to rotation, Eqs. (15) and (16) can again be used, with the boundary conditions:

$$\text{Inner cylinder}: \text{At } r = R_i; \sigma_{rii} = 0$$

$$r = R_c; \sigma_{ric} = 0$$

$$\text{Outer cylinder}: \text{At } r = R_o; \sigma_{roo} = 0$$

$$r = R_c; \sigma_{roc} = 0$$

being used to determine the constants of integration.

2.2.2. Plane strain approach

Perhaps the first application of plane strain to the analysis of electric machine rotors is in [6] with reference to this paper being made by [14–16]. In this approach it is assumed that the axial strain is zero and therefore also that $d\varepsilon_z/dr = 0$. Hence, differentiating Eq. (9) gives:

$$\frac{1}{E}\left(\frac{d\sigma_z}{dr} - \nu\left(\frac{d\sigma_r}{dr} + \frac{d\sigma_\theta}{dr}\right)\right) + \alpha\frac{d\Delta T}{dr} = 0 \tag{23}$$

As before, assuming $d\Delta T/dr = 0$:

$$\frac{d\sigma_z}{dr} = \nu\left(\frac{d\sigma_r}{dr} + \frac{d\sigma_\theta}{dr}\right) \tag{24}$$

and introducing this into Eq. (12) gives:

$$\left(\frac{d\sigma_r}{dr} + \frac{d\sigma_\theta}{dr}\right)(1-\nu) = -(1+\nu)\rho r\omega^2 \tag{25}$$

As with the plane stress case, this equation is integrated, the equilibrium condition (Eq. (6)) reintroduced and a further integration leads to:

$$\sigma_r = A - \frac{B}{r^2} - \frac{3-2\nu}{8(1-\nu)}\rho\omega^2 r^2 \tag{26}$$

$$\sigma_\theta = A + \frac{B}{r^2} - \frac{1+2\nu}{8(1-\nu)}\rho\omega^2 r^2 \tag{27}$$

For the case of no rotation, Eqs. (26) and (27) are identical to Eqs. (15) and (16) so Eqs. (17)–(20) can be used to determine the stresses. However, with rotation present, the stresses and strains will not be the same. In order to determine the circumferential strain due to rotation (and hence the change in interference), the axial stress, σ_z, is required in Eq. (8). To determine this stress, the condition of plane strain is imposed on Eq. (9) to give:

$$\sigma_z = \nu(\sigma_r + \sigma_\theta) - E\Delta T\alpha \tag{28}$$

To determine the shrinkage pressure, Eqs. (26)–(28) are substituted into Eq. (8) for each cylinder at the common radius and then Eq. (21) is used to enforce compatibility between the cylinders. This leads to:

$$S = \frac{R_{ci} - R_{co} + \Delta T[\alpha_i(1+\nu_i)R_{ci} - \alpha_o(1+\nu_o)R_{co}]}{\frac{R_{co}\left\{R_c^2\left(1-\nu_o-2\nu_o^2\right)+R_o^2(1+\nu_o)\right\}}{E_o\left(R_o^2-R_c^2\right)} + \frac{R_{ci}\left\{R_c^2\left(1-\nu_i-2\nu_i^2\right)+R_i^2(1+\nu_i)\right\}}{E_i\left(R_c^2-R_i^2\right)}} \tag{29}$$

The analysis methodology is otherwise identical to that for the plane stress case with stresses due to interference and rotation being found individually and then summed. Again, as [13] observe, it is important to account for the change in interference when analyzing the combined rotation and interference load case. The impact of this can be seen in **Figure 7** for a cylinder

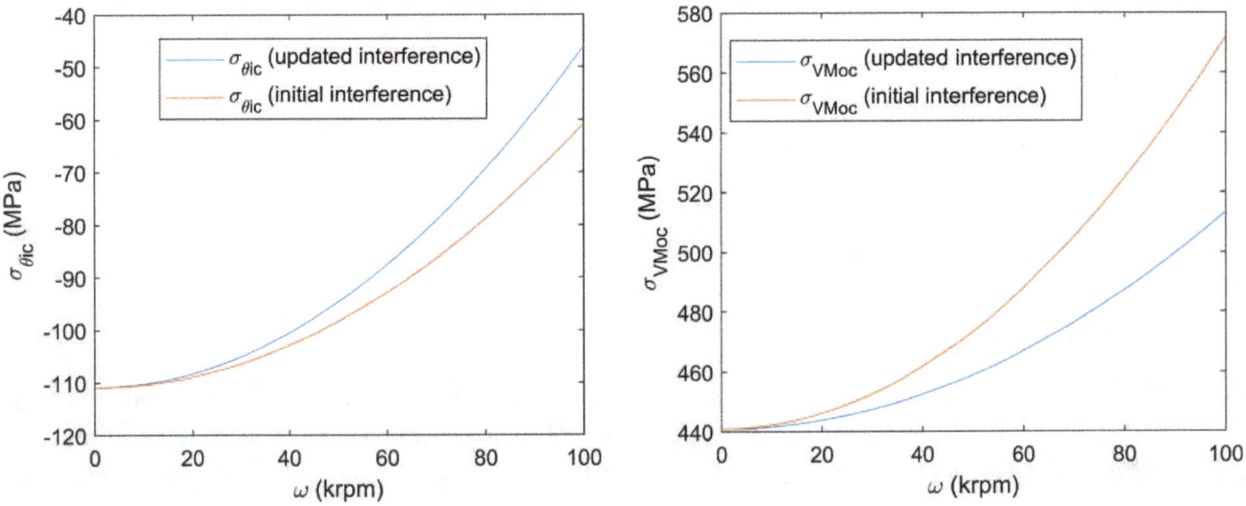

Figure 7. Impact of not updating interference.

Inner radius, R_i	6.35 mm
Common radius, R_c	11.4 mm
Outer radius, R_o	13 mm
Inner Poisson's ratio, ν_i	0.27
Outer Poisson's ratio, ν_o	0.284
Inner density, ρ_i	8400 kg/m^3
Outer density, ρ_o	8220 kg/m^3

Table 1. Properties for axial stress study.

pair with the properties given in **Table 1** and an initial interference of 0.03 mm. It can be seen that if the interference is not updated to account for rotation, the compressive stress at the outer radius of the magnet, $\sigma_{\theta ic}$, is overestimated and similarly, the von Mises stress on the inner radius of the sleeve, σ_{VMoc}, is overestimated as speed increases.

A criticism of the plane strain approach is the requirement for axial strain to be zero. Although a rotor assembly maybe mounted on a solid shaft with end caps to retain the magnet and sleeve, this shaft will be a deformable body. Hence, it will be possible for the rotor components to change in length.

3. Generalized plane strain approach

3.1. Axial stress generation

The generalized plane strain approach is similar to the plain strain approach and Eq. (24) can be applied. Using this in conjunction with Eq. (12), the variation of axial stress with radius is given by:

$$\frac{d\sigma_z}{dr} = \frac{-r\nu_i\rho_i\omega^2}{(1-\nu_i)} \tag{30}$$

Integrating gives the axial stress as:

$$\sigma_z = \frac{\nu_i\rho_i\omega^2}{2(1-\nu_i)}(C - r^2) \tag{31}$$

where C is a constant of integration.

Assuming that an external axial force acts on the cylinders, as indicated in **Figure 6**, and taking the inner cylinder as an example, there must be equilibrium between the internal and external axial forces, expressed as:

$$F_{zi} = 2\pi \int_{R_i}^{R_c} r\sigma_z dr \tag{32}$$

where F_{zi} is the external axial force acting on the inner cylinder. Carrying out the integration allows the constant C to be determined and the axial stress is then given by:

$$\sigma_{zi} = \frac{F_{zi}}{\pi(R_c^2 - R_i^2)} + \frac{\nu_i\rho_i\omega^2}{4(1-\nu_i)}(R_c^2 + R_i^2 - 2r^2) \tag{33}$$

Similarly, for the outer cylinder:

$$\sigma_{zo} = \frac{F_{zo}}{\pi(R_o^2 - R_c^2)} + \frac{\nu_o\rho_o\omega^2}{4(1-\nu_o)}(R_c^2 + R_o^2 - 2r^2) \tag{34}$$

where F_{zo} is the external axial force acting on the outer cylinder.

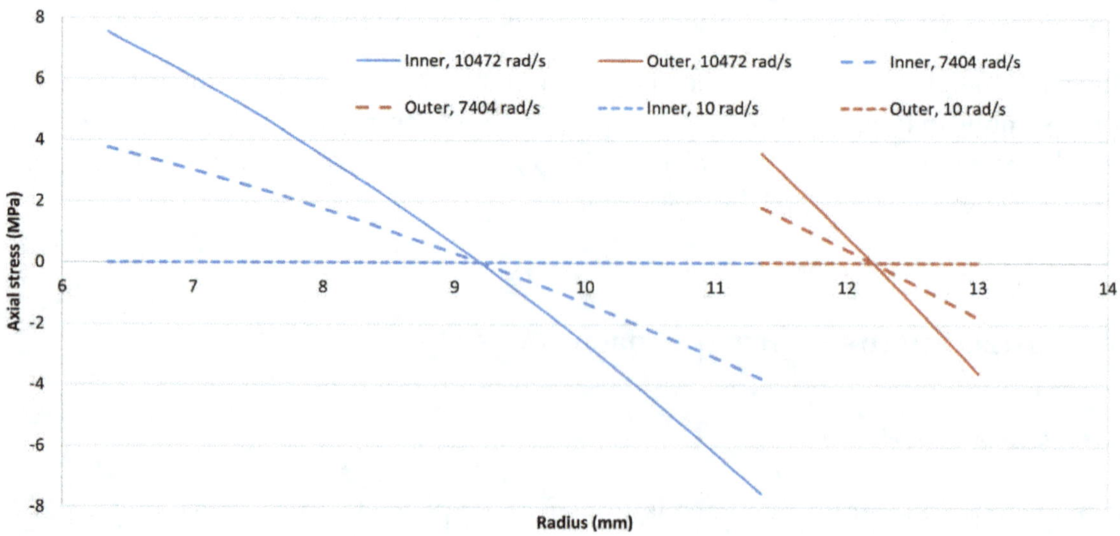

Figure 8. Axial stress generation.

As an aside at this point, it is worth considering the possible impact of axial stress generation. Taking a cylinder pair with the dimensions and material properties shown in **Table 1** and assuming that no axial force is applied ($F_{zi} = F_{zo} = 0$), the axial stress due to rotation alone, over a range of speeds is shown in **Figure 8**. With no rotation applied, no axial stress is generated and the generalized plane strain approach is then identical to the plane stress approach. However, once rotation is introduced, axial stress is generated with peak values at the inner and outer radii. These are the positions where the circumferential strain needs to be calculated, according to Eq. (8), in order to determine the degree of interference and shrinkage pressure. The axial stress will therefore impact on these critical values.

A further point worth noting is that at the end of the cylinder axial stress cannot exist as this would place the material out of equilibrium. On the end face, a state of plane stress must exist. This change in stress state may be responsible for the end failure noted by [17].

3.2. Combining the cylinders

To account for the interaction between the cylinders, two equilibrium and two compatibility conditions need to be considered:

A. In the radial direction there must be force equilibrium at the interface between the cylinders.

B. At the interface there must be circumferential strain compatibility. With reference to **Figure 6**, this is expressed as Eq. (21).

C. Assuming that there is sufficient friction acting between the two cylinders to prevent a differential change in length, there must be axial strain compatibility. This change in length is shown in **Figure 6** as $\varepsilon_z L$ where L is the rotor length. This compatibility condition therefore becomes:

$$\varepsilon_{zi} = \varepsilon_{zo} \tag{35}$$

Where ε_{zi} and ε_{zo} are the axial strains of the inner and outer cylinders respectively.

D. There must be equilibrium of forces in the axial direction between the two cylinders. With reference to **Figure 6**:

$$F_{zi} = -F_{zo} \tag{36}$$

To enforce condition A, the constants in Eqs. (26) and (27) are determined with the shrinkage pressure, S, applied at the common radius for the inner and outer cylinders. A pair of simultaneous equations are then determined by considering conditions D and C together and then D and B together. Details of this analysis are given in [18], the result being:

$$\begin{bmatrix} U & V \\ X & Y \end{bmatrix} \begin{bmatrix} F_{zo} \\ S \end{bmatrix} = \begin{bmatrix} W \\ Z \end{bmatrix} \tag{37}$$

where:

$$U = \frac{1}{\pi}\left[E_i\left(R_i^2 - R_c^2\right) - E_o\left(R_o^2 - R_c^2\right)\right]$$

$$V = -2R_c^2\left[\nu_O E_i\left(R_i^2 - R_c^2\right) - \nu_i E_o\left(R_o^2 - R_c^2\right)\right]$$

$$W = \left(R_o^2 - R_c^2\right)\left(R_i^2 - R_c^2\right)\left\{\frac{\omega^2}{2}\left[\nu_O E_i \rho_o\left(R_c^2 + R_o^2\right) - \nu_i E_o \rho_i\left(R_c^2 + R_i^2\right)\right] + T(\alpha_i - \alpha_o)E_o E_i\right\}$$

$$X = \frac{1}{\pi}\left[\nu_i E_o R_{ci}\left(R_o^2 - R_c^2\right) + \nu_o E_i R_{co}\left(R_c^2 - R_i^2\right)\right]$$

$$Y = -\left\{E_o R_{ci}\left(R_o^2 - R_c^2\right)\left[R_c^2(1 - \nu_i) + R_i^2(1 + \nu_i)\right] + R_{co}E_i\left(R_c^2 - R_i^2\right)\left[R_o^2(1 + \nu_o) + R_c^2(1 - \nu_o)\right]\right\}$$

$$Z = \left(R_c^2 - R_i^2\right)\left(R_o^2 - R_c^2\right)\left\{E_o E_i\left[R_{co} - R_{ci} + T(\alpha_o R_{co} - \alpha_i R_{ci})\right]\right.$$

$$+ \frac{\rho_o R_{co}\omega^2 E_i}{8(1 - \nu_o)}\left[2\left(R_c^2 + 3R_o^2\right) - 4\nu_o\left(R_o^2 + R_c^2\right) - 2\nu_o^2\left(R_o^2 - R_c^2\right)\right]$$

$$\left. - \frac{\rho_i R_{ci}\omega^2 E_o}{8(1 - \nu_i)}\left[2\left(R_c^2 + 3R_i^2\right) - 4\nu_i\left(R_c^2 + R_i^2\right) - 2\nu_i^2\left(R_i^2 - R_c^2\right)\right]\right\}$$

4. Comparison of models

4.1. Basic model parameters

The material properties used in the comparison of the various models are as shown in **Table 1**, these being typical properties for samarium cobalt magnets and an Inconel sleeve. Analyses were carried out for the parameters shown in **Table 2**.

For this study temperature was not treated as a variable so thermal expansion effects were ignored.

A critical result is the circumferential stress on the outer radius of the magnet, $\sigma_{\theta ic}$, at maximum speed. This is shown in **Figure 9** for a magnet inner radius of 6 mm. When the degree of interference and sleeve thickness are low there is insufficient compression generated in the magnets by the sleeve and this stress becomes tensile. This may lead to magnet failure dependent on the precise material used. Or, if magnets are segmented, it would lead to magnet separation and localized over-loading of the sleeve.

Parameter	Minimum	Maximum (or discrete values)
Magnet inner radius, R_i, (mm)		4, 6, 8
Common radius, R_c, (mm)		14
Interference, δ, (mm)	0	0.1
Sleeve thickness (mm)	0	10
Rotor speed (ω) (krpm)	0	100

Table 2. Geometric parameters for comparative study.

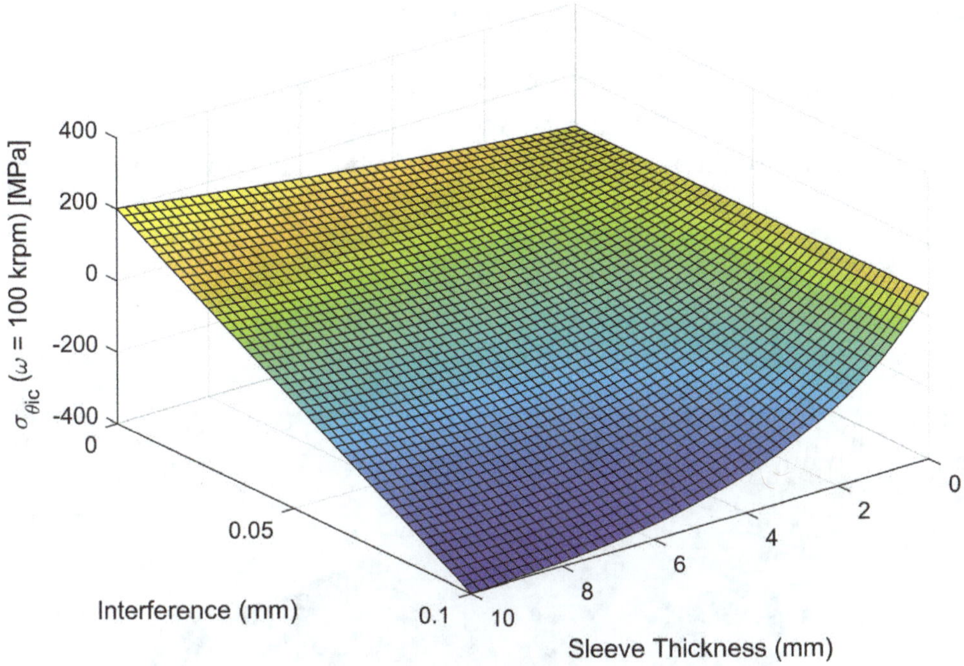

Figure 9. Circumferential stress on the magnet outer radius at speed.

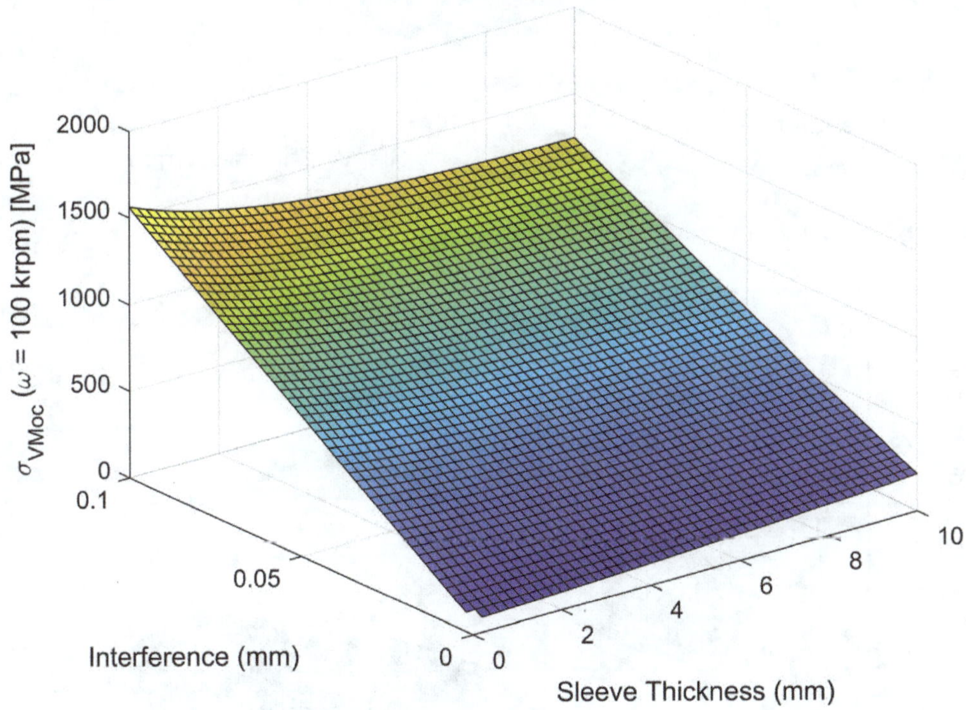

Figure 10. Von Mises stress at the inner radius of the sleeve at speed.

Figure 10 shows the von Mises stress generated at the inner surface of the sleeve, σ_{VMoc}, at zero speed, again with a magnet inner radius of 6 mm. This stress or its equivalent at maximum speed will be critical for the sleeve which is normally a ductile material. This stress becomes larger as interference increases (and hence both radial and circumferential stress magnitudes increase) but reduces as sleeve thickness increases.

4.2. Comparing plane stress and plane strain models

Figure 11 shows the percentage difference in von Mises stress between the plane stress and plane strain theories at maximum speed for a rotor with a magnet inner radius of 4 mm. For smaller levels of interference, this difference is substantial. The percentage difference in circumferential

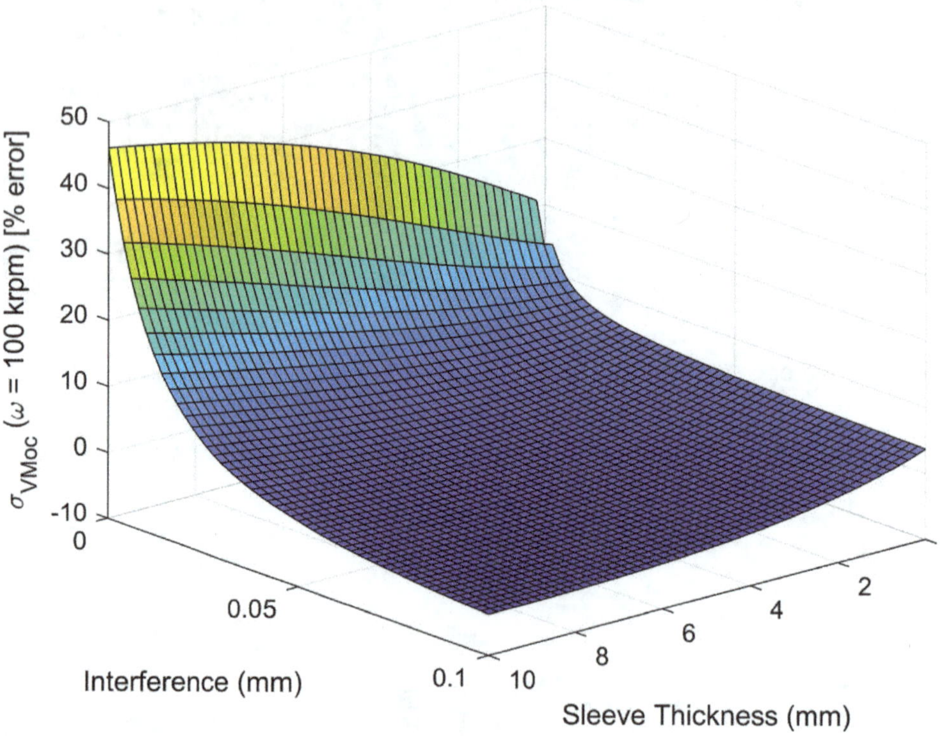

Figure 11. Percentage difference in σ_{VMoc} at maximum speed.

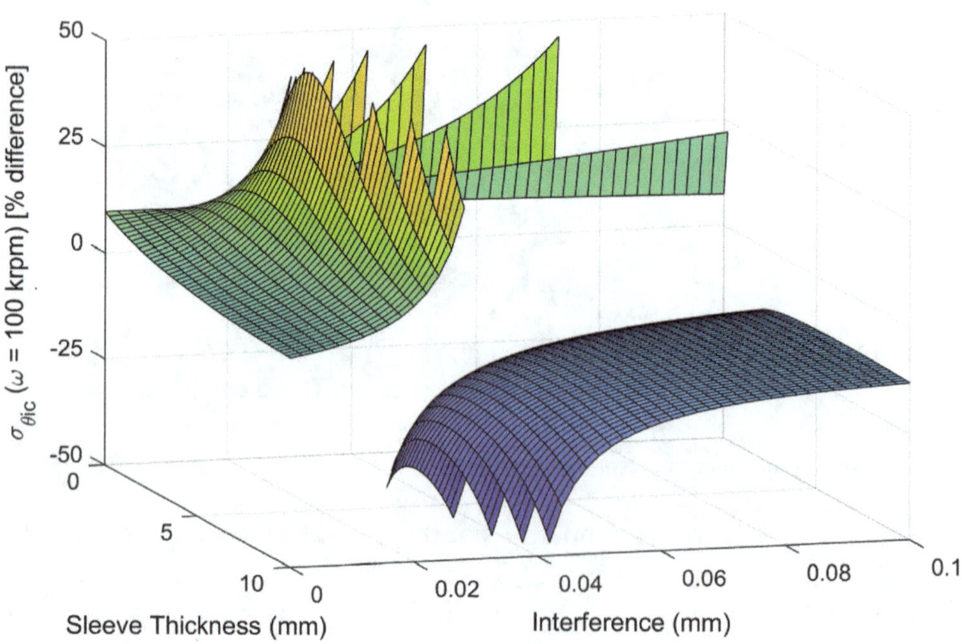

Figure 12. Percentage difference in $\sigma_{\theta ic}$ at maximum speed.

stress in the magnet for the same case is shown in **Figure 12**. Where this stress approaches zero, the results approach infinity and have been omitted from the figure for clarity. In order to optimize rotor design it will be desirable to approach zero circumferential stress in the magnet as maximum speed is reached. Hence, whilst the absolute differences between the two theories are small (because the absolute values are small), they are still critical. These results highlight the need for a theoretical approach that does not incorporate the conflicting approximations of the plane stress and plane strain models.

4.3. Generalized plane strain model

The generalized plane strain model does not include any of the approximations of the plane stress or plane strain theories. For the case of relatively long cylinders with high levels of friction between the components, this theory can be regarded as accurate away from the ends of the rotor. **Figure 13**

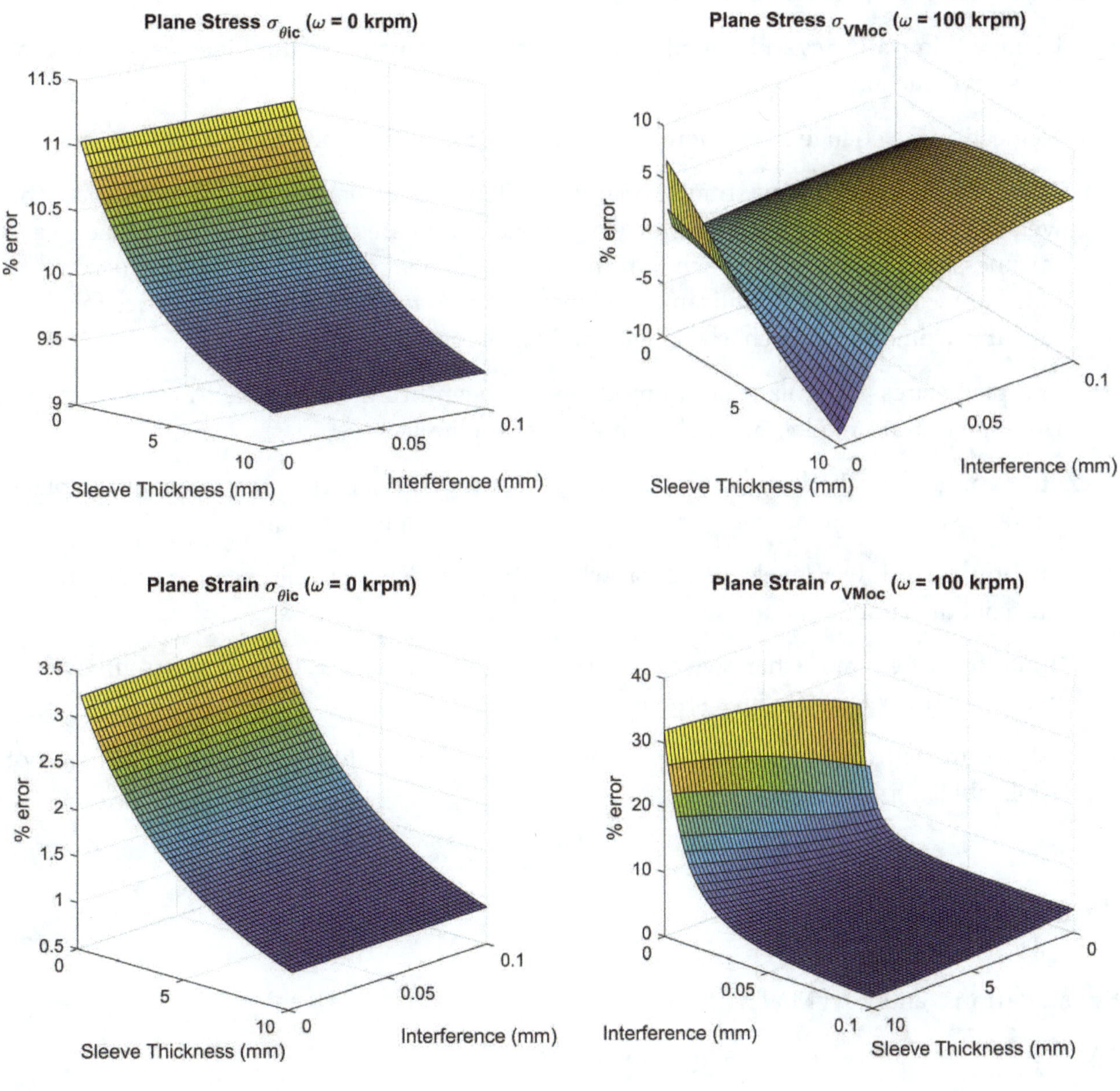

Figure 13. Errors in plane stress and plane strain theories.

shows a selection of results comparing the plane stress and plane strain models to the generalized plane strain model for a rotor with a magnet inner radius of 4 mm. When stationary the error in $\sigma_{\theta ic}$ increases with decreasing sleeve thickness for both the plane stress and plane strain models. This is much more significant for the plane stress model. Looking at the von Mises stress at maximum speed, the plane strain model shows increasing error as the interference is reduced but relatively little dependence on sleeve thickness. The error in the plane stress result is dependent on both sleeve thickness and interference but is never as large as the peak error value for the plane strain theory.

5. Conclusions and further work

The work presented here has shown that:

- Using a thin walled cylinder model for the rotor sleeve will introduce increasing errors as the sleeve thickness increases.
- Neglecting the change in interference due to rotation will introduce significant errors.
- The plane stress and plane strain models will often give similar results for rotors operating well within their design limits. However, when rotors are optimized to minimize sleeve thickness or increase speed such that the circumferential stress in the magnet becomes a limiting factor, there is significant difference between the theories. Both theories incorporate approximations which are not generally true for this application.
- The plane stress and plane strain models show significant differences from the generalized plane strain model, particularly at the design limits.

Despite avoiding the limiting approximations of the other theories, the generalized plane strain model still has limitations. Further development is required to consider:

- The impact of lower levels of friction where the cylinders may slip against each other in the axial direction.
- Situations where more than two cylinders are interfering such as with magnets affixed to a back iron and retained with a sleeve.
- Sleeves made from composite materials where the constitutive equations, 7 to 9, are not applicable in the form given here.

Author details

Simon Barrans* and Levi Mallin

*Address all correspondence to: s.m.barrans@hud.ac.uk

University of Huddersfield, UK

References

[1] Tenconi A, Vaschetto S, Vigliani A. Electrical machines for high-speed applications: Design considerations and tradeoffs. IEEE Transactions on Industrial Electronics. 2014;**61**(6): 3022-3029

[2] Fang H, Qu R, Li J, Zhang P, Fan X. Rotor design for high-speed high-power permanent-magnet synchronous machines. IEEE Transactions on Industry Applications. 2017;**53**(4): 3411-3419

[3] Li S, Sarlioglu B. Assessment of high-speed multi-megawatt electric machines. In: Presented at the Electric Machines & Drives Conference; Coeur d'Alene, ID, USA; 10–13 May, 2015

[4] Gerada D, Mebarki A, Brown NL, Gerada C, Cavagnino A, Boglietti A. High-speed electrical machines: Technologies, trends, and developments. IEEE Transactions on Industrial Electronics. 2014;**61**(6):2946-2959

[5] Lee W, Schubert E, Li Y, Li S, Bobba D, Sarlioglu B. Overview of electric turbocharger and supercharger for downsized internal combustion engines. IEEE Transactions on Transportation Electrification. 2017;**3**(1):36-47

[6] Binder A, Schneider T, Klohr M. Fixation of buried and surface-mounted magnets in high-speed permanent-magnet synchronous machines. IEEE Transactions on Industry Applications. 2006;**42**(4):1031-1037

[7] Hearn EJ. Mechanics of Materials. Vol. 2. Oxford, UK: Pergamon; 1985

[8] Li W, Qiu H, Zhang X, Cao J, Yi R. Analyses on electromagnetic and temperature fields of superhigh-speed permanent-magnet generator with different sleeve materials. IEEE Transactions on Industrial Electronics. 2014;**61**(6):3056-3063

[9] Borisavljevic A, Polinder H, Ferreira JA. Enclosure design for a high-speed permanent magnet rotor. In: Presented at the Power Electronics, Machines and Drives; Brighton, UK; 2010

[10] Zhang F, Du G, Wang T, Liu G, Cao W. Rotor retaining sleeve design for a 1.12-MW high-speed PM machine. IEEE Transactions on Industry Applications. 2015;**51**:3675-3685

[11] Zhu C, Chen L. Rotor strength analysis for stator-permanent magnet machines. In: Presented at the International Conference on Electrical Machines and Systems (ICEMS); Sydney, NSW, Australia; 2017

[12] Cheng W, Xu G, Sun Y, Geng H, Yu L. Optimum design of ultra high speed hybrid rotor of PM machines. In: Presented at the International Conference on Mechatronics and Automation; Tianjin, China; 2014

[13] Burnand G, Araujo DM, Perriard Y. Very-high-speed permanent magnet motors: Mechanical rotor stresses analytical model. In: Presented at the Electric Machines and Drives Conference (IEMDC); Miami, Florida, USA; 2017

[14] Thomas AS, Zhu ZQ, Jewll GW. Comparison of flux switching and surface mounted permanent magnet generators for high-speed applications. In: Presented at the Power Electronics, Machines and Drives (PEMD); 2011

[15] Fernando WU, Gerada C. High speed permanent magnet machine design with minimized stack-length under electromagnetic and mechanical constraints. International Journal of Applied Electromagnetics and Mechanics. 2014;**46**:95-109

[16] Tao Z, Xiaoting Y, Huiping Z, Hongyun J. Strength design on permanent magnet rotor in high speed motor using finite element method. TELKOMNIKA. Indonesian Journal of Electrical Engineering. 2014;**12**:1758-1763

[17] Smith DJB, Mecrow BC, Atkinson GJ, Jack AG, Mehna AAA. Shear stress concentrations in permanent magnet rotor sleeves. In: Presented at the International Conference on Electrical Machines; Rome, Italy; 2010

[18] Barrans S, Al-Ani M, Carter J. Mechanical design of rotors for permanent magnet high-speed electric motors for turbocharger applications. IET Electrical Systems in Transportation. 2017;**7**(4):278-286

Permissions

All chapters in this book were first published in EMSGADSC, by InTech Open; hereby published with permission under the Creative Commons Attribution License or equivalent. Every chapter published in this book has been scrutinized by our experts. Their significance has been extensively debated. The topics covered herein carry significant findings which will fuel the growth of the discipline. They may even be implemented as practical applications or may be referred to as a beginning point for another development.

The contributors of this book come from diverse backgrounds, making this book a truly international effort. This book will bring forth new frontiers with its revolutionizing research information and detailed analysis of the nascent developments around the world.

We would like to thank all the contributing authors for lending their expertise to make the book truly unique. They have played a crucial role in the development of this book. Without their invaluable contributions this book wouldn't have been possible. They have made vital efforts to compile up to date information on the varied aspects of this subject to make this book a valuable addition to the collection of many professionals and students.

This book was conceptualized with the vision of imparting up-to-date information and advanced data in this field. To ensure the same, a matchless editorial board was set up. Every individual on the board went through rigorous rounds of assessment to prove their worth. After which they invested a large part of their time researching and compiling the most relevant data for our readers.

The editorial board has been involved in producing this book since its inception. They have spent rigorous hours researching and exploring the diverse topics which have resulted in the successful publishing of this book. They have passed on their knowledge of decades through this book. To expedite this challenging task, the publisher supported the team at every step. A small team of assistant editors was also appointed to further simplify the editing procedure and attain best results for the readers.

Apart from the editorial board, the designing team has also invested a significant amount of their time in understanding the subject and creating the most relevant covers. They scrutinized every image to scout for the most suitable representation of the subject and create an appropriate cover for the book.

The publishing team has been an ardent support to the editorial, designing and production team. Their endless efforts to recruit the best for this project, has resulted in the accomplishment of this book. They are a veteran in the field of academics and their pool of knowledge is as vast as their experience in printing. Their expertise and guidance has proved useful at every step. Their uncompromising quality standards have made this book an exceptional effort. Their encouragement from time to time has been an inspiration for everyone.

The publisher and the editorial board hope that this book will prove to be a valuable piece of knowledge for researchers, students, practitioners and scholars across the globe.

List of Contributors

Flyur R. Ismagilov, Viacheslav Ye. Vavilov and Valentina V. Ayguzina
Ufa State Aviation Technical University, Ufa, Russia

Nicolae Patrascoiu
System Control, Computer Engineering and Electrical Engineering Department, University of Petrosani, Romania

Sanath Alahakoon and Mark Burridge
School of Engineering and Technology, Central Queensland University, Rockhampton, Australia

Damiano D'Aguanno and Fabrizio Marignetti
Department of Electrical and Information Engineering, University of Cassino and South Lazio, Italy

Francesco Faginoli
Faber Spa, AN, Italy

Ruthber Rodriguez Serrezuela and Jorge Luis Aroca Trujillo
Industrial Engineering, University Corporation of Huila, CORHUILA, Neiva, Republic of Colombia

Roberto Sagaro Zamora
Department of Mechanical Engineering and Design, Universidad de Oriente, Santiago de Cuba, Cuba

Vadim Azhmyakov
Faculty of Basic Sciences, University of Medellin, UdeM, Medellin, Republic of Colombia

Marian Găiceanu
Integrated Energy Conversion Systems and Advanced Control of Complex Processes Research Center, Dunarea de Jos University of Galati, Galati, Romania

Abderrahmène Sallami
Department of Electrical Engineering, Laboratory ACS, Tunisia

Pedro Ponce, Luis Arturo Soriano and Arturo Molina
Escuela de Ingeniería y Ciencias, Tecnologico de Monterrey, Mexico City, Mexico

Manuel Garcia
Escuela Superior de Ingeniería Mecánica y Eléctrica, Instituto Politécnico Nacional, México City, México

Adel El-Shahat
Electrical and Computer Engineering Department, Georgia Southern University, Georgia, USA

Tze-Yee Ho
Department of Electrical Engineering, Feng Chia University, Seatwen, Taichung, Taiwan, R.O.C.

Simon Barrans and Levi Mallin
University of Huddersfield, UK

Index

www.ingramcontent.com/pod-product-compliance
Lightning Source LLC
LaVergne TN
LVHW080931210926
848485LV00004B/771
* 9 7 8 1 6 3 9 8 9 4 3 1 4 *